The Grid

The Grid

The Grid

Core Technologies

Maozhen Li
Brunel University, UK

Mark Baker
University of Portsmouth, UK

John Wiley & Sons, Ltd

Copyright © 2005   John Wiley & Sons Ltd, The Atrium, Southern Gate, Chichester,
                     West Sussex PO19 8SQ, England

                     Telephone   (+44) 1243 779777

Email (for orders and customer service enquiries): cs-books@wiley.co.uk
Visit our Home Page on www.wiley.com

All Rights Reserved. No part of this publication may be reproduced, stored in a retrieval system or transmitted in any form or by any means, electronic, mechanical, photocopying, recording, scanning or otherwise, except under the terms of the Copyright, Designs and Patents Act 1988 or under the terms of a licence issued by the Copyright Licensing Agency Ltd, 90 Tottenham Court Road, London W1T 4LP, UK, without the permission in writing of the Publisher. Requests to the Publisher should be addressed to the Permissions Department, John Wiley & Sons Ltd, The Atrium, Southern Gate, Chichester, West Sussex PO19 8SQ, England, or emailed to permreq@wiley.co.uk, or faxed to (+44) 1243 770620.

Designations used by companies to distinguish their products are often claimed as trademarks. All brand names and product names used in this book are trade names, service marks, trademarks or registered trademarks of their respective owners. The Publisher is not associated with any product or vendor mentioned in this book.

This publication is designed to provide accurate and authoritative information in regard to the subject matter covered. It is sold on the understanding that the Publisher is not engaged in rendering professional services. If professional advice or other expert assistance is required, the services of a competent professional should be sought.

Other Wiley Editorial Offices

John Wiley & Sons Inc., 111 River Street, Hoboken, NJ 07030, USA

Jossey-Bass, 989 Market Street, San Francisco, CA 94103-1741, USA

Wiley-VCH Verlag GmbH, Boschstr. 12, D-69469 Weinheim, Germany

John Wiley & Sons Australia Ltd, 33 Park Road, Milton, Queensland 4064, Australia

John Wiley & Sons (Asia) Pte Ltd, 2 Clementi Loop #02-01, Jin Xing Distripark, Singapore 129809

John Wiley & Sons Canada Ltd, 22 Worcester Road, Etobicoke, Ontario, Canada M9W 1L1

Wiley also publishes its books in a variety of electronic formats. Some content that appears in print may not be available in electronic books.

Library of Congress Cataloging in Publication Data

Li, Maozhen.
  Core technologies / Maozhen Li, Mark Baker.
    p. cm.
  ISBN-13 978-0-470-09417-4 (PB)
  ISBN-10 0-470-09417-6 (PB)
  1. Computational grids (Computer systems)   2. Electronic data processing—Distributed processing.
  I. Baker, Mark. II. Title.
  QA76.9.C58L5 2005
  005.3'6—dc22

                                                                                          2005002378

British Library Cataloguing in Publication Data

A catalogue record for this book is available from the British Library

ISBN-13 978-0-470-09417-4 (PB)
ISBN-10 0-470-09417-6 (PB)

Typeset in 11/13pt Palatino by Integra Software Services Pvt. Ltd, Pondicherry, India
Printed and bound by CPI Antony Rowe, Eastbourne

Contents

About the Authors	xiii
Preface	xv
Acknowledgements	xix
List of Abbreviations	xxi
1 An Introduction to the Grid	**1**
1.1 Introduction	1
1.2 Characterization of the Grid	1
1.3 Grid-Related Standards Bodies	4
1.4 The Architecture of the Grid	5
1.5 References	6
Part One System Infrastructure	**9**
2 OGSA and WSRF	**11**
Learning Objectives	11
Chapter Outline	11
2.1 Introduction	12
2.2 Traditional Paradigms for Distributed Computing	13
2.2.1 Socket programming	14
2.2.2 RPC	15
2.2.3 Java RMI	16
2.2.4 DCOM	18
2.2.5 CORBA	19
2.2.6 A summary on Java RMI, DCOM and CORBA	20
2.3 Web Services	21
2.3.1 SOAP	23
2.3.2 WSDL	24
2.3.3 UDDI	26
2.3.4 WS-Inspection	27
2.3.5 WS-Inspection and UDDI	28
2.3.6 Web services implementations	29
2.3.7 How Web services benefit the Grid	33

	2.4	OGSA	34
		2.4.1 Service instance semantics	35
		2.4.2 Service data semantics	37
		2.4.3 OGSA portTypes	38
		2.4.4 A further discussion on OGSA	40
	2.5	The Globus Toolkit 3 (GT3)	40
		2.5.1 Host environment	41
		2.5.2 Web services engine	42
		2.5.3 Grid services container	42
		2.5.4 GT3 core services	43
		2.5.5 GT3 base services	44
		2.5.6 The GT3 programming model	50
	2.6	OGSA-DAI	53
		2.6.1 OGSA-DAI portTypes	54
		2.6.2 OGSA-DAI functionality	56
		2.6.3 Services interaction in the OGSA-DAI	58
		2.6.4 OGSA-DAI and DAIS	59
	2.7	WSRF	60
		2.7.1 An introduction to WSRF	60
		2.7.2 WSRF and OGSI/GT3	66
		2.7.3 WSRF and OGSA	69
		2.7.4 A summary of WSRF	70
	2.8	Chapter Summary	70
	2.9	Further Reading and Testing	72
	2.10	Key Points	72
	2.11	References	73
3	**The Semantic Grid and Autonomic Computing**	**77**
		Learning Outcomes	77
		Chapter Outline	77
	3.1	Introduction	78
	3.2	Metadata and Ontology in the Semantic Web	79
		3.2.1 RDF	81
		3.2.2 Ontology languages	83
		3.2.3 Ontology editors	87
		3.2.4 A summary of Web ontology languages	88
	3.3	Semantic Web Services	88
		3.3.1 DAML-S	89
		3.3.2 OWL-S	90
	3.4	A Layered Structure of the Semantic Grid	91
	3.5	Semantic Grid Activities	92
		3.5.1 Ontology-based Grid resource matching	93
		3.5.2 Semantic workflow registration and discovery in myGrid	94
		3.5.3 Semantic workflow enactment in Geodise	95
		3.5.4 Semantic service annotation and adaptation in ICENI	98
		3.5.5 PortalLab – A Semantic Grid portal toolkit	99
		3.5.6 Data provenance on the Grid	106
		3.5.7 A summary on the Semantic Grid	107

3.6	Autonomic Computing	108
	3.6.1 What is autonomic computing?	108
	3.6.2 Features of autonomic computing systems	109
	3.6.3 Autonomic computing projects	110
	3.6.4 A vision of autonomic Grid services	113
3.7	Chapter Summary	114
3.8	Further Reading and Testing	115
3.9	Key Points	116
3.10	References	116

Part Two Basic Services 121

4 Grid Security 123

4.1	Introduction	123
4.2	A Brief Security Primer	124
4.3	Cryptography	127
	4.3.1 Introduction	127
	4.3.2 Symmetric cryptosystems	128
	4.3.3 Asymmetric cryptosystems	129
	4.3.4 Digital signatures	130
	4.3.5 Public-key certificate	130
	4.3.6 Certification Authority (CA)	132
	4.3.7 Firewalls	133
4.4	Grid Security	134
	4.4.1 The Grid Security Infrastructure (GSI)	134
	4.4.2 Authorization modes in GSI	136
4.5	Putting it all Together	140
	4.5.1 Getting an e-Science certificate	140
	4.5.2 Managing credentials in Globus	146
	4.5.3 Generate a client proxy	148
	4.5.4 Firewall traversal	148
4.6	Possible Vulnerabilities	149
	4.6.1 Authentication	149
	4.6.2 Proxies	149
	4.6.3 Authorization	150
4.7	Summary	151
4.8	Acknowledgements	151
4.9	Further Reading	151
4.10	References	152

5 Grid Monitoring 153

5.1	Introduction	153
5.2	Grid Monitoring Architecture (GMA)	154
	5.2.1 Consumer	155
	5.2.2 The Directory Service	156
	5.2.3 Producers	157
	5.2.4 Monitoring data	159

5.3	Review Criteria	161
	5.3.1 Scalable wide-area monitoring	161
	5.3.2 Resource monitoring	161
	5.3.3 Cross-API monitoring	161
	5.3.4 Homogeneous data presentation	162
	5.3.5 Information searching	162
	5.3.6 Run-time extensibility	162
	5.3.7 Filtering/fusing of data	163
	5.3.8 Open and standard protocols	163
	5.3.9 Security	163
	5.3.10 Software availability and dependencies	163
	5.3.11 Projects that are active and supported; plus licensing	163
5.4	An Overview of Grid Monitoring Systems	164
	5.4.1 Autopilot	164
	5.4.2 Control and Observation in Distributed Environments (CODE)	168
	5.4.3 GridICE	172
	5.4.4 Grid Portals Information Repository (GPIR)	176
	5.4.5 GridRM	180
	5.4.6 Hawkeye	185
	5.4.7 Java Agents for Monitoring and Management (JAMM)	189
	5.4.8 MapCenter	192
	5.4.9 Monitoring and Discovery Service (MDS3)	196
	5.4.10 Mercury	201
	5.4.11 Network Weather Service	205
	5.4.12 The Relational Grid Monitoring Architecture (R-GMA)	209
	5.4.13 visPerf	214
5.5	Other Monitoring Systems	217
	5.5.1 Ganglia	217
	5.5.2 GridMon	219
	5.5.3 GRM/PROVE	220
	5.5.4 Nagios	221
	5.5.5 NetLogger	222
	5.5.6 SCALEA-G	223
5.6	Summary	225
	5.6.1 Resource categories	225
	5.6.2 Native agents	225
	5.6.3 Architecture	226
	5.6.4 Interoperability	226
	5.6.5 Homogeneous data presentation	226
	5.6.6 Intrusiveness of monitoring	227
	5.6.7 Information searching and retrieval	231
5.7	Chapter Summary	233
5.8	Further Reading and Testing	236
5.9	Key Points	236
5.10	References	236

Part Three Job Management and User Interaction	241
6 Grid Scheduling and Resource Management	**243**
Learning Objectives	243
Chapter Outline	243
6.1 Introduction	244
6.2 Scheduling Paradigms	245
6.2.1 Centralized scheduling	245
6.2.2 Distributed scheduling	246
6.2.3 Hierarchical scheduling	248
6.3 How Scheduling Works	248
6.3.1 Resource discovery	248
6.3.2 Resource selection	251
6.3.3 Schedule generation	251
6.3.4 Job execution	254
6.4 A Review of Condor, SGE, PBS and LSF	254
6.4.1 Condor	254
6.4.2 Sun Grid Engine	269
6.4.3 The Portable Batch System (PBS)	274
6.4.4 LSF	279
6.4.5 A comparison of Condor, SGE, PBS and LSF	288
6.5 Grid Scheduling with QoS	290
6.5.1 AppLeS	291
6.5.2 Scheduling in GrADS	293
6.5.3 Nimrod/G	293
6.5.4 Rescheduling	295
6.5.5 Scheduling with heuristics	296
6.6 Chapter Summary	297
6.7 Further Reading and Testing	298
6.8 Key Points	298
6.9 References	299
7 Workflow Management for the Grid	**301**
Learning Outcomes	301
Chapter Outline	301
7.1 Introduction	302
7.2 The Workflow Management Coalition	303
7.2.1 The workflow enactment service	305
7.2.2 The workflow engine	306
7.2.3 WfMC interfaces	308
7.2.4 Other components in the WfMC reference model	309
7.2.5 A summary of WfMC reference model	310
7.3 Web Services-Oriented Flow Languages	310
7.3.1 XLANG	311
7.3.2 Web services flow language	311
7.3.3 WSCI	313
7.3.4 BPEL4WS	315
7.3.5 BPML	317
7.3.6 A summary of Web services flow languages	318

7.4 Grid Services-Oriented Flow Languages ... 318
   7.4.1 GSFL ... 318
   7.4.2 SWFL ... 321
   7.4.3 GWEL ... 321
   7.4.4 GALE ... 322
   7.4.5 A summary of Grid services flow languages ... 323
7.5 Workflow Management for the Grid ... 323
   7.5.1 Grid workflow management projects ... 323
   7.5.2 A summary of Grid workflow management ... 329
7.6 Chapter Summary ... 330
7.7 Further Reading and Testing ... 331
7.8 Key Points ... 332
7.9 References ... 332

8 Grid Portals ... 335
Learning Outcomes ... 335
Chapter Outline ... 335
8.1 Introduction ... 336
8.2 First-Generation Grid Portals ... 337
   8.2.1 A three-tiered architecture ... 337
   8.2.2 Grid portal services ... 338
   8.2.3 First-generation Grid portal implementations ... 339
   8.2.4 First-generation Grid portal toolkits ... 341
   8.2.5 A summary of the four portal tools ... 348
   8.2.6 A summary of first-generation Grid portals ... 349
8.3 Second-Generation Grid Portals ... 350
   8.3.1 An introduction to portlets ... 350
   8.3.2 Portlet specifications ... 355
   8.3.3 Portal frameworks supporting portlets ... 357
   8.3.4 A Comparison of Jetspeed, WebSphere Portal and GridSphere ... 368
   8.3.5 The development of Grid portals with portlets ... 369
   8.3.6 A summary on second-generation Grid portals ... 371
8.4 Chapter Summary ... 372
8.5 Further Reading and Testing ... 373
8.6 Key Points ... 373
8.7 References ... 374

Part Four Applications ... 377

9 Grid Applications – Case Studies ... 379
Learning Objectives ... 379
Chapter Outline ... 379
9.1 Introduction ... 380
9.2 GT3 Use Cases ... 380
   9.2.1 GT3 in broadcasting ... 381
   9.2.2 GT3 in software reuse ... 382
   9.2.3 A GT3 bioinformatics application ... 387

9.3	OGSA-DAI Use Cases	387
	9.3.1 eDiaMoND	387
	9.3.2 ODD-Genes	388
9.4	Resource Management Case Studies	388
	9.4.1 The UCL Condor pool	388
	9.4.2 SGE use cases	389
9.5	Grid Portal Use Cases	390
	9.5.1 Chiron	390
	9.5.2 GENIUS	390
9.6	Workflow Management – Discovery Net Use Cases	391
	9.6.1 Genome annotation	391
	9.6.2 SARS virus evolution analysis	391
	9.6.3 Urban air pollution monitoring	392
	9.6.4 Geo-hazard modelling	394
9.7	Semantic Grid – myGrid Use Case	394
9.8	Autonomic Computing – AutoMate Use Case	395
9.9	Conclusions	397
9.10	References	398

Glossary 401

Index 419

About the Authors

Dr Maozhen Li is currently Lecturer in Electronics and Computer Engineering, in the School of Engineering and Design at Brunel University, UK. From January 1999 to January 2002, he was Research Associate in the Department of Computer Science, Cardiff University, UK. Dr Li received his PhD degree in 1997, from the Institute of Software, Chinese Academy of Sciences, Beijing, China. His research interests are in the areas of Grid computing, problem-solving environments for large-scale simulations, software agents for semantic information retrieval, multi-modal user interface design and computer support for cooperative work. Since 1997, Dr Li has published 30 research papers in prestigious international journals and conferences.

Dr Mark Baker is a hardworking Reader in Distributed Systems at the University of Portsmouth. He also currently holds visiting chairs at the universities of Reading and Westminster. Mark has resided in the relative safety of academia since leaving the British Merchant, where he was a navigating officer, in the early 1980s. Mark has held posts at various universities, including Cardiff, Edinburgh and Syracuse. He has a number of geek-like interests, which his research group at Portsmouth help him pursue. These include wide-area resource monitoring, messaging systems for parallel and wide-area applications, middleware such as information and security services, as well as performance evaluation and modelling of computer systems.

Mark's non-academic interests include squash (getting too old), DIY (he may one day finish his house off), reading (far too many science fiction books), keeping the garden ship-shape and a beer or two to reduce the pain of the aforementioned activities.

Preface

Grid technologies and the associated applications are currently of unprecedented interest and importance to a variety of communities. This book aims to outline and describe all of the components that are currently needed to create a Grid infrastructure that can support a range of wide-area distributed applications. In this book we take a pragmatic approach to presenting the material; we attempt not only to describe a particular component, but also to give practical examples of how that software may be used in context. We also intend to ensure that the companion Web site has extensive material that can be used by not only novices, but experienced practitioners too, to learn or gather technical material that can help in the process of understanding and using various Grid components and tools.

PURPOSE AND READERSHIP

The purpose of this book is not to convince the reader that one framework, technology or specification is better than another; rather its purpose is to expose the reader to a wide variety of what we call core technologies so that they can determine which is best for their own use.

This book is intended for postgraduate students and researchers from various fields who are interested in learning about the core technologies that make up the Grid today. The material being developed for the companion Web site will supplement the book's content. We intend that the book, along with Web content, will provide sufficient material to allow a complete self-study course of all the components addressed.

The book takes a bottom-up approach, addressing lower-level components first, then mid-level frameworks and systems, and then finally higher-level concepts, concluding by outlining a number of representative Grid applications that provide examples of how the aforementioned frameworks and components are used in practice.

We cover the core technologies currently in Grid environments to a sufficient depth that readers will be prepared to take on research papers and other related literature. In fact, there is often sufficient depth that a reader may use the book as a reference of how to get started with a particular Grid component.

The subject material should be accessible to postgraduates and researchers who have a limited knowledge about the Grid, but technically have some knowledge about distributed systems, and experience in programming with C or Java.

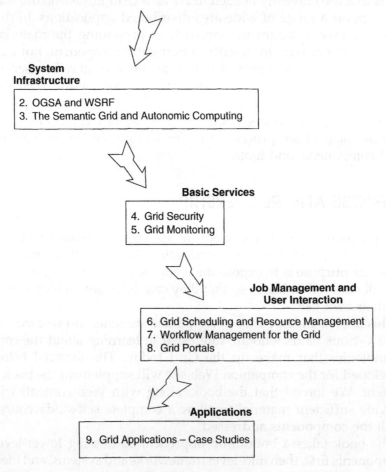

Figure P.1 Organization of the book

ORGANIZATION OF THE BOOK

The organization of the book is shown in Figure P.P.1. We have organized the book into four general parts, which reflect the bottom-up view that we use to address the topics covered. We know that certain topics have been discussed under different parts, but we feel that this should assist the reader label topics more easily and hopefully help them get to grips with the content more easily.

The first section, "system infrastructure", contains the chapters that discuss and outline the current architecture, services and instantiations of the Grid. These chapters provide the underpinning information that the proceeding chapters build on. The second section, "basic services", contains the chapters that describe Grid security and monitoring. Both these chapters explain services that do not actually need to exist to have a Grid environment, but without security and monitoring services it is impossible to have a secure, robust and reliable environment that can be used by higher-level services and applications. The third section we have labelled "Job management and User interaction". At this level users have potentially direct access to tools and utilities that can change their working environment (in the case of a Portal), or manage and schedule their jobs (in the case of workflow and scheduling systems). Finally, the last section of the book is called "Applications"; here we discuss a number of representative Grid-based applications that highlight the technologies and components discussed in the earlier chapters of the book.

Acknowledgements

This first edition of our textbook was prepared during mid–late 2004, when the Grid-based technologies were not only at an embryonic stage, but also in a great state of flux. With any effort, such as writing a book, nothing would really be accomplished in a timely fashion without the aid of a large number of willing helpers and volunteers. The technology landscape that we have been writing about is changing rapidly, so we sought and asked experts in various fields to read through and comment on all parts of the book. We would like to thank the following people for reviewing parts of the book:

- Chapter 2 – OGSA and WSRF: Stephen Pickles and Mark McKeown (Manchester Computing, University of Manchester) and Helen Xiang (DSG, University of Portsmouth).
- Chapter 3 – The Semantic Grid and Autonomic Computing: Rich Boaks (DSG, University of Portsmouth) and Manish Parashar (Rutgers, The State University of New Jersey, USA).
- Chapter 4 – Grid Security: Alistair Mills (Grid Deployment Group, CERN).
- Chapter 5 – Grid Monitoring: A special thank you to Garry Smith (DSG, University of Portsmouth), who provided a lot of detailed content for this chapter, and still managed to write and submit his PhD.
- Chapter 6 – Grid Scheduling and Resource Management: NG1 – Fritz Ferstl (Sun Microsystems), Condor – Todd Tannenbaum (Condor project, University of Wisconsin, USA), LSF – Songnian Zhou (Platform Computing Inc, Canada), PBS – Bob Henderson (Altair Grid Technologies, USA).
- Chapter 7 – Workflow Management for the Grid: Omer Rana (Cardiff University).

- Chapter 8 – Grid Portals: Rob Allan (Daresbury Laboratory).
- Chapter 9 – Grid Applications – Case Studies: Rob Allan (Daresbury Laboratory).

We like to make a special mention of and an acknowledgement to Rob Allan (Daresbury Laboratory, UK), who meticulously reviewed the book as a whole and fed back many useful comments about its presentation and content.

We would like to say a special thanks to Birgit Gruber, our Wiley editor, who worked closely with us through the production of the book, and generally made the effort involved a pleasant one.

COMPANION WEB SITE

We have set up a Web site (coregridtechnologies.org) containing companion material to the book that will assist readers and teachers. The amount of content will grow with time and eventually include:

- Tables and figures from the book in various formats
- Slides of the content
- Notes highlighting various aspects of the content
- Links and references to companion material
- Laboratory exercises and solutions
- Source code for examples
- Potential audio/visual material.

Obviously, from the inception of book to its publication and distribution, the landscape that we describe will have undulated some more, so the book is a snapshot of the technologies during mid–late 2004. We believe that we can overcome some of the gaps that may appear in the book's coverage of material by adding the appropriate content to the companion Web site.

List of Abbreviations

Abbreviation	Expanded form	Context
ACL	Access Control List	
AM	Actuator Manager	CODE
AMUSE	Autonomic Management of Ubiquitous Systems for e-Health	
AppLeS	Application Level Scheduler	
APST	AppLeS Parameter Sweep Template	AppLeS
ASA	Autonomic Storage Architecture	
ASAP	Asynchronous Service Access Protocol	
ASP	Active Server Pages	Microsoft .Net
BLAST	Basic Local Alignment Search Tool	
BPEL4WS	Business Process Execution Language for Web Services	
BPML	Business Process Modelling Language	
CA	Certification Authority	
CCA	Common Component Architecture	
CFD	Computational Fluid Dynamics	
CGI	Common Gateway Interface	
CIM	Common Information Model	
ClassAd	Classified Advertisement	Condor
CMS	Compact Muon Solenoid	
COD	Computing On Demand	Condor
CODE	Control and Observation in Distributed Environments	

CORBA	Common Object Request Broker Architecture	OMG
CPS	Certificate Policy Statement	
CSF	Community Scheduler Framework	
CSR	Certificate Signing Request	
DA	Data Analyser	GridICE
DAG	Directed Acyclic Graph	Condor
DAGMan	Directed Acyclic Graph Manager	Condor
DAIS	Database Access and Integration Services	
DAISGR	DAI Service Group Registry	OGSA-DAI
DAML	DARPA Agent Markup Language	
DAML-S	DAML Services	DAML
DCE	Distributed Computing Environment	
DCOM	Distributed Component Object Model	Microsoft
DCS	Data Collector Service	GridICE
DES	Data Encryption Standard	
DL	Description Logic	
DMTF	Distributed Management Task Force	
DNS	Detection and Notification Service	GridICE
DPML	Discovery Process Markup Language	Discovery Net
DSP	Distributed Systems Group	GridRM
DTD	Document Type Definition	W3C
EC	Event Consumer	JAMM
ECS	Element Construction Set	
EDG	European Data Grid	
EDSO	Engineering Design Search and Optimization	Geodise
EG	Event Gateway	JAMM
EGEE	Enabling Grids for E-science in Europe	
EJB	Enterprise JavaBeans	J2EE
FaCT	Fast Classification of Terminologies	
FIFO	First In First Out	

GA	Genetic Algorithm	
GAC	GPIR Admin Client	GPIR
GALE	Grid Access Language for high-performance computing Environments	
GAR	Grid Archive	GT3
GARA	GARA	
GDS	Grid Data Service	OGSA-DAI
GDSF	Grid Data Service Factory	OGSA-DAI
GDS-Perform	Grid Data Service Perform	OGSA-DAI
GDS-Response	Grid Data Service Response	OGSA-DAI
GEMLCA	Grid Execution Management for Legacy Code Architecture	
Geodise	Grid Enabled Optimization and DesIgn Search for Engineering	
GGF	Global Grid Forum	
GIP	GPIR Information Provider	GPIR
GIS	Grid Information Services	
GMA	Grid Monitoring Architecture	GGF
GPDK	Grid Portal Development Kit	
GPIR	Grid Portals Information Repository	
GRAAP-WG	Grid Resource Allocation Agreement Protocol Working Group	
GrADS	Grid Application Development Software	
GRAM	Globus Resource Allocation Manager	Globus
GridFTP	Grid File Transfer Protocol	Globus
GRIM	Grid Resource Identity Mapper	GT3
GSA	Grid System Agent	PortalLab
GSFL	Grid Services Flow Language	
GSH	Grid Service Handle	OGSA
GSI	Grid Security Infrastructure	Globus
GSR	Grid Service Reference	OGSA
GSSAPI	Generic Security Services Application Programming Interface	GSSAPI

GT2	Globus Toolkit 2	Globus
GT3	Globus Toolkit 3	Globus
GT4	Globus Toolkit 4	Globus
GUSTO	Generic Ultraviolet Sensors Technologies and Observations	
GWEL	Grid Workflow Execution Language	
GWSDL	Grid WSDL	GT3
HAT	Heterogeneous Application Template	AppLeS
HM	Hawkeye Manager	Condor
HMA	Hawkeye Monitoring Agent	Condor
HPSS	High-Performance Storage System	
ICENI	Imperial College e-Science Networked Infrastructure	ICENI
IDL	Interface Definition Language	
IPG	Information Power Grid	
IIOP	Internet-Inter ORB Protocol	CORBA
ISAPI	Internet Server Application Programming Interface	Microsoft .Net
J2EE	Java 2 Enterprise Edition	
J2SE	Java 2 Standard Edition	
JAMM	Java Agents for Monitoring and Management	
JAR	Java Archive	Java
Java CoG	Java Commodity Grid	
JAXB	Java Architecture for XML Binding	J2EE
JAXM	Java API for XML Messaging	J2EE
JAXP	Java API for XML Processing	J2EE
JAXR	Java API for XML Registries	J2EE
JAX-RPC	Java API for XML-Based RPC	J2EE
JCE	Java Cryptography Extension	
JCP	Java Community Process	
JCR	Java Certificate Request	
JNDI	Java Native Directory Interface	
JISGA	Jini-based Service-Oriented Grid Architecture	
JRE	Java Run time Environment	

JRMP	Java Remote Method Protocol	RMI
JSP	Java Server Page	
JSR	Java Specification Requests	JCP
LCG	LHC Computing Grid	
LCID	Legacy Code Interface Description	GEMLCA
LDAP	Lightweight Directory Access Protocol	
LMJFS	Local Managed Job Factory Service	GT3
LSF	Load Sharing Facility	
MAC	Message Authentication Code	
MCA	Machine Check Architecture	
MDS	Monitoring and Discovery Service	Globus
MJS	Managed Job Service	GT3
MMJFS	Master Managed Job Factory Service	GT3
MPI	Message Passing Interface	
MS	Measurement Service	GridICE
MSXML	Microsoft XML Parser	Microsoft .Net
MVC	Model-View-Controller	
N1GE	N1 Grid Engine	
NetLogger	Networked Application Logger	
NMI	NSF Middleware Initiative	
NS	Naming Schema	GridRM
NWS	Network Weather Service	
OASIS	Organization for the Advancement of Structured Information Standards	
OCS	Open Content Syndication	
OGCE	Open Grid Computing Environments	
OGSA	Open Grid Services Architecture	GGF
OGSA-DAI	OGSA Data Integration and Access	GGF
OGSI	Open Grid Services Infrastructure	GGF
OGSI-WG	OGSI Working Group	GGF
OIL	Ontology Inference Layer	

OLE	Object Linking and Embedding	
OMG	Object Management Group	
ONC	Open Network Computing	
ORPC	Object Remote Procedure Call	DCOM
OSF	Open Software Foundation	
OWL	Web Ontology Language	W3C
OWL-S	OWL Services	OWL
P2P	Peer-to-Peer	
PASOA	Provenance-Aware Service-Oriented Architecture	
PBS	Portable Batch System	
PDSR	Portlet Domain Service Repository	PortalLab
PGT3	Platform Globus Toolkit 3.0	LSF
PI	Producer Interface	CODE
PIF	Process Interchange Format	
PII	Portlet Invocation Interface	
PImR	Portlet Implementation Repository	PortalLab
PInR	Portlet Interface Repository	PortalLab
PKI	Public Key Infrastructure	
PMA	Port Manager Agent	JAMM
portType	Port Type	WSDL
PS	Presentation Service	GridICE
PUB	Publisher Service	GridICE
PSL	Process Specification Language	
PSML	Portlet Structure Markup Language	
PVM	Parallel Virtual Machine	
PWG	Portlet Wrapper Generator	PortalLab
RBAC	Role-Based Access Control	
RDF	Resource Description Framework	W3C
RDFS	RDF Schema	W3C
RMI	Remote Method Invocation	Java
RPC	Remote Procedure Call	
RSA	Rivest, Shamir and Adleman	
RSL	Resource Specification Language	Globus
RSS	Really Simple Syndication	
RFT	Reliable File Transfer	GT3

RUS-WG	Resource Usage Services Working Group	
SA	Simulated Annealing	
SARS	Severe Acute Respiratory Syndrome	
SD	Sensor Directory	JAMM
SDDF	Self-Defining Data Format	Pablo
SDE	Service Data Element	OGSA
SGE	Sun Grid Engine	
SGP	Semantic Grid Portlet	PortalLab
SM	Sensor Manager	CODE
SNMP	Simple Network Management Protocol	
SOA	Service-Oriented Architecture	
SPM	Semantic Portlet Matcher	PortalLab
SPR	Semantic Portlet Register	PortalLab
SRB	Storage Resource Broker	
SSL	Secure Sockets Layer	
SWAP	Simple Workflow Access Protocol	WfMC
SWFL	Service Workflow Language	JISGA
TLS	Transport Layer Security	
UA	User Agent	PortalLab
UDDI	Universal Description, Discovery and Integration	W3C
UHE	User-Hosting Environment	GT3
ULM	Universal Logger Message	
URI	Uniform Resource Identifier	
VO	Virtual Organization	GGF
VPCE	The Visual Portal Composition Environment	PortalLab
W3C	World Wide Web Consortium	
WAR	Web Application Repository	
WBEM	Web-Based Enterprise Management	
WBS	Williams–Beuren Syndrome	
WfMC	Workflow Management Coalition	
WFMS	Workflow Management System	
WML	Wireless Markup Language	
WPDL	Workflow Process Definition Language	WfMC
WS	Web Services	W3C

WSCI	Web Services Choreography Interface	
WSFL	Web Services Flow Language	
WSDD	Web Services Deployment Descriptor	WS
WSDL	Web Services Description Language	W3C
WSIF	Web Services Invocation Framework	
WSIL	WS-Inspection Language	
WSML	Web Services Meta Language	Microsoft .Net
WSRF	Web Services Resource Framework	
WSRP	Web Services for Remote Portlets	OASIS
XDR	External Data Representation	
XML	eXtensible Markup Language	W3C
XPDL	XML Process Definition Language	WfMC
XSD	XML Schema Definition	W3C
XSL	eXtensible Stylesheet Language	
XSLT	XSL Transformation	
YAWL	Yet Another Workflow Language	

1
An Introduction to the Grid

1.1 INTRODUCTION

The Grid concepts and technologies are all very new, first expressed by Foster and Kesselman in 1998 [1]. Before this, efforts to orchestrate wide-area distributed resources were known as metacomputing [2]. Even so, whichever date we use to identify when efforts in this area started, compared to general distributed computing, the Grid is a very new discipline and its exact focus and the core components that make up its infrastructure are still being investigated and have yet to be determined. Generally it can be said that the Grid has evolved from a carefully configured infrastructure that supported a limited number of grand challenge applications executing on high-performance hardware between a number of US national centres [3], to what we are aiming at today, which can be seen as a seamless and dynamic virtual environment. In this book we take a step-by-step approach to describe the middleware components that make up this virtual environment which is now called the Grid.

1.2 CHARACTERIZATION OF THE GRID

Before we go any further we need to somehow define and characterize what can be seen as a Grid infrastructure. To start with, let us think about the execution of a distributed application. Here

---

The Grid: Core Technologies Maozhen Li and Mark Baker
© 2005 John Wiley & Sons, Ltd

we usually visualize running such an application "on top" of a software layer called middleware that unifies the resources being used by the application into a single coherent virtual machine. To help understand this view of a distributed application and its accompanying middleware, consider Figure 1.1, which shows the hardware and software components that would be typically found on a PC-based cluster. This view then raises the question, what is the difference between a distributed system and the Grid? Obviously the Grid is a type of distributed system, but this does not really answer the question. So, perhaps we should try and establish "What is a Grid?"

In 1998, Ian Foster and Carl Kesselman provided an initial definition in their book *The Grid: Blueprint for a New Computing Infrastructure* [1]: "A computational grid is a hardware and software infrastructure that provides dependable, consistent, pervasive, and inexpensive access to high-end computational capabilities." This particular definition stems from the earlier roots of the Grid, that of interconnecting high-performance facilities at various US laboratories and universities.

Since this early definition there have been a number of other attempts to define what a Grid is. For example, "A grid is a software framework providing layers of services to access and manage distributed hardware and software resources" [4] or a "widely

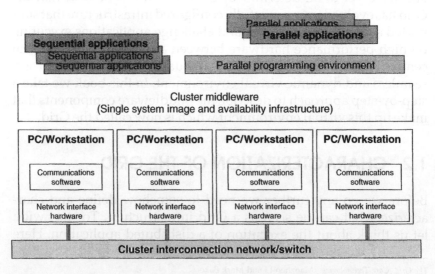

Figure 1.1 The hardware and software components of a typical cluster

1.2 CHARACTERIZATION OF THE GRID

distributed network of high-performance computers, stored data, instruments, and collaboration environments shared across institutional boundaries" [5]. In 2001, Foster, Kesselman and Tuecke refined their definition of a Grid to "coordinated resource sharing and problem solving in dynamic, multi-institutional virtual organizations" [6]. This latest definition is the one most commonly used today to abstractly define a Grid.

Foster later produced a checklist [7] that could be used to help understand exactly what can be identified as a Grid system. He suggested that the checklist should have three parts to it. (The first part to check off is that there is coordinated resource sharing with no centralized point of control that the users reside within different administrative domains.) If this is not true, it is probably the case that this is not a Grid system. The second part to check off is the use of standard, open, general-purpose protocols and interfaces. If this is not the case it is unlikely that system components will be able to communicate or interoperate, and it is likely that we are dealing with an application-specific system, and not the Grid. The final part to check off is that of delivering non-trivial qualities of service. Here we are considering how the components that make up a Grid can be used in a coordinated way to deliver combined services, which are appreciably greater than the sum of the individual components. These services may be associated with throughput, response time, meantime between failure, security or many other facets.

From a commercial view point, IBM define a grid as "a standards-based application/resource sharing architecture that makes it possible for heterogeneous systems and applications to share, compute and storage resources transparently" [8].

So, overall, we can say that the Grid is about resource sharing; this includes computers, storage, sensors and networks. Sharing is obviously always conditional and based on factors like trust, resource-based policies, negotiation and how payment should be considered. The Grid also includes coordinated problem solving, which is beyond simple client–server paradigm, where we may be interested in combinations of distributed data analysis, computation and collaboration. The Grid also involves dynamic, multi-institutional Virtual Organizations (VOs), where these new communities overlay classical organization structures, and these virtual organizations may be large or small, static or dynamic. The LHC Computing Grid Project at CERN [9] is a classic example of where VOs are being used in anger.

1.3 GRID-RELATED STANDARDS BODIES

For Grid-related technologies, tools and utilities to be taken up widely by the community at large, it is vital that developers design their software to conform to the relevant standards. For the Grid community, the most important standards organizations are the Global Grid Forum (GGF) [10], which is the primary standards setting organization for the Grid, and OASIS [11], a not-for-profit consortium that drives the development, convergence and adoption of e-business standards, which is having an increasing influence on Grid standards. Other bodies that are involved with related standards efforts are the Distributed Management Task Force (DMTF) [12], here there are overlaps and on-going collaborative efforts with the management standards, the Common Information Model (CIM) [13] and the Web-Based Enterprise Management (WBEM) [14]. In addition, the World Wide Web Consortium (W3C) [15] is also active in setting Web services standards, particularly those that relate to XML.

The GGF produces four document types related to standards that are defined as:

- *Informational*: These are used to inform the community about a useful idea or set of ideas, for example GFD.7 (A Grid Monitoring Architecture), GFD.8 (A Simple Case Study of a Grid Performance System) and GFD.11 (Grid Scheduling Dictionary of Terms and Keywords). There are currently eighteen Informational documents from a range of working groups.
- *Experimental*: These are used to inform the community about a useful experiment, testbed or implementation of an idea or set of ideas, for example GFD.5 (Advanced Reservation API), GFD.21 (GridFTP Protocol Improvements) and GFD.24 (GSS-API Extensions). There are currently three Experimental documents.
- *Community practice*: These are to inform the community of common practice or process, with the objective to influence the community, for example GFD.1 (GGF Document Series), GFD.3 (GGF Management) and GFD.16 (GGF Certificate Policy Model). There are currently four Common Practice documents.
- *Recommendations*: These are used to document a specification, analogous to an Internet Standards track document, for example GFD.15 (Open Grid Services Infrastructure), GFD.20 (GridFTP:

Protocol Extensions to FTP for the Grid) and GFD.23 (A Hierarchy of Network Performance Characteristics for Grid Applications and Services). There are currently four Recommendation documents.

1.4 THE ARCHITECTURE OF THE GRID

Perhaps the most important standard that has emerged recently is the Open Grid Services Architecture (OGSA), which was developed by the GGF. OGSA is an Informational specification that aims to define a common, standard and open architecture for Grid-based applications. The goal of OGSA is to standardize almost all the services that a grid application may use, for example job and resource management services, communications and security. OGSA specifies a Service-Oriented Architecture (SOA) for the Grid that realizes a model of a computing system as a set of distributed computing patterns realized using Web services as the underlying technology. Basically, the OGSA standard defines service interfaces and identifies the protocols for invoking these services.

OGSA was first announced at GGF4 in February 2002. In March 2004, at GGF10, it was declared as the GGF's flagship architecture. The OGSA document, first released at GGF11 in June 2004, explains the OGSA Working Group's current thinking on the required capabilities and was released in order to stimulate further discussion. Instantiations of OGSA depend on emerging specifications (e.g. WS-RF and WS-Notification). Currently the OGSA document does not contain sufficient information to develop an actual implementation of an OSGA-based system. A comprehensive analysis of OGSA was undertaken by Gannon *et al.*, and is well worth reading [16].

There are many standards involved in building a service-oriented Grid architecture, which form the basic building blocks that allow applications execute service requests. The Web services-based standards and specifications include:

- Program-to-program interaction (SOAP, WSDL and UDDI);
- Data sharing (eXtensible Markup Language – XML);
- Messaging (SOAP and WS-Addressing);
- Reliable messaging (WS-ReliableMessaging);

- Managing workload (WS-Management);
- Transaction-handling (WS-Coordination and WS-AtomicTransaction);
- Managing resources (WS-RF or Web Services Resource Framework);
- Establishing security (WS-Security, WS-SecureConversation, WS-Trust and WS-Federation);
- Handling metadata (WSDL, UDDI and WS-Policy);
- Building and integrating Web Services architecture over a Grid (see OGSA);
- Overlaying business process flow (Business Process Execution Language for Web Services – BPEL4WS);
- Triggering process flow events (WS-Notification).

As the aforementioned list indicates, developing a solid and concrete instantiation of OGSA is currently difficult as there is a moving target – as the choice of which standard or specification will emerge and/or become popular is unknown. This is causing the Grid community a dilemma as to exactly what route to use to develop their middleware. For example, WS-GAF [17] and WS-I [18] are being mooted as possible alternative routes to WS-RF [19].

Later in this book (Chapters 2 and 3), we describe in depth what is briefly outlined here in Sections 1.2–1.4.

1.5 REFERENCES

[1] Ian Foster and Carl Kesselman (eds), *The Grid: Blueprint for a New Computing Infrastructure*, 1st edition, Morgan Kaufmann Publishers, San Francisco, USA (1 November 1998), ISBN: 1558604758.
[2] Smarr, L. and Catlett, C., Metacomputing, *Communication of the ACM*, 35, 1992, pp. 44–52, ISSN: 0001-0782.
[3] De Roure, D., Baker, M.A., Jennings, N. and Shadbolt, N., The Evolution of the Grid, in *Grid Computing: Making the Global Infrastructure a Reality*, Fran Berman, Anthony J.G. Hey and Geoffrey Fox (eds), pp. 65–100, John Wiley & Sons, Chichester, England (8 April 2003), ISBN: 0470853190.
[4] CCA, http://www.extreme.indiana.edu/ccat/glossary.html.
[5] IPG, http://www.ipg.nasa.gov/ipgflat/aboutipg/glossary.html.
[6] Foster, I., Kesselman, C. and Tuecke, S., The Anatomy of the Grid: Enabling Scalable Virtual Organizations, *International Journal of Supercomputer Applications*, 15(3), 2001.
[7] Grid Checklist, http://www.gridtoday.com/02/0722/100136.html.

1.5 REFERENCES

[8] IBM Grid Computing, http://www-1.ibm.com/grid/grid_literature.shtml.
[9] LCG, http://lcg.web.cern.ch/LCG/.
[10] GGF, http://www.ggf.org.
[11] OASIS, http://www.oasis-open.org.
[12] DMTF, http://www.dmtf.org.
[13] CIM, http://www.dmtf.org/standards/cim.
[14] WBEM, http://www.dmtf.org/standards/wbem.
[15] W3C, http://www.w3.org.
[16] Gannon, D., Chiu, K., Govindaraju, M. and Slominski, A., A Revised Analysis of the Open Grid Services Infrastructure, *Journal of Computing and Informatics*, 21, 2002, 321–332, http://www.extreme.indiana.edu/~aslom/papers/ogsa_analysis4.pdf.
[17] WS-GAF, http://www.neresc.ac.uk/ws-gaf.
[18] WS-I, http://www.ws-i.org.
[19] WS-RF, http://www.globus.org/wsrf.

Part One
System Infrastructure

2
OGSA and WSRF

LEARNING OBJECTIVES

In this chapter we will study the Open Grid Services Architecture (OGSA) and the Web Services Resource Framework (WSRF). From this chapter you will learn:

- What is OGSA, and what role it will play with the Grid?
- What is the Open Grid Services Infrastructure (OGSI)?
- What are Web services technologies?
- Traditional paradigms for constructing Client/Server applications.
- What is WSRF and what impact will WSRF have on OGSA and OGSI?

CHAPTER OUTLINE

2.1 Introduction
2.2 Traditional Paradigms for Distributed Computing
2.3 Web Services
2.4 OGSA
2.5 The Globus Toolkit 3 (GT3)

The Grid: Core Technologies Maozhen Li and Mark Baker
© 2005 John Wiley & Sons, Ltd

2.6 OGSA-DAI
2.7 WSRF
2.8 Chapter Summary
2.9 Further Reading and Testing

2.1 INTRODUCTION

The Grid couples disparate and distributed heterogeneous software and hardware resources to provide a uniform computing environment for scientists and engineers to solve data and computation-intensive problems. Because of the heterogeneity of the Grid, the Global Grid Forum (GGF) [1] has been organized as a working body for designing standards for the Grid.

Globus [2] Toolkit 2 (GT2) and earlier versions have been widely used for building pre-OGSA oriented Grid systems. However, Grid systems based on Globus at this stage are heterogeneous in nature because these Grid systems are developed with heterogeneous protocols, which make it hard for them to interoperate. With the parallel development of GT2, Web services [3], as promoted by IBM, Microsoft, Sun Microsystems and many other Information Technology (IT) players, are emerging as a promising computing platform for building distributed business related applications in a heterogeneous environment.

At the GGF4 meeting in February 2002, the Globus team and IBM proposed a first OGSA specification [4] to merge the efforts of Globus and Web services. OGSA was proposed as the architecture for building the next generation of service-oriented Grid systems in a standard way. A working group in GGF has also been organized, called OGSA-WG [5], to work on the design of the OGSA specification. This was an important step and represented a significant milestone in the evolution of the Grid. OGSA is based on Web services, which use standard protocols such as XML and HTTP for building service-oriented distributed systems. OGSA introduces the concept of Grid services, which are Web services with some extensions to meet the specific need of the Grid. OGSA defines various aspects related to Grid services, e.g. what kind of features a Grid service should have and the life cycle management of Grid services. However, OGSA merely defines what interfaces are needed, but does not specify how these interfaces should be implemented. Another working group in GGF has been organized, called OGSI-WG [6], to work on OGSI, a technical specification for the implementation of Grid services as proposed in the OGSA specification in the context of Web services. Based on the OGSI technical specification, Globus Toolkit Version 3 (GT3) has been implemented and released as a toolkit for building OGSA compliant service-oriented Grid systems.

Standard Web services are persistent and stateless; OGSI compliant Grid services, however, can be transient and are stateful. The Web services community has recently criticized the work on the extension of standard Web services in OGSI mainly because the OGSI specification is too heavy with everything in one specification, and it does not work well with existing Web services and XML tooling. In January 2004, the Globus Alliance and IBM in conjunction with HP introduced the WSRF [7] to resolve this issue. WSRF is emerging as a promising standard for modelling stateful resources with Web services.

This chapter is organized as follows. In Section 2.2, we give a review on traditional paradigms for building distributed client/server applications. In Section 2.3, we present Web services and describe their core technologies. In Section 2.4, we introduce OGSA and describe the concepts of Grid services in the context of Web services. In Section 2.5, we present GT3 which has been widely deployed for building service-oriented Grid systems. In Section 2.6, we present OGSA-DAI which defines Grid data services for data access and integration on the Grid. In Section 2.7, we present WSRF and its concepts. The impacts of WSRF on OGSI and OGSA will also be discussed in this section. In Section 2.8 we conclude this chapter, and in Section 2.9 we give further readings and testing.

2.2 TRADITIONAL PARADIGMS FOR DISTRIBUTED COMPUTING

In this section, we review traditional computing paradigms for building distributed client/server applications. Figure 2.1 shows a simplistic sketch of the possible traditional client/server architecture using a variety of communication techniques such as sockets, Remote Procedure Calls (RPC) [8], Java Remote Method Invocation

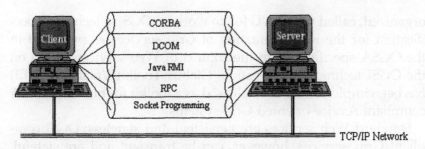

Figure 2.1 Traditional paradigms for distributed computing

(RMI) [9], Distributed Component Object Model (DCOM) [10] and Common Object Request Broker Architecture (CORBA) [11]. In the following sections, we give a brief overview of each technique.

2.2.1 Socket programming

Sockets provide a low-level API for writing distributed client/server applications. Before a client communicates with a server, a socket endpoint needs to be created. The transport protocol chosen for communications can be either TCP or UDP in the TCP/IP protocol stack. The client also needs to specify the hostname and port number that the server process is listening on. The standard socket API is well-defined, however the implementation is language dependant. So, this means socket-based programs can be written in any language, but the socket APIs will vary with each language use. Typically, the socket client and server will be implemented in the same language and use the same socket package, but can run on different operating systems (i.e. in the Java case).

As mentioned above, socket programming is a low-level communication technique, but has the advantage of a low latency and high-bandwidth mechanism for transferring large amount of data compared with other paradigms. However, sockets are designed for the client/server paradigm, and today many applications have multiple components interacting in complex ways, which means that application development can be an onerous and time-consuming task. This is due to the need for the developer to explicitly create, maintain, manipulate and close multiple sockets.

2.2.2 RPC

RPC is another mechanism that can be used to construct distributed client/server applications. RPC can use either TCP or UDP for its transport protocol. RPC relies heavily on an Interface Definition Language (IDL) interface to describe the remote procedures executing on the server-side. From an RPC IDL interface, an RPC compiler can automatically generate a client-side stub and a server-side skeleton. With the help of the stub and skeleton, RPC hides the low-level communication and provides a high-level communication abstraction for a client to directly call a remote procedure as if the procedure were local. RPC itself is a specification and implementations such as Open Network Computing (ONC) RPC [12] from Sun Microsystems and Distributed Computing Environment (DCE) RPC [13] from the Open Software Foundation (OSF) can be used directly for implementing RPC-based client/server applications.

RPC is not restricted to any specific language, but most implementations are in C. An RPC client and server have to be implemented in the same language and use the same RPC package. When communicating with a server, a client needs to specify the hostname or the IP address of the server. Figure 2.2 shows the data-flow control in an RPC-based client/server application.

Compared with socket programming, RPC is arguably easier to use for implementing distributed applications. However, RPC

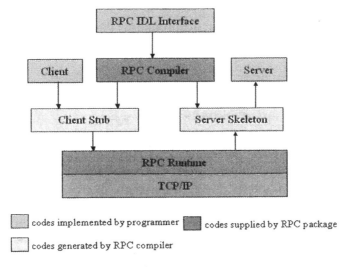

Figure 2.2 Data-flow control in an RPC application

only supports synchronous communication (call/wait) between the client and server; here the client has to wait until it receives a response from the server. In addition, RPC is not object-oriented. The steps to implement and run a client/server application with RPC are:

- Write an RPC interface in RPC IDL;
- Use an RPC compiler to compile the interface to generate a client-side stub and a server-side skeleton;
- Implement the server;
- Implement the client;
- Compile all the code with a RPC library;
- Start the server;
- Start the client with the IP address of the server.

2.2.3 Java RMI

The Java RMI is an object-oriented mechanism from Sun Microsystems for building distributed client/server applications. Java RMI is an RPC implementation in Java. Similar to RPC, Java RMI hides the low-level communications between client and server by using a client-side stub and a server-side skeleton (which is not needed in Java 1.2 or later) that are automatically generated from a class that extends *java.rmi.UnicastRemoteObject* and implements an RMI *Remote* interface.

At run time there are three interacting entities involved in an RMI application. These are:

- A client that invokes a method on a remote object.
- A server that runs the remote object which is an ordinary object in the address space of the server process.
- The object registry (*rmiregistry*), which is a name server that relates objects with names. Remote objects need to be registered with the registry. Once an object has been registered, the registry can be used to obtain access to a remote object using the name of that object.

Java RMI itself is both a specification and an implementation. Java RMI is restricted to the Java language in that an RMI client and server have to be implemented in Java, but they can run on

2.2 TRADITIONAL PARADIGMS FOR DISTRIBUTED COMPUTING

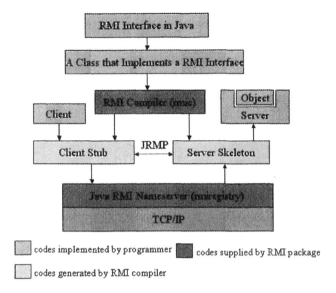

Figure 2.3 Data-flow control in a Java RMI application

different operating systems in distributed locations. When communicating with a server, an RMI client has to specify the server's hostname (or IP address) and use the Java Remote Method Protocol (JRMP) to invoke the remote object on the server. Figure 2.3 shows the data-flow control in a Java RMI client/server application.

Java RMI uses an object-oriented approach, compared to the procedural one that RPC uses. A client can pass an object as a parameter to a remote object. Unlike RPC which needs an IDL interface, a Java RMI interface is written in Java. RMI has good support for marshalling, which is a process of passing parameters from client to a remote object, i.e. a *Serializable* Java object can be passed as a parameter. The main drawbacks of Java RMI are its limitation to the Java language, its proprietary invocation protocol-JRMP, and it only supports synchronous communications.

The steps to implement and run a Java RMI client/server application are:

- Write an RMI interface;
- Write an RMI object to implement the interface;
- Use RMI compiler (*rmic*) to compile the RMI object to generate a client-side stub and an server-side skeleton;
- Write an RMI server to register the RMI object;

- Write an RMI client;
- Use Java compiler (*javac*) to compile all the Java source codes;
- Start the RMI name server (*rmiregistry*);
- Start the RMI server;
- Start the RMI client.

2.2.4 DCOM

The Component Object Model (COM) is a binary standard for building Microsoft-based component applications, which is independent of the implementation language. DCOM is an extension to COM for distributed client/server applications. Similar to RPC, DCOM hides the low-level communication by automatically generating a client-side stub (called proxy in DCOM) and a server-side skeleton (called stub in DCOM) using Microsoft's Interface Definition Language (MIDL) interface. DCOM uses a protocol called the Object Remote Procedure Call (ORPC) to invoke remote COM components. The ORPC is layered on top of the OSF DCE RPC specification. Figure 2.4 shows the data-flow control in a client/server application with DCOM.

DCOM is language independent; clients and DCOM components can be implemented in different languages. Although DCOM is available on non-Microsoft platforms, it has only achieved broad popularity on Windows. Another drawback of DCOM is that it

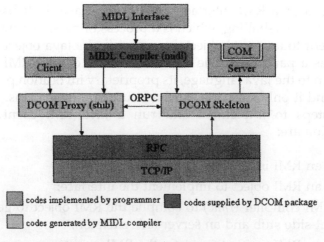

Figure 2.4 Data-flow control in a DCOM application

only supports synchronous communications. The steps to implement and run a DCOM client/server application are:

- Write an MIDL interface;
- Use an interface compiler (*midl*) to compile the interface to generate a client-side stub and a server-side skeleton;
- Write the COM component to implement the interface;
- Write a DCOM client;
- Compile all the codes;
- Register the COM component with a DCOM server;
- Start the DCOM server;
- Start the DCOM client.

2.2.5 CORBA

CORBA is an object-oriented middleware infrastructure from Object Management Group (OMG) [14] for building distributed client/server applications. Similar to Java RMI and DCOM, CORBA hides the low-level communication between the client and server by automatically generating a client-side stub and a server-side skeleton through an Interface Definition Language (IDL) interface. CORBA uses Internet-Inter ORB Protocol (IIOP) to invoke remote CORBA objects. The Object Request Broker (ORB) is the core of CORBA; it performs data marshaling and unmarshalling between CORBA clients and objects. Figure 2.5 shows the dataflow control in a client/server application using the CORBA.

Compared with Java RMI and DCOM, CORBA is independent of location, a particular platform or programming language. CORBA supports both synchronous and asynchronous communications. CORBA has an advanced directory service called COSNaming, which provides the mechanisms to allow the transparent location of objects. However, CORBA itself is only an OMG specification. There are many CORBA products available that can be used to build CORBA applications. The steps to implement and run a CORBA client/server application are:

- Write a CORBA IDL interface;
- Use an IDL compiler to compile the interface to generate a client-side stub and a server-side skeleton;

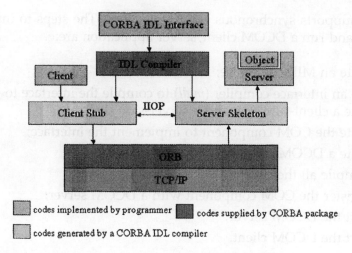

Figure 2.5 Data-flow control in a CORBA application

- Write a CORBA object to implement the interface;
- Write a CORBA server to register the CORBA object;
- Write a CORBA client;
- Compile all the source codes;
- Start a CORBA name server;
- Start the CORBA server;
- Start the CORBA client.

2.2.6 A summary on Java RMI, DCOM and CORBA

Java RMI, DCOM and CORBA have all been around for some time and represent the most popular distributed, object-oriented middleware which can be used to rapidly develop distributed client/server applications. Although they differ in their specific implementations and features [15], they all share the following features:

- An interface is needed for invoking a remote object or a component.
- The complexity of low-level communications is hidden from the users by automatically generating a client-side stub and a server-side skeleton via the interface definition.

- They use proprietary communication protocols – e.g. Java RMI uses JRMP, DCOM uses ORPC and CORBA uses IIOP – to invoke remote objects or components.
- The interface definition is in binary format. It is difficult for client applications to make a query on an interface, such as to find out what kinds of methods are defined, inputs/outputs of each method to make a better use of the methods.
- Clients and objects are tightly coupled with their interfaces. For example, changing a part of the client means the other parts, such as the server, also need modification.

In summary, middleware such as Java RMI, DCOM and CORBA are not based on open standards, which makes it difficult for them to be ubiquitously taken up in heterogeneous environments. Ideally, what is needed is an open standards-based middleware infrastructure for building and integrating applications in heterogeneous environments, and Web services are emerging as such an infrastructure.

2.3 WEB SERVICES

Web services are emerging as a promising infrastructure for building distributed applications. Web services are based on a Service-Oriented Architecture (SOA) in which clients are service requestors and servers are service providers. Web services differ from other approaches such as Java RMI, CORBA and DCOM in their focus on simple open standards such as XML and HTTP, which have wide industry support and a chance of becoming truly ubiquitous. Web services provide a stratum on top of other mechanisms, as shown in Figure 2.6. We define a Web service as given below.

> Essentially, a Web service is a loosely coupled, encapsulated, platform and programming language neutral, composable server-side component that can be described, published, discovered and invoked over an internal network or on the Internet.

The explanation of the definition is given below:

- *Loosely coupled*: A Web service implementation is free to change without unduly impacting the service client as long as the service interface remains the same.

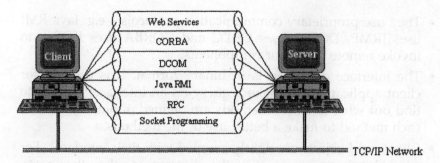

Figure 2.6 Paradigms including Web services for building distributed applications

- *Encapsulated*: The implementation of a Web service is completely invisible to the client of a service.
- *Platform and programming language neutral*: A Web service can be implemented in any language and deployed on any platform.
- *Composable*: A Web service can be composed from a number of deployed services.
- *Server-side component*: A Web service can range in scope from a complete application to a subroutine that runs on a server.
- *Described*: An XML-based interface is used to describe the functionality and capabilities that a Web service can provide.
- *Published*: A Web service can be registered with a service registry that can be accessed on an intranet or on the Internet.
- *Discovered*: A Web service client can discover a service by searching a service registry and match their service requirements.
- *Invoked*: A Web service can be bound to by a service client via standard transport protocols such as HTTP or FTP.
- *Internal network or the Internet*: A Web service can be made available strictly within an organization or it can be offered across the firewall, available to any consumer connected to the Internet.

The core standards of Web services, as defined by W3C consortium, are SOAP [16], Web Services Description Language (WSDL) [17] and the Universal Description, Discovery and Integration (UDDI) [18]. Another standard for service discovery is the Web Services Inspection (WS-Inspection) specification [19] defined by IBM and Microsoft. The specification defines WS-Inspection Language (WSIL) for service description and discovery.

2.3.1 SOAP

SOAP is a simple and lightweight communication protocol for clients and servers to exchange messages in an XML format over a transport-level protocol, which is normally HTTP. From Figure 2.7 we can see that a SOAP message is encapsulated in an envelope that consists of the following four parts:

- Various namespaces are used by the SOAP message, typically these include xmlns:SOAP-ENV (SOAP Envelope), xmlns:xsi (XML Schema for Instance) and xmlns:xsd (XML Schema Definition).
- A set of encoding rules for expressing instances of application-defined data types.
- An optional header for carrying auxiliary information for authentication, transactions and payments.
- The Body is the main payload of the message. When an RPC call is used in the SOAP message, the Body has a single element that contains the method name, arguments and a Uniform Resource Identifier (URI) of the service target address. In addition, the fault entry can be used to explain a failure.

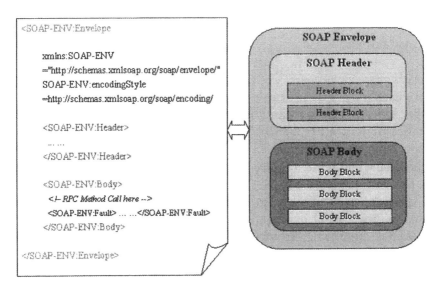

Figure 2.7 The structure of a SOAP message

SOAP is independent of the underlying transport protocol, so SOAP messages can be carried over many transport-level protocols such as HTTP, FTP, SMTP or more sophisticated protocols such as Java RMI JRMP or CORBA IIOP. HTTP is the most commonly used protocol because it can normally pass firewalls. Since XML is a universal standard, clients and servers built on different platforms can communicate with SOAP.

2.3.2 WSDL

WSDL is an XML-based specification that is used to completely describe a Web service, e.g. what a service can do, where it resides and how to invoke it. A WSDL interface is similar to a CORBA IDL or a DCOM MIDL interface, but with richer semantics to describe a service. WSDL defines services as a set of network endpoints or *ports* using an RPC-based mechanism or a document-oriented message exchange for the communication between a service requestor and provider. An RPC-oriented operation is one in which the SOAP messages contain parameters and return values, and a document-oriented operation is one in which the SOAP messages contain XML documents. The communication in RPC-based message exchanging is synchronous, but the communication in Document-oriented message exchanging is often asynchronous.

The common elements in WSDL, as shown in Figure 2.8, are explained below.

Data types
The *data types* part encloses data type definitions that are relevant for message exchanging. For maximum interoperability and platform neutrality, WSDL uses XML XSD as the default data type. This part is *extensible*, meaning that it can contain arbitrary subsidiary elements to allow general data types to be constructed.

The XSD namespace can be used to define the data types in a message regardless of whether or not the resulting message exchanging format is actually XML, or the resulting XSD schema validates the particular wire format.

<message>
The <message> element defines the data elements of an operation in a service. Each message can consist of one or more parts. The

2.3 WEB SERVICES

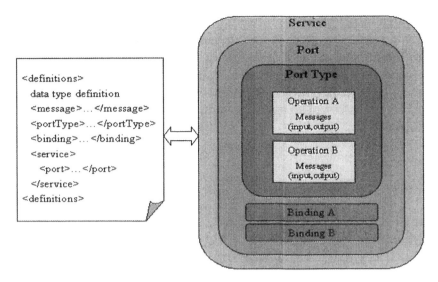

Figure 2.8 The structure of a WSDL document

parts are similar to the parameters of a function or method call in a traditional programming language.

<portType>
<portType> is the core part of a WSDL document. Similar to a Java interface or a C++ class, it defines a set of abstract operations provided by a service. Each operation uses messages defined in the *<message>* element to describe its inputs and outputs.

<binding>
<binding> identifies a concrete protocol and data format for the operations and messages defined by a particular *<portType>*. There may be an arbitrary number of bindings for a given portType, i.e. a binding can be document-oriented or use RPC. SOAP over HTTP is the most commonly used mechanism for transmitting messages between a service client and a service itself.

<port>
A *<port>* defines an individual service endpoint by specifying a single address for a binding.

<service>
A *<service>* is a set of related ports. Ports within a service have the following relationship:

- None of the ports communicate with each other.
- If a service has several ports that share a <portType>, but employ different bindings or addresses, these are alternative ports where the port provides semantically equivalent behaviour. This allows a consumer of a WSDL document to choose particular port(s) to communicate with, based on some criteria (such as a protocol or distance).

2.3.3 UDDI

The UDDI is an industry standard for service registration (publication) and discovery. A service provider uses UDDI to advertise the services that it is making available. A client uses UDDI to find the appropriate service(s) for its purposes. A UDDI registry is similar to a CORBA trader service, or it can be thought of as a Domain Name Server (DNS) service for business applications. A UDDI registry has two kinds of players: businesses that want to publish a service, and clients who want to obtain services of a certain kind, and then use them via some binding process. Data in UDDI can be organized in the following ways:

- *White pages*: This includes general information about a service provider, such as its name, contact information and other identifiers.
- *Yellow pages*: This information describes a Web service using different categorizations (taxonomies) and allows others to discover a Web service based on its categorization (such as car manufacturing or car sales business).
- *Green pages*: Green pages have technical information about a Web service, usually with a reference to an external WSDL document of the service, enabling the client to know how to interact with the service.

UDDI is layered over SOAP, which means that a client uses SOAP to access a UDDI registry. A UDDI registry exposes a set of APIs in the form of SOAP-based Web services. The API contains Inquiry and Publishing APIs for services discovery and service publication.

2.3.4 WS-Inspection

WS-Inspection is similar in scope to UDDI; it is a complementary rather than a competitive technology. It allows service description information to be distributed to any location using a simple extensible XML document format. WS-Inspection does not concern itself with business entity information (whereas UDDI does). It works under the assumption that a service client is aware of the services provided by the service provider.

The WS-Inspection specification mainly provides the following two functions:

- It defines an XML format for listing references to existing service descriptions.
- It defines a set of conventions so that it is easy to locate WS-Inspection documents.

In WS-Inspection, Web services are described in WS-Inspection documents. A WS-Inspection document provides a means for aggregating references to pre-existing service description documents which have been authored in arbitrary number of formats such as WSDL, UDDI or plain HTML. A WS-Inspection document is generally made available at the point-of-offering for the services that are referenced within the document. Within a WS-Inspection document, a single service can have more than one reference to a service description. A service description is usually a URL that points to a WSDL document; occasionally, a service description can be a reference to an entry within a UDDI registry. With WS-Inspection, a service provider creates a WS-Inspection document and makes the document network accessible. Service requestors use standard Web-based access mechanisms (e.g. HTTP GET) to retrieve this document and discover what services the provider is advertising. Figure 2.9 shows an example of WS-Inspection document.

This example contains a reference to two service descriptions and a single reference to another WS-Inspection document. The first <service> element contains one service description, which is a reference to a WSDL document. The second <service> element also contains one service description reference to a business service entry in a UDDI registry. The UDDI service key identifies one unique business service. The <link> element is used to reference

```xml
<?xml version="1.0"?>
<inspection xmlns="http://schemas.xmlsoap.org/ws/2001/10/inspection/">
  <service>
    <description
        referencedNamespace="http://schemas.xmlsoap.org/wsdl/"
        location="http://example.com/exampleservice.wsdl"/>
  </service>

  <service>
    <description
        referencedNamespace="urn:uddi-org:api">
        <wsiluddi:serviceDescription location=
                 "http://example.com/uddi/inquiryapi">
            <wsiluddi:serviceKey>
                  2946BB0-BC28-11D5-A432-0004AC49CC1E
            </wsiluddi:serviceKey>
        </wsiluddi:serviceDescription>
    </description>
  </service>

  <link referencedNamespace=
            "http://schemas.xmlsoap.org/ws/2001/10/inspection/"
            location="http://example.com/tools/toolservices.wsil"/>
</inspection>
```

Figure 2.9 An example WS-Inspection document

a collection of service descriptions. In this case, it is referencing another WS-Inspection document.

WS-Inspection supports a completely distributed model for providing service-related information; the service descriptions may be stored at any location, and requests to retrieve the information are generally made directly to the entities that are offering the services.

2.3.5 WS-Inspection and UDDI

As mentioned in Nagy and Ballinger [20], the UDDI and WS-Inspection specifications address different sets of issues with service registration and discovery, which are characterized by different trade-offs. UDDI provides a high degree of functionality, but it comes at the cost of increased complexity. The WS-Inspection specification provides less functionality in order to maintain a low overhead. With this in mind, the two specifications should be viewed as complementary technologies, to be used either together or separately depending upon the situation. For example, a UDDI

2.3 WEB SERVICES

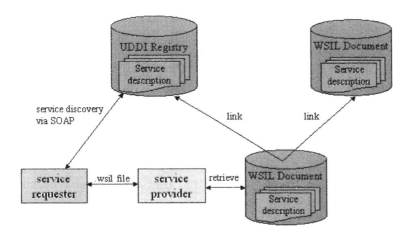

Figure 2.10 Service discovery with UDDI and WS-Inspection

registry could be populated by the results from robot crawling WS-Inspection documents on the Web. Likewise, a UDDI registry may itself be discovered when a requestor retrieves a WS-Inspection document, which refers to an entry in the UDDI registry as shown in Figure 2.10. In environments where the additional functionality afforded by UDDI is not required and where constraints do not allow for its deployment, the WS-Inspection mechanism may provide all of the capabilities that are needed. In situations where data needs to be centrally managed, a UDDI solution alone may provide the best solution.

2.3.6 Web services implementations

Web services are based on a set of specifications. Currently there are many Web services implementations that can be used to build distributed applications. There are three aspects that need to be considered when using Web services.

- A *programming model* specifies how to write client codes to access Web services, how to write service implementations, how to handle other parts of the SOAP specification, such as headers and attachments.
- A *deployment model* is the framework used to deploy a service and provide a Web service deployment descriptor (a wsdd file) to map the implementation of the service to SOAP messages.

- A *SOAP Engine* receives SOAP messages and invokes Web service implementations.

In the following section, we describe three frameworks for implementing Web services applications – J2EE, .Net and Apache Axis.

2.3.6.1 J2EE

J2EE [21] is standard for developing, building and deploying Java-based applications. It can be used to build traditional Web sites, software components, or packaged applications. J2EE has recently been extended to include support for building XML-based Web services as well. J2EE provides the following APIs for Web services:

- The Java API for XML Processing (JAXP) – processes XML documents using various parsers.
- The Java Architecture for XML Binding (JAXB) – processes XML documents using schema-derived JavaBeans component classes.
- The Java API for XML-based RPC (JAX-RPC) – a standard for RPC. It provides APIs for XML RPC invocation and uses base-level protocol bindings with SOAP/HTTP, but is not limited to HTTP.
- The Java API for XML Messaging (JAXM) and SOAP with Attachments API for Java (SAAJ) – send SOAP messages over the Web in a standard way.
- The Java API for XML Registries (JAXR) – provides a standard way to interact with business UDDI registries.

Figure 2.11 shows data-control flow for a client to invoke a Web service with J2EE JAX-RPC.

2.3.6.2 Apache Axis

Apache Axis [22] is a SOAP engine that can be used to exchange SOAP messages between clients and services. It also provides support for WSDL operations, e.g. Java2WSDL can be used to generate a WSDL document from a Java interface, and WSDL2Java can be used to generate a client-side stub and a server-side skeleton based on the WSDL document. Axis does not provide support for service discovery and publication. UDDI4Java [23] from IBM can be used

2.3 WEB SERVICES

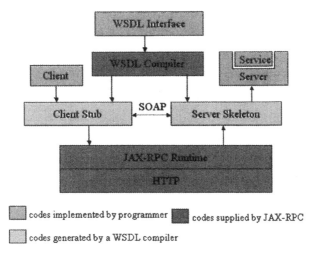

Figure 2.11 Data-flow control in invoking a Web service with J2EE JAX-RPC

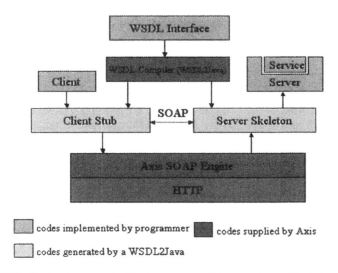

Figure 2.12 Data-flow control in invoking a Web service with Axis

together with Axis for this purpose. Figure 2.12 shows data-flow control for a client to invoke a Web service with Axis.

A Web service application with Axis can be implemented and started as follows:

- Write a Java interface;
- Use Java2WSDL to compile the interface to generate a WSDL interface;

- Use WSDL2Java to compile the WSDL interface to generate a client-side stub and a server-side skeleton;
- Write a service to implement the WSDL interface;
- Write a client;
- Compile all the codes with *javac* compiler;
- Write a Web service deployment descriptor (a wsdd file) to deploy the service in Jakarta Tomcat Web server [24];
- Start Tomcat;
- Start the client to invoke the service.

2.3.6.3 Microsoft .Net

.Net is a Microsoft Platform [25] for building Web services applications. Similar to a J2EE Web service, a .NET Web service supports the WSDL specification and uses a WSDL document to describe itself. However, an XML namespace has to be used within a WSDL file to uniquely identify the Web service's endpoint.

.Net provides a client-side component that lets a client invoke a Web service described by WSDL. It also provides a server-side component that maps Web service operations to a COM-object method call as described by the WSDL interface and a Web Services Meta Language (WSML) file, which is needed for Microsoft's implementation of SOAP. Web services can be published using DISCO [26] files or via a UDDI registry. DISCO is a Microsoft publishing/discovery technology built into .NET.

.Net provides a UDDI software development kit (SDK) for discovery of Web services. With regard to the invocation of Web services, .Net provides three choices:

- Use the built-in .Net SOAP message classes.
- Construct a Web service listener manually, using for example, Microsoft XML Parser (MSXML), Active Server Pages (ASP) or Internet Server Application Programming Interface (ISAPI).
- Use the Microsoft SOAP Toolkit 2.0 to build a Web service listener that communicates with a service implemented with COM. The toolkit can generate a client-side stub from a WSDL interface that can be used by a client to communicate with the service.

2.3.7 How Web services benefit the Grid

Web services are emerging as an XML-based open standard for building distributed applications in a heterogeneous computing environment. Web services are independent of platforms, programming languages and locations. Web services can be described, published and dynamically discovered and bound to WSDL, a rich interface description language. The technologies associated with Web services provide a promising platform for integrating services provided by heterogeneous systems. Figure 2.13 shows the architecture of Web services and how it makes use of its core technologies. First, a service provider publishes its services into a UDDI registry with SOAP. Then a service requestor (client) searches the registry to find services of interest. Finally the client requests a service by binding to the service.

The Grid can benefit from the Web services framework by taking advantage of the following factors:

- The Grid requires the support for the dynamic discovery and composition of Grid services in heterogeneous environments; this necessitates mechanisms for registering and discovering interface definitions and endpoint implementation descriptions; for dynamically generating proxies based on (potentially multiple) bindings of specific interfaces. WSDL supports this requirement by providing a standard mechanism for defining interface

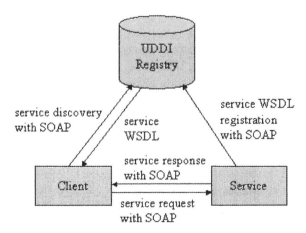

Figure 2.13 Web services core technologies

definitions separately from their embodiment within a particular binding (transport protocol and data encoding format).
- Web services technologies are based on internationally recognized standards. The widespread adoption means that a framework based on Web services will be able to exploit numerous tools and extended services, such as WSDL processors that can generate bindings for a variety of environments, e.g. Web Services Invocation Framework (WSIF) [27], workflow systems that utilize WSDL, and hosting environments for Web services (e.g. Microsoft .NET and Apache Axis).

2.4 OGSA

OGSA is a *de facto* standard for building the next generation of service-oriented Grid systems. The GGF is currently coordinating a worldwide effort to complete the OGSA specification. OGSA is based on Web services technologies, but with some extensions. OGSA extends Web services by introducing interfaces and conventions in three main areas.

- First, there is a dynamic and potentially transient nature of services in a Grid environment, in which particular service instances may come and go as work is dispatched, as resources are configured and provisioned, and as system state changes. Therefore, Grid services need interfaces to manage their creation, destruction and life cycle management.
- Second, there is state. Grid services can have attributes and data associated with them. This is similar in concept to the traditional structure of objects in object-oriented programming. Objects have behaviour and data. Likewise, Web services needed to be extended to support state data associated with Grid services.
- Third, clients can subscribe their interests in services. Once there is any change in a service, the clients are notified. This is a call-back operation from services to clients.

As shown in Figure 2.14, Grid applications can be built from OGSA compliant services. Services in OGSA are composed of two parts, OGSA platform and core services. The OGSA platform services are Grid-based services related to user authentication and authorization, fault tolerance, job submission, monitoring and data

2.4 OGSA

Figure 2.14 Building OGSA compliant Grid applications with OGSI

access. The core services in OGSA mainly include service creation, destruction, life cycle management, service registration, discovery and notification. OGSA introduces Grid service interfaces such as *GridService, Factory, Registration, HandleResolver* and *Notification* to support its core services. OGSA introduces the concepts of service instance and service data associated with each Grid service to support transient and stateful Grid services. In addition, the notification model in OGSA allows services to notify subscribed clients about the events they are interested in.

OGSA defines various aspects related to a Grid service, e.g. the features of Grid services, and what interfaces are needed; but it does not specify how these interfaces should be implemented. That is the task of OGSI, which is a technical specification to specify how to implement the core Grid services as defined in OGSA in the context of Web services, specifically WSDL. The OGSI specifies exactly what needs to be implemented to conform to OGSA. Therefore, a Grid services can be defined as an OGSI compliant Web service. An OGSA compliant service can be defined as any OGSI compliant service whose interface has been defined by OGSA to be a standard OGSA service interface.

2.4.1 Service instance semantics

While standard Web services are persistent, Grid services can be transient. OGSA provides a soft-service management by introducing the concept of Grid service instances. A Grid service instance is an instantiation of a Grid service that can be dynamically created and explicitly destroyed. A Grid service that can create a service instance is called a service factory, a persistent service itself. A client can request a factory to create many service instances and multiple clients can access the same service instance.

As shown in Figure 2.15, a user job submission can involve one or more Grid service instances, which are created from corresponding Grid service factories running on three nodes. The implementations of the three services are independent of location, platform and programming language.

A Grid Service Handle (GSH), a globally unique URI that distinguishes a specific Grid service instance from all other Grid service instances, identifies each Grid service instance. However, Grid services may be upgraded during their lifetime, for example to support a new protocol version or to add alternative protocols. Thus, the GSH carries no protocol- or instance-specific information such as a network address and supported protocol bindings. Instead, this information is encapsulated, along with all other instance-specific information required to interact with a specific service instance, into a single abstraction called a Grid Service Reference (GSR). Unlike a GSH, which is invariant, the GSR(s) for a Grid service instance can change over that service's lifetime. A GSR has an explicit expiration time, or may become invalid at any time during a service's lifetime, and OGSA defines mapping mechanisms for obtaining an updated GSR. The GSR format is specific

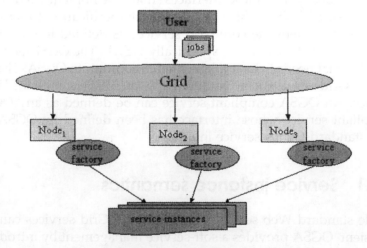

Figure 2.15 A user job submission involves multiple Grid service instances

2.4 OGSA

to the binding mechanism used by the client to communicate with the Grid service instance. For example, if the client uses a SOAP binding, the GSR assumes that an annotated WSDL document format will be used.

2.4.2 Service data semantics

In OGSA, apart from methods, each Grid service instance is also associated with *service data*, which is a collection of XML elements encapsulated as Service Data Elements (SDE). Service data are used to describe information about a service instance and their run-time states. Unlike standard Web services, which are stateless, Grid services are stateful and can be introspected. A client can use the standard FindServiceData() method defined in the *GridService* portType for querying and retrieving service data associated with a Grid service registered in a registry, i.e. the service type; if it is a service instance, the GSH of the service instance; the location of a service factory; and the run-time states. Figure 2.16 shows a hierarchical view in terms of service factory, service instances and service data.

A service factory can create many service instances, of which each has a Service Data Set. A Service Data Set can contain zero or multiple SDEs. Each SDE can be of a different type. As shown in Figure 2.16, the first instance has two "type A" SDEs and one

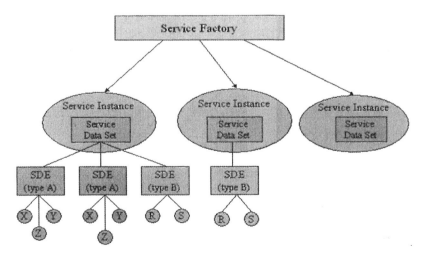

Figure 2.16 A hierarchical view of service factory, service data and service data elements

"type B" SDE. The second instance has only one "type A" SDE. The third instance has no SDEs at all (even so, it does have an *empty* Service Data Set). Notice how SDEs of the same type always contain the same information ("type A" has data X, Y, Z; "type B" has data R and S). The SDEs are the ones that actually contain the data (X, Y, Z, R, S).

2.4.3 OGSA portTypes

OGSA provides the following interfaces, which are extended WSDL portTypes, to define Grid services. In OGSA, the *GridService* interface must be implemented by all Grid services, while the other interfaces are optional. OGSA supports the following interfaces.

GridService portType
A Grid service must implement the *GridService* portType as it serves as the base interface in OGSA. This portType is analogous to the base Object class within object-oriented programming languages such as C++ or Java, in that it encapsulates the root behaviour of the component model. The three methods encapsulated by the *GridService* portType are FindServiceData(), SetTerminationTime() and Destroy() that are used for service discovery, introspection and soft-state life cycle management.

Factory portType
A factory is a persistent Grid service that implements the *Factory* portType. It can be used to create transient Grid service instances with its createService() method.

HandleResolver portType
A Grid service that implements the *HandleResolver* portType can be used to resolve a GSH to a GSR using its FindbyHandle() method.

Registration portType
A registry is a Grid service that implements the *Registration* portType to support service discovery by maintaining collections of GSHs and their associated policies. Clients can query a registry to discover services' availability, properties and policies. Two elements define a registry service – the registration interface, which allows a service instance to register a GSH with the registry service, and a set of associated service data, that contains information about the registered GSH and the run-time states of the service instance.

2.4 OGSA

RegisterService() and UnRegisterService() are the two methods defined in the portType for service registration and unregistration.

NotificationSource / NotificationSink portType

The OGSA notification model allows interested parties to subscribe to service data elements and receive notification events when their values are modified. A Grid service that implements the *NotificationSource* portType is called a notification source. A Grid service that implements the *NotificationSink* portType is called a notification sink. To subscribe notification to a particular Grid service, a notification sink invokes a notification source using the SubscribeToNotificationTopic() method in the *NotificationSource* interface, giving it the service GSH of the notification sink and the topics interested. A notification source will use the DeliverNotification() method in the *NotificationSink* interface to send a stream of notification messages to the sink, while the sink sends periodic messages to notify the source that it is still interested in receiving notifications. To ensure reliable delivery, a user can implement this behaviour by defining an appropriate protocol binding for the service.

As shown in Figure 2.17, a Grid service must implement the *GridService* interface, and may implement other interfaces such as *Factory*, *Registration*, *HandleResolver* and *NotificationSource/*

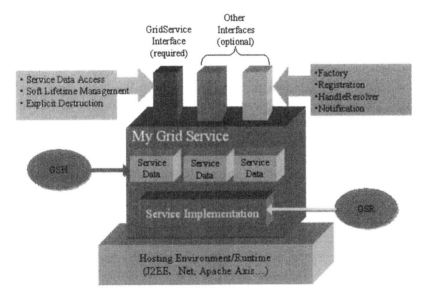

Figure 2.17 The structure of a Grid service in OGSA

NotificationSink. OGSA defines standard mechanisms for service creation, destruction, life cycle management, service registration, discovery and service notification. A Grid service can be a persistent service, or a transient service instance. Each Grid service has a unique GSH and one or more GSRs to refer to its implementation, which is independent of location, platform and programming language. A Grid service can be deployed in environments hosted by J2EE, .Net or Apache Axis.

2.4.4 A further discussion on OGSA

There is a long way for OGSA to go before there is complete architecture specification where all of the desired properties of the Grid are addressed. OGSA will have to be refined and adjusted iteratively, but this is a natural and healthy process, and the first very important step has been taken. It is felt that this step has been taken in the right direction and we hope that OGSA will be successful in its evolution into the open standards-based architecture that it sets out to define.

Kunszt [28] points out some issues remain to be resolved in the future development of OGSA with respect to aspects such as availability, robustness, scalability, measurability, interoperability, compatibility, service discovery, manageability and changeability. Dialani *et al*. [29] propose a transparent fault tolerance mechanism for Web services that may be used to support Grid services' fault tolerance. A service may become overloaded on a node at certain times if large numbers of users concurrently request a service instance to be created. Zhang *et al*. [30] propose the concept of a Grid mobility service that we feel may be a solution to the service-overloading problem. A service could be a mobile code that can move from node to node in a Grid environment. When necessary, a service can move to another more lightly loaded node to spawn service instances. However, practical work needs to be done to serve as proof-of-concept.

2.5 THE GLOBUS TOOLKIT 3 (GT3)

OGSI provides a technical specification for implementing Grid services defined in the OGSA specification. Currently OGSI implementations such as GT3 [31], MS.NETGrid [32], OGSI.NET [33],

2.5 THE GLOBUS TOOLKIT 3 (GT3)

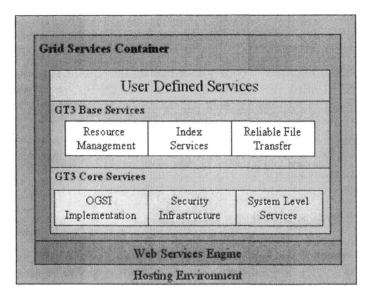

Figure 2.18 The GT3 structure

OGSI::Lite [34], PyOGSI [35] have been released. In this section, we introduce GT3 and describe the services provided by the GT3 framework.

As shown in Figure 2.18, a Grid services container in GT3 runs on top of a Web services engine such as the Apache Axis for service management. The functionality of each part in the structure is described below.

2.5.1 Host environment

A hosting environment is a specific execution environment that defines not only the programming model and language, but also the development and debugging tools that can be used to implement Grid services. It also defines how the implementation of a Grid service meets its obligations with respect to Grid service semantics. GT3 supports the following four Java hosting environments:

- *Embedded*: A library allowing an OGSI-hosting environment to be embedded in any existing J2SE applications.
- *Standalone*: A lightweight J2SE server that hosts Grid services.

- **J2EE Web container**: An OGSI hosting environment inside a Web server that can be hosted by any Java Servlet-compliant engine, such as the Jakarta Tomcat.
- **J2EE EJB container**: A code generator to allow exposure of stateful J2EE Entity and Session JavaBeans as OGSI compliant Grid services.

Depending on security, reliability, scalability and performance requirements, any of these hosting environments can be chosen as a target environment for implementing Grid services.

2.5.2 Web services engine

A Web services engine is responsible for SOAP message exchange between clients and services. GT3 currently uses the Apache Axis as its SOAP engine, which manages SOAP message exchange.

2.5.3 Grid services container

A Grid services container runs on top of a Web services engine, and provides a run-time environment for hosting various services. The idea of using a container in GT3 is borrowed from the Enterprise JavaBeans (EJB) model, which uses containers to host various application or components with business logic. A GT3 container can be deployed into a range of hosting environments in order to overcome the heterogeneity of today's Grid deployments. For example, a Grid service could be implemented as an enterprise B2B application serving a large number of concurrent users, as well as a lightweight entry point into a Grid scheduling system for batch submissions. If a service is developed to comply with a container interface contract, it can be deployed in all environments supported by the container. Compared with Web services, there are three major functional areas covered by a Grid service container:

- Lightweight service introspection and discovery supporting both pull and push information flows.
- Dynamic deployment and soft-state management of stateful service instances that can be globally referenced using an extensible resolution scheme.

2.5 THE GLOBUS TOOLKIT 3 (GT3)

- A transport independent Grid Security Infrastructure (GSI) [36] supporting credential delegation; message signing and encryption; as well as authorization.

2.5.4 GT3 core services

The GT3 Core implements the interfaces and behaviour defined by OGSI. GT3 core services are focused on the implementation of the OGSI specification. Apart from that, security and system level services are also part of the core services.

2.5.4.1 OGSI implementation

OGSI is a technical specification that can be used for implementing Grid services as defined by OGSA. The OGSI implementation in GT3 is a set of primitives implementing the standard OGSI interfaces also called portTypes, such as: *GridService, Factory, Notification* (source/sink/subscription), *HandleResolver* and *ServiceGroup* (entry/registration). Grid services inherit these interfaces but often implement or extend them. The implementation of the *GridService* interface in GT3 is essentially the base container implementation, and the *Factory* interface implements most of the state management of Grid services in the GT3 container. These two implementations are hence fundamental parts of GT3 and not likely to be replaced by other implementations. The implementations of the other OGSI interfaces in GT3 should, however, be seen more as a reference implementation that could be replaced by more robust implementations provided by users.

2.5.4.2 Security infrastructure

Two levels of security are provided in GT3: transport-level security uses the *HTTPG* protocol to enable GSI over HTTP; message-level security offers both GSI Secure Conversation and GSI XML Signature by using the WS-Security [37], XML Encryption [38] and XML Signature [39]. GT3 has made a number of improvements to GSI, which can be found in Welch *et al.* [40].

2.5.4.3 System level services

System level services are OGSI compliant Grid services that are generic enough to be used by all other services. Currently GT3 contains three system level services:

- The *Admin Service* is used to "ping" a hosting environment and to try and cleanly shutdown a container.
- The *Logging Management Service* allows the user to modify log filters and to group existing log producers into more easily manageable units at run time.
- The *Management Service* provides an interface for monitoring the current status and loading a Grid service container. It also allows users to activate and deactivate Grid service instances.

2.5.5 GT3 base services

Based on GT3 core services, base services in GT3 are mainly focused on resource management, information services and reliable file transfer.

2.5.5.1 Resource management in GT3

The Globus Resource Allocation Manager (GRAM) [41] is the lowest level of Globus resource management architecture. To invoke a job via GRAM, a client uses the Resource Specification Language (RSL) to describe the job to be run, specifying such details as the name of the executable; the working directory, where input and output should be stored; and the queue in which it should run. While GT3 offers a number of other services (e.g. for file movement and job monitoring), GRAM is the most complicated service in GT3 from a security perspective because it provides a secure, remote, dynamic instantiation of processes, involving both secured interaction with a remote process and the local operating system.

The GT3 GRAM model
In GT3, the RSL is marked up in XML. GT3 allows users to run jobs remotely in a secure fashion, using a set of WSDL documents

2.5 THE GLOBUS TOOLKIT 3 (GT3)

and client interfaces for submitting, monitoring and terminating a job. A job submission ultimately results in the creation of a Managed Job Service (MJS). An MJS is a Grid service that acts as an interface to its associated job, instantiating it and then allowing it to be controlled and monitored with standard Grid and Web services mechanisms. An MJS is created by invoking an operation on an MJS factory service. While conceptually one MJS factory service can be managed by each user account, this approach is not ideal in practice because it involves resource consumption by factories that sit idle when the user is not using the resource. Thus, GT3 introduces the Master Managed Job Factory Service (MMJFS), which is leveraged from GRAM in previous Globus toolkit versions. One MMJFS runs in a non-privileged account, and invokes a Local Managed Job Factory Service (LMJFS) for users to create an MJS in their accounts as needed. A service called a Proxy Router routes incoming requests from a user to either that user's LMJFS, if present, or the MMJFS, if an LMJFS is not present for the user making the request. Figure 2.19 shows the relation between MMJFS, LMJFS, and MJS. Each LMJFS runs in a User-Hosting Environment (UHE) and is valid for the lifetime of the UHE. An LMJFS can create one or more MJSs as needed.

Each active user account has a hosting environment running for its use, with an LMJFS and one or more MJS instances running in that hosting environment. This approach allows for the creation of

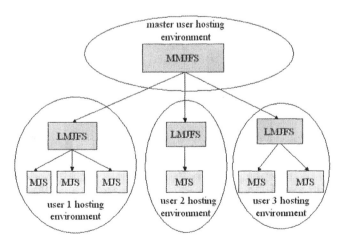

Figure 2.19 MMJFS, LMJFS, and MJS in GT3

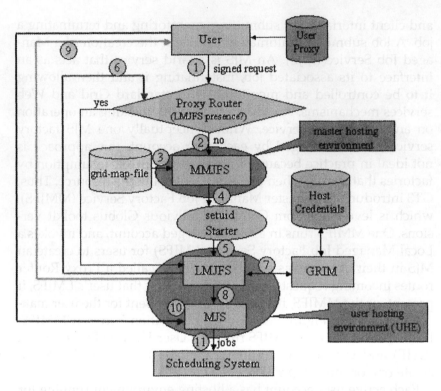

Figure 2.20 The data flow of submitting jobs in GT3

multiple services in a lightweight manner. Figure 2.20 shows the data flow of submitting user jobs in GT3, which are:

1. The user forms a job description in RSL and signs it with their GSI credentials. This signed request is sent to the target resource on which process initiation is desired.
2. The Proxy Router service accepts the request and either routes it to an LMJFS, if present (skip to step 6), or to the MMJFS otherwise.
3. The MMJFS verifies the signature of the request and establishes the identity of the user. Then it determines the local account in which the job should be executed based on the user's identity and the *grid-map-file*, which is a local configuration file containing the mappings from GSI identities to local identities.

2.5 THE GLOBUS TOOLKIT 3 (GT3)

4. The MMJFS invokes the *setuid Starter* process to start an LMJFS for the requestor.

5. The *setuid Starter* is a privileged program (typically setuid-root) that has the sole function of starting a preconfigured LMJFS for a user.

6. When an LMJFS starts, it needs to acquire credentials and register itself with the Proxy Router. To register, the LMJFS sends a message to the Proxy Router. This informs the Proxy Router of the existence of the LMJFS so that it can route future requests for job initiation to it.

7. The LMJFS invokes the Grid Resource Identity Mapper (GRIM) to acquire a set of credentials. GRIM is a privileged program (typically setuid-root) that accesses the local host credentials and from them generates a set of GSI proxy credentials for the LMJFS. This proxy credential has embedded in it the user's Grid identity, local account name and policy to help the client verify that the LMJFS is appropriate for its needs.

8. The LMJFS receives the signed job request. The LMJFS verifies the signature on the request to make sure it has not been tampered with and to verify the requestor is authorized to access the local user account in which the LMJFS is running. Once these verifications are complete, the LMJFS creates and invokes an MJS with the job initiation request.

9. LMJFS returns the address (GSH) of the MJS to the user.

10. The user connects to the MJS to initiate a job submission. The requestor and MJS perform mutual authentication using the credentials acquired from GRIM. The MJS verifies whether the requestor is authorized to initiate processes in the local account. The requestor authorizes the MJS with a GRIM credential issued from an appropriate host credential containing a Grid identity of the user. This approach allows the client to verify that the MJS is communicating with it running not only on the right host but also in an appropriate account. The user then delegates GSI credentials to the MJS to dispatch the job for running.

11. Jobs can be dispatched to a scheduling system such as Condor [42], Sun Grid Engine [43], LSF [44] or PBS [45].

Benefits of GT3 GRAM model
The GRAM model described in this section has the following benefits from a privileged security perspective:

- *Effective resource usage*: In GT3, users with different user accounts can share an MMJFS service or use an LMJFS invoked by the MMJFS for job submission. LMJFS is only valid for the lifetime of a user-hosting environment. It can be released once a user has completed using it. This helps effective resource usage, which results in performance improvement in services management.
- *No privileged services*: Network services, since they accept and process external communications are likely to be compromised by logic errors, buffer overflows, and similar. Removing privileges from these services can reduce the impact of compromises by minimizing the privileges gained.
- *Minimal privileged code*: Privileged code is confined to two programs, GRIM and the setuid Starter. The design of these programs allows them to be audited effectively and should reduce the chances of using them maliciously to gain privilege authorization.
- *Client-side authorization*: GRIM allows the client to verify not only the resource on which an MJS is running but also the account in which it is executing. Thus, a client can act to prevent spoofing of addresses or social engineering tricks that might mislead the user into connecting to, and more importantly delegating credentials to, an MJS other than the one they intended to.

2.5.5.2 The GT3 Index Service

The Index Service [46] uses an extensible framework for managing static and dynamic data for GT3-based Grid systems. It provides the following functions:

- Dynamic service data creation and management via Service Data Provider components.
- Aggregation of service data from multiple instances.
- Registration of Grid service instances.

Figure 2.21 shows the structure of the GT3 Index Service. The Index Service combines Service Data Provider, Service Data Aggregation, and Registry components. Service Data Provider components

2.5 THE GLOBUS TOOLKIT 3 (GT3)

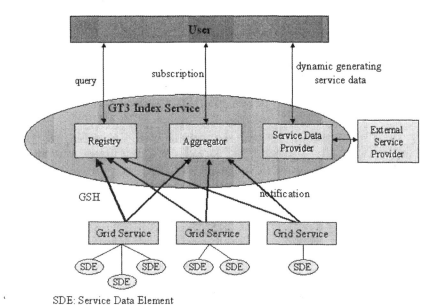

SDE: Service Data Element

Figure 2.21 The structure of the GT3 Index Service

provide a standard mechanism for the dynamic generation of service data via external programs, which can be the GT3 core service providers or be user-created custom providers. Service Data Aggregation components provide a mechanism for handling subscription, notification, and update of service data, which can be aggregated from multiple service instances. Finally, Registry components maintain a set of available GSHs of Grid service instances whose service data can be queried with the FindServiceData() method provided in the *GridService* interface.

The GT3 Index Service provides a GUI-based browser that allows users to view the details of the services available in a registry. In addition, it also provides *ogsi_find_service_data* command, a command-line interface for querying the service data available.

2.5.5.3 Reliable File Transfer in GT3

The GT3 Reliable File Transfer (RFT) [47] service is that provides interfaces for controlling and monitoring third-party file transfers using GridFTP [48]. The client controlling the transfer is hosted inside of a Grid service. The GT3 RFT guarantees that file transfers will be reliable. For example, if a file transfer is interrupted (due to a network failure, for example), it can restart the file transfer from the moment it failed, instead of starting all over again.

2.5.6 The GT3 programming model

The programming model of GT3 core is just the same as Web services computing model using loosely coupled client-side stubs and server-side skeletons (also called stubs in GT3). However, GT3 programming has the following characteristics:

- GT3 uses GWSDL to define a Grid service interface. A GWSDL is a WSDL with all the OGSI-specific types, messages, portTypes, and namespaces. GWSDL uses the <gwsdl:portType> tag instead of the <portType> tag in standard WSDL.
- The Web Services Deployment Descriptor (WSDD) contains information related to deploying a Grid service such as name of the Grid service, name of a Grid service instance, and the base class of the Grid service instance.
- GT3 packages all the compiled Java codes plus related files such as the WSDD file of the Grid service as a GAR file for service deployment. GAR is a special kind of JAR.
- A Grid service can be a transient service that can be dynamically created and explicitly destroyed.
- A Grid service is a stateful service associated with service data.
- Grid services can notify subscribed clients for events of interest.

Figure 2.22 shows the data-flow control in implementing GT3 applications. The steps to implement a GT3 application with Apache Axis are given below.

- Write a Grid service interface, which can be defined in Java, WSDL or GWSDL.
- Write a service to implement the Grid service interface.
- Write a WSDD file for the deployment of the Grid service factory.
- Use a GT3 builder script to compile the Grid service interface file and implementation file and package them plus other related files such as the stubs generated from the interface and the WSDD file into a GAR file.

2.5 THE GLOBUS TOOLKIT 3 (GT3)

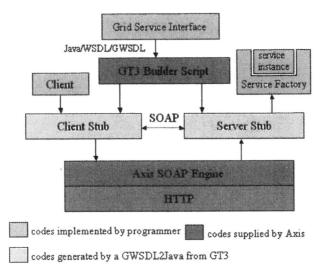

Figure 2.22 Data-flow control in implementing GT3 applications

- Use Apache Ant [49] to deploy the GAR file in a Grid service container to publish the service.
- Write a client to request the factory to create an instance of the service. The client first gets the GSH of the factory and then the GSR of the factory. Then it uses the GSR of the factory to create an instance and obtains the GSR of the created instance. Finally the client uses the GSR of the instance to access the service instance.
- Start a Grid service container.
- Start the client to request a service.

2.5.6.1 Server-side components in GT3

As shown in Figure 2.23, the major architectural components of the server-side frameworks include the following:

- *The Web service engine*: This engine is used to deal with normal Web service behaviour, SOAP message processing, JAX-RPC handler processing, and Web service configuration.
- *The Grid service container*: GT3 provides a container for the creation, destruction, and life cycle management of stateful Grid services.

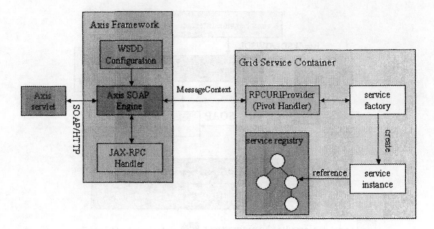

Figure 2.23 Service-side components in GT3

GT3 uses Apache Axis as the SOAP engine, which can be deployed in a Tomcat Web server running as a Java Servlet container. The SOAP engine is responsible for SOAP request/response serialization and de-serialization, JAX-RPC handler invocation, and Grid service configuration. The GT3 container provides a pivot handler to the Axis framework to pass the request messages to the Grid service container. Once a Grid service factory creates a service instance, the framework creates a unique GSH for that instance, which will be registered with the container registry. This registry holds all of the stateful service instances and is contacted by other components and handlers to perform services:

- Identify services and invoke service methods;
- Get/set service properties (such as instance GSH and GSR);
- Activate/deactivate services;
- Resolve GSHs to GSRs.

2.5.6.2 Client-side components in GT3

As shown in Figure 2.24, GT3 uses the normal JAX-RPC client-side programming model for Grid service clients. In addition, GT3 provides a number of helper classes at the client-side to hide the details of the OGSI client-side programming model.

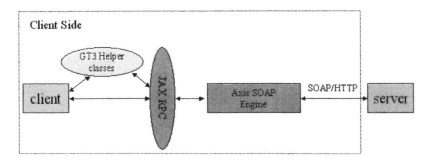

Figure 2.24 Client-side components in GT3

2.5.6.3 A summary of GT3

GT3 includes an implementation of OGSI and has been employed for building Grid applications in the context of OGSA. GT3 has core and base services. The core services are concerned with creation, destruction, and life cycle management of stateful Grid services. The base services are concerned with secure job submission, information services and reliable data transfer. To submit a job that needs to run in parallel, there are two choices with GT3.

- Deploy a service factory on multiple nodes and run multiple service instances in parallel.
- Use MMJFS and LMJFS to submit a user job to a backend scheduling system such as Condor, Sun Grid Engine, PBS or LSF to run jobs in parallel.

2.6 OGSA-DAI

While OGSI specifies the issues related to Grid services such as creation and destruction, registration and discovery, notification, however, the concept of Grid services is general. OGSI does not specifically specify how to manage data services on the Grid. As shown in Figure 2.25, sitting on top of OGSI, the Open Grid Services Architecture Data Access and Integration (OGSA-DAI) [50] is a middleware technology that can be used to easily access and integrate data from different data sources such as relational databases, XML databases, and file systems on the Grid. The OGSA-DAI is both a specification and an implementation. As a specification, it defines the services and interfaces that are needed

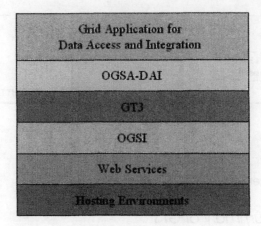

Figure 2.25 The position of OGSA-DAI in OGSA

for data access and integration on the Grid. The interfaces are extended WSDL portTypes based on the OGSI specification. The OGSA-DAI implementation has been released which implements the interfaces defined in the OGSA-DAI framework based on GT3. The aim of the OGSA-DAI is to allow external data resources, such as databases, to be incorporated within the OGSA framework, and hence accessible via standard Grid services interfaces. By using the OGSA-DAI, heterogeneous disparate data resources can be accessed uniformly. It supports the registration/discovery of databases and the interaction with those databases.

2.6.1 OGSA-DAI portTypes

The OGSA-DAI provides the following interfaces (portTypes) to define Grid services for data access and integration. A Grid service in the OGSA-DAI is called a Grid Data Service (GDS). A GDS must implement the *GDSPortType* interface. The OGSA-DAI provides the following interfaces.

GDSPortType portType
The *GDSPortType* portType supports data access, integration and delivery and will be implemented by all GDSs. It extends three portTypes. One is *GridService* portType specified in OGSI, the other two are *GridDataPerform* (GDP) portType and *GridDataTransport* portType defined by the OGSA-DAI.

2.6 OGSA-DAI

GridDataPerform portType

The *GridDataPerform* portType provides methods for clients to access data sources and retrieve results. As shown in Figure 2.26, it supports a document-oriented interface for database queries in which a query request is submitted using Grid Data Service Perform (GDS-Perform) documents to specify the operations on the data sources and a response is returned using Grid Data Service Response (GDS-Response) documents containing the results operations. The nature of the query document submitted to the GDS and the subsequent result document depends on the type of the data resource that the Grid service is configured to represent. For example, a relational database may accept SQL queries while an XML database may accept XPath queries. Using a document to describe a request allows the request to be analysed and optimized.

GridDataTransport portType

The *GridDataTransport* portType provides supports for data transfer between OGSA-DAI services, and between OGSA-DAI clients and the OGSA-DAI services. It allows data to be pushed or pulled. The portType is derived from *GridService* portType in OGSI and provides the following methods:

- PutFully() – transfer a complete set of data.
- GetFully() – receive a complete set of data.
- PutBlock() – transfer a block of data which is part of a larger batch of data using the operation to specify the index of the block.
- GetBlock() – receive a block of data which is part of a larger batch of data using the operation to specify the index of the block.

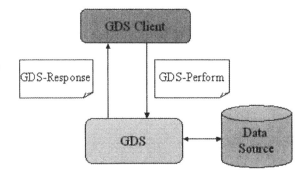

Figure 2.26 Access data in the OGSA-DAI

GridDataServiceFactory portType
The *GridDataServiceFactory* portType is used to implement a Grid Data Service Factory (GDSF), which is a persistent Grid data service used to create GDSs. The portType extends three OGSI portTypes – *GridService, Factory* and *NotificationSource*.

DAIServiceGroupRegistry portType
The *DAIServiceGroupRegistry* portType is used to implement a DAIServiceGroupRegistry (DAISGR). A DAISGR service can be used to register any services that implement one or more OGSA-DAI portTypes. The DAIServiceGroupRegistry portType extends four OGSI portTypes – *GridService, ServiceGroup, ServiceGroup Registration* and *NotificationSource*.

2.6.2 OGSA-DAI functionality

Based on the OGSA-DAI portTypes introduced, the OGSA-DAI provides the following functionality for data access and integration on the Grid.

2.6.2.1 The lifetime management of GDS instances

A GDS instance is a transient service. It is created by a GDSF associated with a data source. A client can access a data source through the interaction with the GDS. A GDS instance can be dynamically created and explicitly destroyed.

2.6.2.2 Service registration/unregistration

A OGSA-DAI service can register itself with a DAISGR via the ServiceGroupRegistration::Add() method. Apart from that, a DAISGR can also be registered in another DAISGR. By querying a DAISGR, clients can discover OGSA-DAI services that offer particular services or capabilities or manage particular data sources. Registered services in a DAISGR can also be unregistered via the ServiceGroupRegistration::Remove() method.

2.6.2.3 Service discovery

A client can query a DAISGR via the GridService:: FindServiceData() method to discover an OGSA-DAI service meeting its requirements.

2.6 OGSA-DAI

A client can then query the OGSA-DAI service directly, and also query the OGSI ServiceGroupEntry service managing the OGSA-DAI service's registration. There are three purposes for a client to query a DAISGR:

- To discover a GDSF to create GDS instances for specific applications.
- To query an OGSA-DAI service instance to retrieve its state information.
- To query the OGSI ServiceGroupEntry associated with an OGSA-DAI service to retrieve the OGSA-DAI service's registration.

2.6.2.4 Service notification

Clients can subscribe to a DAISGR for notifications of events which happen in the DAISGR, i.e., changes in the state of the DAISGR, the registration of a new service or the unregistration of an existing service. Figure 2.27 shows the steps used for notification.

1. A client uses the NotificationSource::Subscribe() method to subscribe to a DAISGR to specify the events of interest for notifications. In the subscription, it specifies the location of a notification sink service which implements the *NotificationSink* interface.
2. The DAISGR creates a notification subscription service which implements the *NotificationSubscription* interface to manage the subscription.

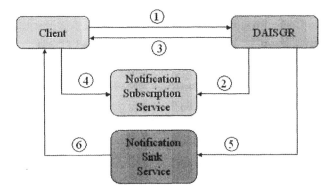

Figure 2.27 Service notification in OGSA-DAI

3. The DAISGR informs the client of the identity of this NotificationSubscription service.
4. The client queries the notification subscription service via GridService::FindServiceData() method to manage its subscription, e.g. its lifetime management.
5. Once the DAISGR has some changes in its state, it will notify the notification sink service via the NotificationSink::DeliverNotification() method.
6. The notification sink service will send the notification messages to the client.

2.6.3 Services interaction in the OGSA-DAI

Now we give a whole view on the interactions between OGSA-DAI services as shown in Figure 2.28.

1. Start a Grid service container, which reads a server-config.wsdd file. The server-conf.wsdd file allows the Grid service container to access information on services to be deployed and to map between service names and the associated classes.
2. The Grid service container creates a persistent DAISGR based on the GSH specified in the server-config.wsdd file.
3. The Grid service container creates a persistent GDSF based on the GSH specified in the server-config.wsdd file.
4. The GDSF registers itself in the DAISGR with the ServiceGroupRegistration::Add() method.
5. The client queries the DAISGR using the GridService::FindServiceData() method. The client selects a registered GDSF.
6. The DAISGR returns the GSH of a selected GDSF.
7. The client can query the service data elements of the GDSF to retrieve its configuration information.
8. The client calls the GDSF Factory::createService() method to create a GDS instance.
9. The GDSF creates the GDS instance.
10. The GDSF returns the GSH of the GDS instance to the client.
11. The client queries the service data elements of the newly created GDS instance using the GridService::FindServiceData()

2.6 OGSA-DAI

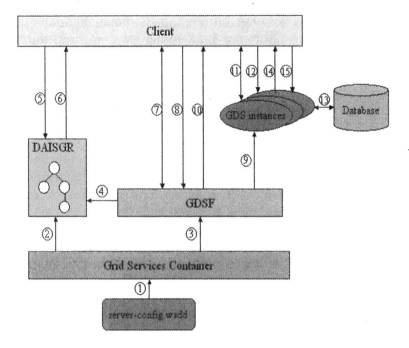

Figure 2.28 Interactions between services in OGSA-DAI

method to establish its configuration and the schema that describes the GDS-Perform documents that may be submitted through the GridDataPerform::perform() method.

12. The client submits a GDS-Perform document to the GDS instance.
13. The GDS instance accesses a database to get some data, and generates a GDS-Response document.
14. The GDS instance returns the GDS-Response document to the client.
15. The client destroys the GDS instance via GridService::Destroy() method.

2.6.4 OGSA-DAI and DAIS

The Database Access and Integration Services (DAIS) working group [51] within the GGF seeks to promote OGSA compliant standards for Grid database services, initially focusing on providing consistent access to existing, autonomously managed databases. It does not seek to develop new data storage systems, but rather to make such systems more readily usable individually or collectively within a Grid framework.

OGSA-DAI is a collaborative programme of work involving the Universities of Edinburgh, Manchester and Newcastle, with industrial participation by IBM and Oracle. Its principal objective is to produce open-source database access and integration middleware that meets the needs of the UK e-Science community for developing Grid-related applications. Its scope includes the definition and development of generic Grid data services providing access to, and integration of data, held in relational database management systems, as well as semi-structured data held in XML repositories. OGSA-DAI also represents a significant contribution on behalf of the UK e-Science Core Programme to extend the Grid model to include database interoperability. OGSA-DAI works closely with the DAIS working group and it is intended that the software will serve as a reference implementation for DAIS standards.

2.7 WSRF

In this section, we present the Web Services Resource Framework (WSRF). We first introduce WSRF and its key concepts. We then describe the impact of WSRF on OGSA and OGSI.

2.7.1 An introduction to WSRF

The WSRF is a set of WS specifications. It introduces WS-Resource to model and manage state information in the context of Web services. A stateful resource associated with a Web service is referred as a WS-Resource. WS specifications are briefly described below.

WS-ResourceLifetime
The WS-ResourceLifetime specification is an initial draft [52] that defines mechanisms for service requestors to request Web services to destroy associated WS-Resources immediately or after certain time. It also defines the means by which a WS-Resource may have a self-destruct action after the expiration of a period of time.

2.7 WSRF

WS-ResourceProperties
The WS-ResourceProperties specification is an initial draft [53] that defines the means by which the definition of the properties of a WS-Resource may be declared as part of a Web service interface. The declaration of the WS-Resource's properties represents a projection of or a *view* on the WS-Resource's state. This projection represents a WS-Resource as an implicit stateful resource type that defines a basis for access to the WS-Resource properties through Web service interfaces. The term "implied" will be explained in Section 2.7.1.2.

This specification also defines a standard set of message exchanges that allow a requestor to query or update the property values of the implied WS-Resource. The set of properties defined in the WS-Resource projection, and associated with the Web service interface, defines the constraints on the valid contents of these message exchanges.

WS-Notification
The WS-Notification specification is an initial draft [54] that defines mechanisms for event subscription and notification using a topic-based publish/subscribe pattern. It defines the following contents:

- Standard message exchanges to be implemented by service providers that wish to participate in notifications.
- Standard message exchanges for a notification broker service provider. The notification broker interface (*NotificationBroker*) defines a standard set of message exchanges to describe a message broker, providing an intermediary between Publishers and Subscribers on a collection of Topics.
- Operational requirements expected of service providers and requestors that participate in notifications, and an XML model that describes topics.

The WS-Notification has been split into three specific specifications, WS-BaseNotification [55], WS-BrokeredNotification [56] and WS-Topics [57]. The three specifications are also the initial drafts at the time of writing. The WS-BaseNotification is the base specification on which the other WS-Notification specification documents depend on. It defines the Web services interfaces for NotificationProducers and NotificationConsumers to exchange messages. The WS-BrokeredNotification defines the Web services interfaces for NotificationBrokers, which play the roles of both NotificationProducers and NotificationConsumers as defined in the WS-BaseNotification. A NotificationBroker can deliver notification messages to NotificationConsumers and subscribe to notifications distributed by NotificationProducers. In addition, a NotificationBroker must support hierarchical topics, and the ConcreteTopicPath topic expression dialects defined in WS-Topics. The WS-Topics defines a mechanism to organize and categorize items of interest for subscription known as "topics". It provides a convenient means by which subscribers can reason about WS-Base notifications of interest. WS-Topics defines three topic expression dialects that can be used as subscription expressions. It further specifies an XML model for describing metadata associated with topics.

WS-BaseFaults

The WS-BaseFaults specification is an initial draft [58] that defines an XML Schema for base faults, along with rules to specify how these faults types are used and extended by Web services. The goal of WS-BaseFaults is to standardize the terminology, concepts, XML types, and WSDL usage of a base fault type for Web services interfaces. Specifying Web services fault messages in a common way enhances support for problem determination and fault management. It is easier for requestors to understand faults if the fault information from various interfaces is consistent.

WS-ServiceGroup

To build higher-level services, we often need primitives for managing collections of services. Archetypal examples are registries and index services. The WS-ServiceGroup specification is an initial draft [59] that defines a means by which Web services and WS-Resources can be aggregated or grouped together. A ServiceGroup is a WS-Resource that represents a collection of Web services. The individual services represented within the ServiceGroup are the ServiceGroup's members or its membership. ServiceGroup membership rules, membership constraints and classifications are expressed using the resource property model as defined in the WS-ResourceProperties specification.

WS-RenewableReferences

No draft has been released yet at the time of writing. The specification will standardize mechanisms for Web services to renew endpoint references when the current reference becomes invalid. It replaces the concepts of GSH and GSR in the OGSI 1.0 specification.

2.7.1.1 The WS-Resource concept

A Web service, which can be viewed as a stateless message processor, can be associated with stateful resources (WS-Resource). A WS-Resource has the following characteristics:

- It is a stateful resource that can be used as a data context for a Web service to exchange messages.
- It can be created, identified, and destroyed. A WS-Resource can have lots of identifiers within the same Web service or within different Web services.
- A stateful WS-Resource type can be associated with a Web service interface definition to allow well-formed queries against the WS-Resource via its service interface, and the status of the stateful WS-Resource can be queried and modified via service message exchanges.

The WSRF does not attempt to define the message exchange used to request the creation of new WS-Resources. Instead, it simply notes that new WS-Resources may be created by the use of a WS-Resource factory pattern as defined in the WS-ResourceLifetime specification. A WS-Resource factory is any Web service capable of bringing one or more WS-Resources into existence. The response message of a factory operation typically contains at least one endpoint reference that refers to the newly created WS-Resource.

2.7.1.2 The implied WS-Resource pattern

The term *implied WS-Resource pattern* is used to describe a specific kind of relationship between a Web service and one or more stateful WS-Resources.

- The term *implied* means that when a client accesses a Web service, the Web service will return a WS-Addressing [60] endpoint reference used to refer to the WS-Resources associated with the

Web service. Each WS-Resource has an identifier (ID) for managing its state. The IDs of the WS-Resources to be accessed by the client will be automatically encapsulated in the endpoint reference and returned to the client. A WS-Resource ID is only used by a Web service as an implicit input to locate a specific WS-Resource; it is opaque to the client.

- The term *pattern* is used to indicate that the relationship between Web services and stateful WS-Resources is codified by a set of conventions on existing Web services technologies, in particular XML, WSDL and WS-Addressing.

WS-Addressing plays a key role in the WSRF. The concept of endpoint references is the core of the WS-Addressing specification. A WS-Addressing endpoint reference is an XML serialization of a network-wide *pointer* to a Web service. A WS-Addressing endpoint reference contains, among others, two important parts. One refers to the network transport-specific address of the Web service, e.g. a URL in the case of HTTP. This is the same address that would appear within a *port* element in a WSDL description of the Web service. The other part is reference properties that contain an ID to uniquely identify a WS-Resource. Figure 2.29 gives an example to explain how the concept of endpoint references uses the implied WS-Resource pattern in accessing a WS-Resource via a Web service.

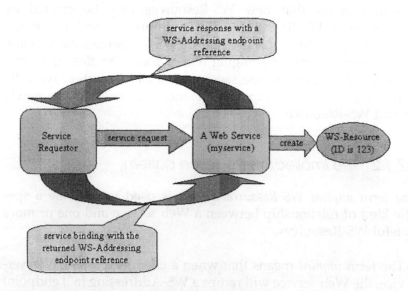

Figure 2.29 Accessing a stateful WS-Resource via a Web service

2.7 WSRF

As shown in Figure 2.29, a service requestor sends a request to a Web service called *myservice* that will cause the service to create a WS-Resource whose ID is 123. The service generates a response with a WS-Addressing endpoint reference and sends the response to the requestor. The endpoint reference is similar to the one as shown in Figure 2.30. It contains a URL of the service (wsa:Address) and the resourceID (resourceID is one property of the wsa:ReferenceProperties).

The service requestor binds to the service with the endpoint reference using SOAP. The WS-Addressing specification requires that each *ReferenceProperties* element in the endpoint reference must appear as a header block in the SOAP message sent to the service identified by the endpoint reference. The SOAP message combined with the ReferenceProperties in Figure 2.30 is similar to the one shown in Figure 2.31. wsa:To specifies where to locate the service, and resourceID specifies which WS-Resource the service will use for the request.

```
<wsa:EndpointReference>
  <wsa:Address>
      http://SomeURL/myservice
  </wsa:Address>
  <wsa:ReferenceProperties>
      <resourceID>123</resourceID>
  </wsaReferenceProperties>
</wsa:EndpointReference>
```

Figure 2.30 A WS-Addressing endpoint reference sample

```
<soap:Envelope>
  <soap:Header>
    ...
       <wsa:To>http://SomeURL/myservice</wsa:To>
       <resourceID>123</resourceID>
    ...
  </soapHeader>
  <soap:Body>
   ... some message here ...
  </soap:Body>
</soap:Envelope>
```

Figure 2.31 A SOAP message for binding to a Web service associated with a WS-Resource

As shown in Figure 2.30, the *ReferenceProperties* in a WS-Addressing endpoint reference of a Web service is opaque to the service requestor, which does not need to do anything about it. The *ReferenceProperties* such as the WS-Resource ID is only used by a Web service as an implicit input to locate a specific WS-Resource associated with the Web service.

2.7.2 WSRF and OGSI/GT3 (61, 62)

A huge effort from the Grid community went into the OGSI specification, implementations, and development and deployment of OGSI-based services. WSRF effectively declared that OGSI was obsolescent.

IBM realized that OGSI was too far from standard Web services to be acceptable to the Web services community. This is what led to WSRF. Some of the key differences are: WSRF does not break WSDL; one can expect better tooling support for WSRF; WSRF is less object-oriented; WSRF is more mix and match. In OGSI, one talked to service instances about their service data. In WSRF, one talks to the service about its resources and their properties. Many of the OGSI ideas and patterns do go through in WSRF, once you substitute "resource" for "service instance". A good thing about WSRF is that it permits multiple service interfaces to the same stateful resource. The equivalent thing in OGSI would be multiple service interfaces for the same service instance. However, assertions that the differences between OGSI and WSRF are "syntactic sugar", and that it is a straightforward refactoring process to go from OGSI implementations to WSRF implementations or OGSI services to WSRF services, are unproven.

2.7.2.1 A comparison of WSRF and OGSI

Advantages of WSRF over OGSI
While the definition of WSRF has been motivated primarily by the desire to integrate recent developments in Web services architecture, in particular WS-Addressing, its design also addresses three criticisms of OGSI v1.0 from the Web services community:

- OGSI is a heavyweight specification with too much definition in one specification. WSRF partitions the OGSI functionality into a set of specifications.

- OGSI does not work well with existing Web services tooling. OGSI uses XML Schema extensively, for example there is frequent use of xsd:any, attributes. It also uses a lot of "document-oriented" WSDL operations. These features cause problems with, for example, JAX-RPC. WSRF somewhat reduces the use of XML Schema.
- OGSI models a stateful resource as a Web service that encapsulates the state of the resource, with the identity and life cycle of the service and resource state coupled. WSRF re-articulates the underlying OGSI architecture to make an explicit distinction between a stateless Web service and the stateful resources acted upon by that service.

Advantages of OGSI over WSRF
There are some advantages of OGSI compared with WSRF.

- With the features of Object-Oriented Paradigm (OOP), OGSI has strong notions of extensibility through inheritance, which are absent in WSRF; one can imagine designing frameworks for developing services that rely heavily on OGSI inheritance.
- Similarly, the criticism that OGSI is heavyweight also has its flip side: in OGSI, we can count on an "OGSI compliant Grid services" providing certain portTypes and behaviour; this is not the case in WSRF where everything is optional. Indeed, one runs into language difficulties: what does WSRF-compliance mean?

2.7.2.2 Modelling stateful resources in WSRF and OGSI

Both WSRF and OGSI are concerned with the extension of standard Web services with stateful information, but with different approaches. As shown in Figure 2.32, WSRF separates associated resources from Web services. It uses a WS-Addressing endpoint reference to associate Web services with stateful WS-Resources by using *ReferenceProperties* in which resources IDs can be specified.

OGSI models stateful resources with service data elements which are tightly coupled with Grid services or services instances, as shown in Figure 2.33.

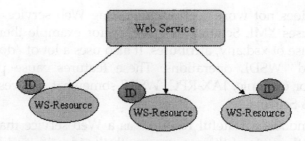

Figure 2.32 Associating Web services with stateful resources (WS-Resources) in WSRF

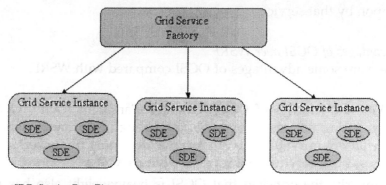

SDE: Service Data Element

Figure 2.33 Associating stateful Grid services with service data in OGSI

2.7.2.3 The impacts of WSRF on OGSI/GT3

As shown in Figure 2.34, WSRF retains essentially all of OGSI concepts, and introduces some changes to OGSI messages and their associated semantics. It is expected that the effort required to modify an OGSI-based system or specification to use WSRF will be small. Services implemented using OGSI-based middleware, such as the GT3, are likely to require some changes to exploit WSRF-based tools, but it is expected that the changes should be modest due to the similarities between WSRF and OGSI. More generally, the fact that WSRF is based on mainstream Web services standards and has been embraced by major vendors means that we can expect to see its integration into commercial Web services products, enabling a much richer choice of products upon which WSRF compliant services can be built.

2.7 WSRF

OGSI	WSRF
Grid Service Reference	WS-Addressing Endpoint Reference
Grid Service Handle	WS-RenewableReferences
Handle Resolver	WS-RenewableReferences
Service Data	WS-ResourceProperties
Grid Service Lifetime Management	WS-ResourceLifetime
Notification	WS-Notification
Factory	Implied WS-Resource Pattern
Service Group	WS-SeviceGroup
Base Fault Type	WS-BaseFaults

Figure 2.34 The functionality in OGSI and WSRF

OGSI is being replaced by WSRF as it is converging the efforts put by the Web services and Grid communities. It is expected that the Globus Toolkit 4 (GT4), which is based on WSRF, will be released in early 2005. However, GT3 and GT4 will not interoperate.

2.7.3 WSRF and OGSA

Based on standard Web services, OGSA attempts to define the core standard services required to build service-oriented Grid applications, through the OGSA working group at GGF. But it does not say anything about the implementation of Grid services in the context of Web services. OGSA is an architecture, so all the WSRF ideas should go through regardless of the plumbing into OGSA. However, the OGSA-WG documents now name WSRF as one of the enabling technologies and will use WSRF language where convenient.

Since WSRF promotes a standard way to model Web services as stateful resources, WSRF can be used as an infrastructure for developing Grid services in the context of OGSA. OGSA can be enriched by WSRF as shown in Figure 2.35. OGSA platform services are Grid base services that are used for job submission, user authentication and authorization, information services, data replication and transfer, as well as data access and integration. WSRF services

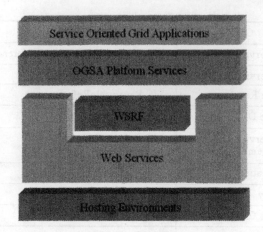

Figure 2.35 A view of OGSA on top of WSRF

are Grid core services that are used for the creation, destruction, and life cycle management of stateful Web services resources.

2.7.4 A summary of WSRF

WSRF promotes a standard way to model Web services with stateful resources. It is hoped that WSRF will provide a convergence point for the Web services and Grid communities. While the work of OGSI is undertaken primarily by the Grid community, the work on WSRF is being undertaken by joint efforts both from Web services vendors like IBM, HP, Fujitsu, SAP AG, Sonic Software and Grid researchers from the Globus Alliance, Indiana University, Lawrence Berkeley National Laboratory, University of Manchester, University of Virginia; WSRF is emerging as a promising standard for building the next generation of interoperable service-oriented Grid applications.

2.8 CHAPTER SUMMARY

In this chapter, we have studied OGSA, its structure, functionality, how it is used and why the Grid community needs it. We started the chapter by providing an overview of traditional distributed middleware techniques such as socket programming, RPC paradigm, Java RMI, DCOM and CORBA. Then we introduced

2.8 CHAPTER SUMMARY

Web services and the benefits that they can bring us. It is clear that Web services are the path ahead as they provide a simple and open standards-based middleware for distributed computing. OGSA uses a range of Web services technologies to provide an infrastructure that is appropriate for Grid applications. Since in the distributed world of the Grid the same problems arise as on the Internet, concerning the description and discovery of services, OGSA certainly takes an approach to embrace these technologies.

OGSA is the *de facto* standard for building service-oriented Grid services and applications. OGSA introduces a couple of new interfaces (portTypes) to meet the Grid needs in the context of WSDL. While Web services are persistent and stateless, Grid services can be transient and stateful. To meet this end, OGSA introduces the concept of GSH/GSR and service data element associated with each Grid service. While OGSA defines what kinds of services should be a part in the architecture, it does not say anything about the implementation of the services. It is the OGSI that technically specifies how the core services in OGSA can be implemented in terms of Web services. The core services include service creation and destruction, service registration and discovery, and service notification. GT3 is a reference implementation of OGSI and has been widely deployed for building OGSA compliant Grid applications. Since the concept of services in OGSA is more general, the OGSA-DAI is specifically designed to provide supports for implementing OGSA compliant Grid services for data access and integration on the Grid.

Many working groups within the GGF are working with the OGSA-WG to define new services which may be integrated with the architecture, i.e., the DAIS working group seeks to promote OGSA compliant standards for Grid database services, the Grid Economic Services Architecture Working Group (GESA-WG) [63] seeks to define standards for charging Grid services, the Resource Usage Services Working Group (RUS-WG) [64] working group seeks to define resource usage as Grid services, the Grid Resource Allocation Agreement Protocol Working Group (GRAAP-WG) [65] seeks to address the protocol between a Super-Scheduler (Grid Level Scheduler) and local schedulers necessary to reserve and allocate resources in the Grid as a building block for this service.

Apart from OGSA, we have also studied WSRF, which is emerging as a promising standard for modelling stateful resources with Web services. WSRF is replacing OGSI and is converging the efforts put from both the Web services and Grid communities. The future Grid middleware toolkits such as GT4 will be based on WSRF to provide the core Grid services as defined by OGSA.

2.9 FURTHER READING AND TESTING

In this chapter, we focused on the theoretical part of OGSA, i.e. how it extends the Web services model, and what kind of services should be a part of the architecture. In terms of the implementation of OGSA core services, we described the OGSI compliant GT3 including its structure, services, and programming model. We introduced OGSA-DAI and discussed how data from different data sources such as relational and XML databases can be accessed and integrated. However, we did not give details related to programming, i.e., how to program Web services, GT3 or OGSA-DAI. Information related to writing Java RMI, DCOM and CORBA applications can be found in McCarty and Dorion [66]. The Apache Axis Web site provides detailed information on how to install Axis and write Web services applications with the Axis framework. Ferreira *et al.* [67] describe how to install GT3 and manage GT3 services. The GT3 Tutorial [68] explains how to program GT3 services. Information related to OGSA-DAI installation and programming can be found in the OGSA-DAI Web site.

2.10 KEY POINTS

- Web services are emerging as a promising platform for building distributed applications in heterogeneous computing environments.
- The core standards in Web services are WSDL for service description, SOAP for message exchanging, UDDI/WS-Inspection for service publication and discovery.
- Standard Web services are stateless and persistent services.
- OGSA is the *de facto* standard for building service-oriented Grid systems.
- OGSA defines the features of Grid services in the context of Web services. It introduces the concept of Grid service instance to provide transient Grid services and uses service data to be associated with Grid services to provide stateful Grid services.

- OGSI is a technical implementation specification to define the interfaces of the core Grid services in the context of OGSA. The GT3 includes an implementation of OGSI.
- OGSA-DAI is an OGSA compliant middleware technology that can be used specifically for data access and integration over the Grid.
- WSRF is emerging as a promising standard for modelling stateful resources with Web services.
- The work on WSRF is a joint effort put by the Web services and the Grid communities. WSRF will bring the two parties further closer.
- WSRF is replacing OGSI and will have little impact on OGSA.
- WSRF can be used as an infrastructure for the construction of the core Grid services in the context of OGSA.
- WSRF is being taken by the Grid community, e.g. it is expected that WSRF-based GT4 will be released in early 2005.

2.11 REFERENCES

[1] GGF, http://www.ggf.org.
[2] Globus, http://www.globus.org.
[3] Web services, http://www.w3.org/2002/ws/.
[4] Foster, I., Kesselman, C., Nick, J. and Tuecke, S. (June 2002). *The Physiology of the Grid: An Open Grid Services Architecture for Distributed Systems Integration*, http://www.globus.org/research/papers/ogsa.pdf.
[5] OGSA-WG, http://www.Gridforum.org/ogsa-wg.
[6] OGSI-WG, http://www.Gridforum.org/ogsi-wg/.
[7] Czajkowski, K., Ferguson, D.F., Foster, I., Frey, J., Graham, S., Sedukhin, I., Snelling, D., Tuecke, S. and Vambenepe, W. (March 2004). *The WS-Resource Framework*, Version 1.0, http://www-106.ibm.com/developerworks/library/ws-resource/ws-wsrf.pdf.
[8] Birrell, A.D. and Nelson, B.J. (1984). Implementing Remote Procedure Calls. *ACM Transactions on Computer Systems*, 2(1): 39–59.
[9] Java RMI, http://java.sun.com/products/jdk/rmi/.
[10] DCOM, http://www.microsoft.com/com/tech/DCOM.asp.
[11] CORBA, http://www.corba.org.
[12] Sun Microsystems (June 1991). *Open Network Computing: RPC Programming*.
[13] OSF DCE 1.0 Application Development Reference, 2 December 1991.
[14] OMG, http://www.omg.org.
[15] Gopalan, S.R. (September 1998). A Detailed Comparison of CORBA, DCOM, and Java/RMI, Object Management Group (OMG), White Paper.
[16] SOAP, http://www.w3.org/TR/soap/.

[17] Christensen, E., Curbera, F., Meredith, G. and Weerawarana, S. (2001). Web Services Description Language (WSDL) 1.1, *W3C Note 15*, http://www.w3.org/TR/wsdl.
[18] UDDI, *Universal Description, Discovery and Integration*, http://www.uddi.org.
[19] Brittenham, P. (2001). *An Overview of the Web Services Inspection Language*, http://www.ibm.com/developerworks/webservices/library/ws-wsilover.
[20] Nagy, William A. and Ballinger, Keith. (November 2001). *The WS-Inspection and UDDI Relationship*, http://www-106.ibm.com/developerworks/webservices/library/ws-wsiluddi.html.
[21] J2EE, http://java.sun.com/j2ee.
[22] Apache Axis, http://ws.apache.org/axis/.
[23] UDDI4J, http://www-124.ibm.com/developerworks/oss/uddi4j/.
[24] Jakarta Tomcat, http://jakarta.apache.org/tomcat/.
[25] Microsoft .Net, http://www.microsoft.com/net/.
[26] Microsoft DISCO, http://msdn.microsoft.com/library/default.asp?url=/library/en-us/cptools/html/cpgrfwebservicesdiscoverytooldiscoexe.asp.
[27] Mukhi, N. (2001). *Web Service Invocation Sans SOAP*, http://www.ibm.com/developerworks/library/ws-wsif.html.
[28] Kunszt, Peter Z. (April 2002). *The Open Grid Services Architecture – A Summary and Evaluation*, http://edms.cern.ch/file/350096/1/OGSAreview.pdf.
[29] Dialani, V., Miles, S., Moreau, L., Roure, D.D. and Luck, M. (August 2002). *Transparent Fault Tolerance for Web Services Based Architectures*. Proceedings of 8th International Europar Conference (EURO-PAR '02), Paderborn, Germany. Lecture Notes in Computer Science, Springer-Verlag.
[30] Zhang, W., Zhang, J., Ma, D., Wang, B. and Chen, Y. (2004). *Key Technique Research on Grid Mobile Service*. Proceedings of the 2nd International Conference on Information Technology for Application (ICITA 2004), Harbin, China.
[31] Sandholm, T. and Gawor, J. (July 2003). *The Globus Toolkit 3 Core – A Grid Service Container Framework*, http://www-unix.globus.org/toolkit/3.0/ogsa/docs/gt3_core.pdf.
[32] MS.NETGrid, http://www.epcc.ed.ac.uk/ogsanet/.
[33] Wasson, G., Beekwilder, N., Morgan, M., Humphrey, M. (2004). *OGSI.NET: OGSI-Compliance on the .NET Framework*. Proceedings of the 4th IEEE/ACM International Symposium on Cluster Computing and the Grid (CCGrid 2004). Chicago, Illinois. CS Press.
[34] OGSI::Lite, http://www.sve.man.ac.uk/Research/AtoZ/ILCT.
[35] PyOGSI, http://dsd.lbl.gov/gtg/projects/pyOGSI/.
[36] Butler, R., Engert, D., Foster, I., Kesselman, C., Tuecke, S., Volmer, J. and Welch, V. (2000). Design and Deployment of a National-Scale Authentication Infrastructure. *IEEE Computer*, 33(12): 60–66.
[37] IBM, Microsoft and VeriSign. (2002). *Web Services Security Language (WS-Security)*.
[38] XML Encryption, http://www.w3.org/Encryption/2001/.
[39] XML Signature, http://www.w3.org/TR/xmldsig-core/.
[40] Welch, V., Siebenlist, F., Foster, I., Bresnahan, J., Czajkowski, K., Gawor, J., Kesselman, C., Meder, S., Pearlman, L. and Tuecke, S. (2003). *Security for Grid Services*, http://www.globus.org/security/GSI3/GT3-Security-HPDC.pdf.

2.11 REFERENCES

[41] Czajkowski, K., Foster, I., Karonis, N., Kesselman, C., Martin, S., Smith, W. and Tuecke, S. (1998). *A Resource Management Architecture for Metacomputing Systems*. Proceedings of IPPS/SPDP '98 Workshop on Job Scheduling Strategies for Parallel Processing, pp. 62–82, Orlando, FL, USA. Lecture Notes in Computer Science, Springer-Verlag.
[42] Condor, http://www.cs.wisc.edu/condor/.
[43] Sun Grid Engine, http://wwws.sun.com/software/Gridware/.
[44] LSF, http://www.platform.com/products/LSF/.
[45] PBS, http://www.openpbs.org/.
[46] GT3 Index Service, http://www.globus.org/ogsa/releases/final/docs/infosvcs/indexsvc_overview.html.
[47] GT3 RFT, http://www-unix.globus.org/toolkit/docs/3.2/rft/key/index.html.
[48] GridFTP, http://www.globus.org/dataGrid/Gridftp.html.
[49] Apache Ant, http://ant.apache.org/.
[50] OGSA-DAI, http://www.ogsadai.org.uk/.
[51] DAIS, http://www.Gridforum.org/6_DATA/dais.htm.
[52] Frey, J., Graham, S., Czajkowski, K., Ferguson, D.F., Foster, I., Leymann, F., Maguire, T., Nagaratnam, N., Nally, M., Storey, T., Sedukhin, I., Snelling, D., Tuecke, S., Vambenepe, W. and Weerawarana, S. (March 2004). Web Services Resource Lifetime, Version 1.1, http://www-106.ibm.com/developerworks/library/ws-resource/ws-resourcelifetime.pdf.
[53] Graham, S., Czajkowski, K., Ferguson, D.F., Foster, I., Frey, J., Leymann, F., Maguire, T., Nagaratnam, N., Nally, M., Storey, T., Sedukhin, I., Snelling, D., Tuecke, S., Vambenepe, W. and Weerawarana, S. (March 2004). Web Services Resource Properties, Version 1.1, http://www-106.ibm.com/developerworks/library/ws-resource/ws-resourceproperties.pdf.
[54] Web Services Notification, http://www-106.ibm.com/developerworks/library/specification/ws-notification/.
[55] Graham, S., Niblett, P., Chappell, D., Lewis, A., Nagaratnam, N., Parikh, J., Patil, S., Samdarshi, S., Sedukhin, I., Snelling, D., Tuecke, S., Vambenepe, W. and Weihl, B. (March 2004). Web Services Base Notification, Version 1.0, ftp://www6.software.ibm.com/software/developer/library/ws-notification/WS-BaseN.pdf.
[56] Graham, S., Niblett, P., Chappell, D., Lewis, A., Nagaratnam, N., Parikh, J., Patil, S., Samdarshi, S., Sedukhin, I., Snelling, D., Tuecke, S., Vambenepe, W. and Weihl, B. (March 2004). Web Services Brokered Notification, Version 1.0, ftp://www6.software.ibm.com/software/developer/library/ws-notification/WS-BrokeredN.pdf.
[57] Graham, S., Niblett, P., Chappell, D., Lewis, A., Nagaratnam, N., Parikh, J., Patil, S., Samdarshi, S., Sedukhin, I., Snelling, D., Tuecke, S., Vambenepe, W. and Weihl, B. (2004) Web Services Topics, Version 1.0, ftp://www6.software.ibm.com/software/developer/library/ws-notification/WS-Topics.pdf.
[58] Tuecke, S., Czajkowski, K., Frey, J., Foster, I., Graham, S., Maguire, T., Sedukhin, I., Snelling, D. and Vambenepe, W. (March 2004). Web Services Base Faults, Version 1.0, http://www-106.ibm.com/developerworks/library/ws-resource/ws-basefaults.pdf.
[59] Graham, S., Maguire, T., Frey, J., Nagaratnam, N., Sedukhin, I., Snelling, D., Czajkowski, K., Tuecke, S. and Vambenepe, W. (March 2004).

WS-ServiceGroup Specification, Version 1.0, http://www-106.ibm.com/developerworks/library/ws-resource/.
[60] WS-Addressing, http://www.w3.org/Submission/2004/SUBM-ws-addressing-20040810/.
[61] Foster, I., Frey, J., Graham, S., Tuecke, S., Czajkowski, K., Ferguson, D., Leymann, F., Nally, M., Sedukhin, I., Snelling, D., Storey, T., Vambenepe, W. and Weerawarana, S. (3 March 2004). Modelling Stateful Resources with Web Services, Version 1.1, http://www-106.ibm.com/developerworks/library/ws-resource/ws-modelingresources.pdf.
[62] Czajkowski, K., Ferguson, D., Foster, I., Frey, J., Graham, S., Maguire, T., Snelling, D. and Tuecke, S. (12 February 2004). From Open Grid Services Infrastructure to WS-Resource Framework: Refactoring & Evolution, Version 1.0, http://www-106.ibm.com/developerworks/library/ws-resource/gr-ogsitowsrf.html.
[63] GESA-WG, http://forge.gridforum.org/projects/gesa-wg.
[64] RUS-WG, http://forge.gridforum.org/projects/rus-wg.
[65] GRAAP-WG, http://forge.gridforum.org/projects/graap-wg.
[66] McCarty, B. and Dorion, L.C. (1999). *Java Distributed Objects*. Sams, Indianapolis, Indiana.
[67] Ferreira, L., Jacob, B., Slevin, S., Brown, M., Sundararajan, S., Lepesant, J. and Bank, J. The Globus Toolkit 3.0 Quick Start, *IBM Redbook*, http://www.redbooks.ibm.com/redpapers/pdfs/redp3697.pdf.
[68] The GT3 Tutorial, http://gdp.globus.org/gt3-tutorial/.

3

The Semantic Grid and Autonomic Computing

LEARNING OUTCOMES

In this chapter, we will study the Semantic Grid and autonomic computing. From this chapter, you will learn:

- What the Semantic Grid is about.
- The technologies involved in the development of the Semantic Grid.
- The state-of-the-art development of the Semantic Grid.
- What autonomic computing is about.
- Features of autonomic computing.
- How to apply autonomic computing techniques to Grid services.

CHAPTER OUTLINE

3.1 Introduction
3.2 Metadata and Ontology in the Semantic Web
3.3 Semantic Web Services
3.4 A Layered Structure of the Semantic Grid

The Grid: Core Technologies   Maozhen Li and Mark Baker
© 2005 John Wiley & Sons, Ltd

3.5 Semantic Grid Activities

3.6 Autonomic Computing

3.7 Chapter Summary

3.8 Further Reading and Testing

3.1 INTRODUCTION

The concept of the Semantic Grid [1] is evolved through the concurrent development of the Semantic Web and the Grid. The Semantic Web can be defined as "an extension of the current Web in which information is given well-defined meaning, better enabling computers and people to work in cooperation" [2]. The aim of the Semantic Web is to augment unstructured Web content so that it may be machine-interpretable information to improve the potential capabilities of Web applications. The aim of the Semantic Grid is to explore the use of Semantic Web technologies to enrich the Grid with semantics. The relationship between the Grid, the Semantic Web and the Semantic Grid is shown in Figure 3.1. The Semantic Grid is layered on top of the Semantic Web and the Grid. It is the application of Semantic Web technologies to the Grid. Metadata and ontologies play a critical role in the development of the Semantic Web. Metadata can be viewed as data that is used to describe data. Data can be annotated with metadata to specify its origin or its history. In the Semantic Grid, for example, Grid services can be annotated with metadata associated with an ontology for automatic service discovery. An ontology is a specification of a conceptualization [3]. We will explain metadata and ontology in Section 3.2.

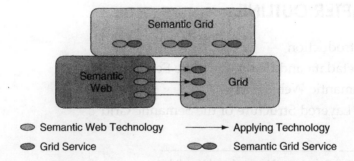

Figure 3.1 The Semantic Web, Grid and Semantic Grid

The Grid is complex in nature because it tries to couple distributed and heterogeneous resources such as data, computers, operating systems, database systems, applications and special devices, which may run across multiple virtual organizations to provide a uniform platform for technical computing. The complexity of managing a large computing system, such as the Grid, has led researchers to consider management techniques that are based on strategies that have evolved in biological systems to deal with complexity, heterogeneity and uncertainty. The approach is referred to autonomic computing [4]. An autonomic computing system is one that has the capabilities of being self-healing, self-configuring, self-optimizing and self-protecting.

This chapter is organized as follows. In Section 3.2, we introduce the ontological languages involved in the development of the Semantic Web. In Section 3.3, we describe how to enrich standard Web services with semantics to provide Semantic Web services. In Section 3.4, we present a layered structure of the Semantic Grid. In Section 3.5, we review the state-of-the-art development of the Semantic Grid. In Section 3.6, we introduce autonomic computing and explain what kinds of benefits it could bring to the Grid. We conclude this chapter in Section 3.7. Finally, in Section 3.8, we provide further readings.

3.2 METADATA AND ONTOLOGY IN THE SEMANTIC WEB

The Semantic Web provides a common framework that allows data to be shared and reused across applications, enterprises and community boundaries. It is a collaborative effort led by W3C [5] with participation from a large number of researchers and industrial partners. The key point of the Semantic Web is to convert the current structure of the Web as a distributed data storage, which is interpretable only by human beings, into a structure of information storage that can be understood by computer-based entities. In order to convert data into information, metadata has to be added into context. The metadata contains the semantics, the explanation of the data to which it refers. Metadata and ontology are critical to the development of the Semantic Web.

Now we give a simple example to show how to use metadata and ontologies to match a service with semantic meanings.

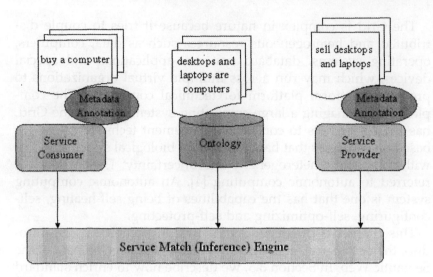

Figure 3.2 Metadata and ontology in semantic service matching

As shown in Figure 3.2, a service consumer is buying a computer. The service request information can be annotated with metadata (perhaps encoded as XML) to describe the service request, e.g. a preferable computer configuration and price. A quote service provided by a vendor selling desktops and laptops can also be annotated with metadata to describe the service. When the service-matching engine receives the two metadata sets related to the service request and quote service, the engine will access the ontology which defines that desktops and laptops are computers. Then the engine will make an inference whether the quote service can satisfy the service request or not.

Metadata and ontologies play a critical role in the development of the Semantic Web. An ontology is a specification of a conceptualization. In this context, specification refers to an explicit representation by some syntactic means. In contrast to schema languages such as XML Schema, ontologies try to capture the semantics of a domain by using knowledge representation primitives, allowing a computer to fully or partially understand the relationships between concepts in a domain. Ontologies provide a common vocabulary for a domain and define the meaning of the terms and the relationships between them. Ontology is referred to as the shared understanding of some domain of interest, which is often conceived as a set of classes (concepts), relations, functions, axioms

3.2 METADATA AND ONTOLOGY IN THE SEMANTIC WEB

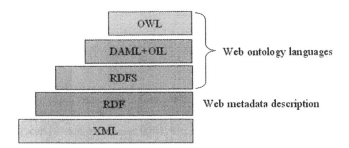

Figure 3.3 The layered structure of the Semantic Web

and instances. Concepts in the ontology are usually organized in taxonomies [6].

In the following sections, we introduce Resource Description Framework (RDF) [7] which is the foundation of the Semantic Web. We also present, as shown in Figure 3.3, RDF-based Web ontology languages such as RDF Schema (RDFS) [8], DAML+OIL [9, 10] and Web Ontology Language (OWL) [11].

3.2.1 RDF

The goal of the Semantic Web is to augment unstructured content of the Web into structured machine-understandable content to improve the efficiency in its access and information discovery. The effective use of metadata among Web applications, however, requires conventions about syntax, structure and semantics. Individual resource description communities define the semantics or meaning, of metadata that address their particular needs. Syntax, which is the systematic arrangement of data elements for machine processing, facilitates the exchange and use of metadata among multiple applications. Structure can be thought of as a formal constraint on the syntax for the consistent representation of semantics.

The RDF, developed under the auspices of the W3C, is an infrastructure that facilitates the encoding, exchange and reuse of structured metadata. The RDF infrastructure enables metadata interoperability through the design of mechanisms that support common conventions of semantics, syntax and structure. RDF does not stipulate semantics for each resource description community, but rather provides the ability for these communities to define metadata elements as needed. RDF uses XML as a common syntax for the exchange and processing of metadata. The XML syntax provides vendor independence, user extensibility, validation, human readability and the ability to represent complex structures.

3.2.1.1 RDF development efforts

RDF is the result of a number of metadata communities bringing together their needs to provide a robust and flexible architecture for supporting metadata for the Web. While the development of RDF as a general metadata framework, and as such, a simple knowledge representation mechanism for the Web, was heavily inspired by the PICS specification [12], no one individual or organization invented RDF. RDF is a collaborative design effort. RDF drew upon the XML design as well as proposals related to XML data submitted by Microsoft's XML Data [13] and Netscape's Meta Content Framework [14]. Other metadata efforts, such as the Dublin Core [15] and the Warwick Framework [16], have also influenced the design of RDF.

3.2.1.2 The RDF data model

As shown in Figure 3.4, an RDF data model contains resources, properties and the values of properties. In RDF, a resource is uniquely identifiable by a Uniform Resource Identifier (URI). The properties associated with resources are identified by property-types which have corresponding values. In RDF, *values* may be

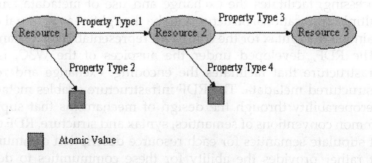

Figure 3.4 The RDF data model

3.2 METADATA AND ONTOLOGY IN THE SEMANTIC WEB

atomic in nature (text strings, numbers, etc.) or other resources, which in turn may have their own properties. RDF is represented as a directed graph in which resources are identified as nodes, property types are defined as directed label arcs, and string values are quoted.

Now let us see how to apply the RDF model for representing RDF statements.

RDF Statement 1: The author of this paper (someURI/thispaper) is John Smith.

Figure 3.5 shows the graph representation of the RDF statement 1. In this example, the RDF resource is *someURI/thispaper* whose property is *author*. The value of the property is *John Smith*.

RDF Statement 2: The author of this paper (someURI/thispaper) is another URI whose name is John Smith.

Figure 3.6 shows the graph representation of the RDF statement 2. In this example, the RDF resource is *someURI/thispaper* whose property is *author*. The value of the property is *another URI* (resource) whose property is *name* and the value of the property is *John Smith*. The RDF statement 2 can be described in XML as shown in Figure 3.7.

3.2.2 Ontology languages

In this section, we outline some representative ontology languages which are based on RDF. These ontology languages can be used to build ontologies on the Web.

Figure 3.5 The graph representation of the RDF statement 1

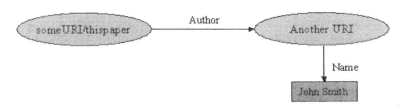

Figure 3.6 The graph representation of the RDF statement 2

```
<rdf:RDF>
  xmlns = "..."
  xmlns:rdf = "..."
  <rdf:Description about = "someURI/thispaper">
    <authored-by>
      <rdf:Description Resource = "anotherURI">
        <name>John Smith</name>
      </rdfDescription>
    </authored-by>
  </rdf:Description>
</rdf:RDF>
```

Figure 3.7 The XML description of the second RDF statement

3.2.2.1 RDFS

RDF itself is a composable and extensible standard for building RDF data models. However, the modelling primitives offered by RDF are very limited in supporting the definition of a specific vocabulary for a data model. RDF does not provide a way to specify resource and property types, i.e. it cannot express the classes to which a resource and its associated properties belong. The RDFS specification, which is built on top of RDF, defines further modelling primitives such as class (*rdfs:Class*), subclass relationship (*subClassOf, subPropertyOf*), domain and range restrictions for property, and sub-property (*rdfs:ConstraintProperty* and *rdfs:ContainerMembershipProperty*). A resource (*rdfs:Resource*) is the base class for modelling primitives defined in RDFS. RDFS define the valid properties in a given RDF description, as well as any characteristics or restrictions of the property-type values themselves.

3.2.2.2 DAML+OIL

RDFS is still a very limited ontology language, e.g. RDFS does not support the definition of properties, the equivalence and disjoint characteristics of classes. DAML+OIL is intended to extend the expressive power of RDFS, and to enable effective automated reasoning.

DAML+OIL is an ontology language designed for the Web, which is built upon XML and RDF, and adds the familiar ontological primitives of object-oriented and frame-based systems [17], as well as the formal rigour of an expressive Description Logic (DL)

[18, 19]. The logical basis of DAML+OIL means that reasoning services can be provided both to support ontology design and to make Web data more accessible to automated processes.

DAML+OIL evolved from a merger of DARPA Agent Markup Language's (DAML) initial ontology language (DAML−ONT) [20], an earlier DAML ontology language, and the Ontology Inference Layer (OIL) [21], an ontology language that couples modelling primitives commonly used in frame-based ontologies, with a simple and well-defined semantics of an expressive DL. DAML+OIL is modelled through an object-oriented approach, and the structure of the domain is described in terms of classes and properties. DAML+OIL classes can be names (URIs) or expressions and a variety of constructors are provided for building class expressions. The axioms supported by DAML+OIL make it possible to assert subsumption or equivalence with respect to classes or properties, the disjoint characteristics of classes, the equivalence or non-equivalence of individuals and various properties of properties. Classes can be combined using conjunction, separation and negation. Within properties both universal and existential quantification are allowed, as well as more exact cardinality constraints. Range and domain restrictions are allowed in the definition of properties, which themselves can be arranged in hierarchies.

In summary, DAML+OIL has the following features:

- DAML+OIL has well-defined semantics and clear properties via an underlying mapping to an expressive DL. The DL gives DAML+OIL the ability and flexibility to compose classes and slots to form new expressions. With the support of DL, an ontology expressed in DAML+OIL can be automatically reasoned by a DL reasoning system such as the FaCT system [22, 23].
- DAML+OIL supports the full range of XML Schema data types. It is tightly integrated with RDFS, e.g. RDFS is used to express DAML+OIL's machine-readable specification, and provides a serialization for DAML+OIL.
- A layered architecture for easy manipulation of the language.
- The DAML+OIL axioms are significantly more extensive than the axioms for either RDF or RDFS.

While the dependence on RDFS has some advantages in terms of the reuse of existing RDFS infrastructure and the portability of DAML+OIL ontologies, using RDFS to completely define the structure of DAML+OIL has proved quite difficult as, unlike XML, RDFS is not designed for the precise specification of syntactic structure [24].

3.2.2.3 OWL

The OWL facilitates greater machine interpretation of Web content than that supported by XML, RDF and RDFS, by providing additional vocabulary along with a formal semantics. OWL is derived from DAML+OIL, which provided a starting point for the W3C Web Ontology Working Group [25] in defining OWL, the language that is aimed to be the standardized and broadly accepted ontology language of the Semantic Web. The OWL Use Cases and Requirements Document [26] provides more details on ontologies, it provides the motivation for a Web Ontology Language in terms of six use cases, and formulates design goals, requirements and objectives for OWL.

OWL has three increasingly expressive sub-languages: OWL Lite, OWL DL (Description Logic) and OWL Full.

- *OWL Lite* supports a classification hierarchy and simple constraints, e.g. while it supports cardinality constraints, it only permits cardinality values of 0 or 1. *OWL Lite* is easy to use and implement.
- *OWL DL* supports the maximum expressiveness while retaining computational *completeness* (all conclusions are guaranteed to be computable) and *decidability* (all computations will finish in finite time). *OWL DL* includes all OWL language constructs, but they can be used only under certain restrictions, e.g. while a class may be a subclass of many classes, a class cannot be an instance of another class.
- *OWL Full* uses all the OWL languages primitives and allows the combination of these primitives in arbitrary ways with RDF and RDFS. It supports maximum expressiveness and the syntactic freedom of RDF with no computational guarantees, e.g. a class in *OWL Full* can be treated simultaneously as a collection of individuals and as an individual in its own right. *OWL Full* allows an ontology to augment the meaning of the pre-defined (RDF or OWL) vocabulary. It is unlikely that any reasoning

software will be able to support complete reasoning for every feature of *OWL Full*.

The advantage of *OWL Full* is that it is completely compatible with RDF both syntactically and semantically: any legal RDF document is also a legal *OWL Full* document; and any valid RDF/RDFS conclusion is also a valid *OWL Full* conclusion.

Antoniou and Harmelen [27] provide a good review of OWL. They suggest that when using OWL, developers should consider which sub-language best suits their needs. The selection of *OWL Lite* depends on the extent to which users require the more-expressive constructs provided by *OWL DL* and *OWL Full*. The choice between *OWL DL* and *OWL Full* mainly depends on the extent to which users require the meta-modelling facilities of RDFS, e.g. defining classes of classes or attaching properties to classes. When using *OWL Full* instead of *OWL DL*, reasoning support is less predictable since complete *OWL Full* implementations will be unlikely. There are strict notions of upward compatibility between these three sub-languages:

- Every legal *OWL Lite* ontology is a legal *OWL DL* ontology.
- Every legal *OWL DL* ontology is a legal *OWL Full* ontology.
- Every valid *OWL Lite* conclusion is a valid *OWL DL* conclusion.
- Every valid *OWL DL* conclusion is a valid *OWL Full* conclusion.

3.2.3 Ontology editors

In this section, we briefly introduce three representative ontology editors that support RDFS, DAML+OIL or OWL. These editors are software tools that can be used to build ontologies. A more detailed survey on ontology editors can be found in Denny [28].

3.2.3.1 OntoEdit

OntoEdit [29, 30] provides a graphical environment for the development and maintenance of ontologies. It supports F-Logic [31], RDFS and DAML+OIL. Ontologies in OntoEdit can be exported to object-relational database schema and Document Type Definitions (DTDs).

3.2.3.2 OilEd

OilEd [32] is an ontology editor allowing the user to build ontologies using DAML+OIL. Basic functionality in OilEd includes the definition and description of classes, slots, individuals and axioms within an ontology. OilEd provides a graphical user interface for editing ontologies.

3.2.3.3 Protégé

Protégé [33, 34] is an extensible, platform-independent and graphical environment for creating and editing ontologies and knowledge bases. Protégé supports DAML+OIL, and it provides beta-level support for editing Semantic Web ontologies in OWL.

3.2.4 A summary of Web ontology languages

So far we have reviewed RDF, RDFS, DAML+OIL and OWL, which are ontology languages to build ontologies for the Semantic Web. The aim of the Semantic Web is to augment the unstructured Web content as structured information and to improve the efficiency of Web information discovery and machine-readability. RDF lays the foundation for the conversion, in that structured information can be expressed with RDF-based metadata. Ontology languages such as RDFS, DAML+OIL and OWL can be used to construct metadata ontologies for a more expressive and structured information on the Web. Both DAML+OIL and OWL try to overcome the limitations of RDFS. However, they are based on RDFS and attempt to be compatible with it, to reuse the effort already invested into RDF and RDFS. Derived from DAML+OIL, OWL is an emerging standard ontology language for the Semantic Web.

3.3 SEMANTIC WEB SERVICES

As we have studied in Chapter 2, Web services are emerging as a promising computing platform for heterogeneous distributed systems. The three core standards in Web services are WSDL for service description, SOAP for message exchange and UDDI for service registration and discovery. A feature of Web services is their support for services composition. It is desirable and necessary for a Web service to automatically find another service in the composition process, which requires that Web services should be enriched with semantics.

One overarching characteristic of the Web services infrastructure is its lack of semantic support. It relies exclusively on XML for interoperation, but that guarantees only syntactic interoperability. Expressing message content in XML lets Web services parse each other's messages, but it does not facilitate the understanding of the messages' content. In addition, in service registration and discovery, UDDI itself does not provide any support for semantic description of a Web service. Web services should have semantic meanings so that services can be matched semantically instead of syntactically. In this section, we introduce DAML-S and OWL-S that can be used to reach this goal.

3.3.1 DAML-S

DAML-S [35] is both a language and an ontology for describing Web services. It attempts to close the gap between the Semantic Web and Web services. As an ontology, it uses DAML+OIL-based constructs to describe Web services; as a language, DAML-S supports the description of specific Web services that users or other services can discover and invoke using standards such as WSDL and SOAP. DAML-S uses semantic annotations and ontologies to relate each Web service's description to a description of its operational domain. The DAML-S ontology describes a set of classes and properties, specific to the description of Web services.

As a DAML+OIL ontology, DAML-S has all the benefits of being capable of utilizing any content described in DAML+OIL. DAML-S has a well-defined semantics and allows the definition of service content vocabulary in terms of objects and their complex relationships, including class, subclass relations and cardinality restrictions. The DAML-S ontology consists of three parts, as shown in Figure 3.8, and described as follows.

- *ServiceProfile*: This is like the Yellow Pages entry for a service. It relates and builds upon the type of content found in UDDI, describing properties of a service necessary for automatic

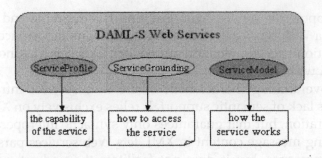

Figure 3.8 DAML-S Web services

discovery, such as what the services offers, and its inputs, outputs and its side effects (preconditions and effects).

- *ServiceModel*: Describes a service's process model, e.g. the control flow and data flow involved in using the service. It is the process model that provides a declarative description of the properties of the Web-accessible programs we wish to reason about. The ServiceModel is designed to allow the automated composition and execution of services.
- *ServiceGrounding*: Connects the process model description to communication-level protocols and message descriptions in WSDL.

A DAML-S-matching engine has also been implemented that allows services to advertise with DAML-S as well as with a UDDI registry so that these services can be discovered by using a UDDI keyword search.

3.3.2 OWL-S

OWL-S [36] is derived from DAML-S; it uses OWL as the ontology language to semantically describe Web services. OWL-S describes the properties, capabilities and process model of a Web service. It allows Web services to be described and discovered, to interoperate, and be composed in an unambiguous, computer-interpretable form. OWL-S elements can be mapped to a WSDL specification, in order to support automatic invocation and execution of a Web service.

3.4 A LAYERED STRUCTURE OF THE SEMANTIC GRID

As we have studied in Chapter 2, OGSA is the *de facto* standard for building service-oriented Grid applications. From a service-oriented point of view, the Semantic Grid can be divided into four service layers – base services, data services, information services and knowledge services. The layered structure is shown in Figure 3.9.

Base services
This layer is primarily concerned with large-scale pooling of computational resources. The base services provided by this layer are related to resource discovery, allocation and monitoring, user authentication, task scheduling or co-scheduling and fault tolerance.

Data services
This layer mainly provides computationally intensive analysis of large-scale-shared data sets or databases, which could range in size from hundreds of terabytes to petabytes, across widely distributed scientific communities. The services provided by this layer are related to data storage, metadata management, data replication and data transfer.

Information services
This layer allows uniform access to heterogeneous information sources and provides commonly used services running on distributed computational resources. Uniform access to information

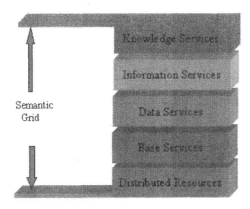

Figure 3.9 A layered structure of the Semantic Grid

sources relies on metadata to describe information and to help with integration of heterogeneous resources. The granularity of the offered services can vary from subroutine or method calls to complete applications. Hence, in scientific computing, services can include the availability of specialized numerical solvers, such as a matrix or partial differential equation solver, to complete scientific codes for applications such as weather forecasting and molecular or fluid dynamics. In commercial computing, services can be statistical routines based on existing libraries or predictive services that offer coarse-grained functionality, such as database profiling or visualization. Services in this layer can, therefore, be offered by individual providers or by corporations; they may be specialized for specific applications, such as genomic databases or general purpose, such as numerical libraries.

Knowledge services
This layer focuses on knowledge representation and extraction. It provides services that can be used to search for patterns in existing data repositories, and the management of information services, e.g. it can provide knowledge discovery from a huge amount of data using a variety of data-mining mechanisms. It can provide semantic meaning of information services aggregated from the information services layer. This layer is domain-oriented such as bioinformatics, and usually uses domain knowledge built with its own ontology.

It is intended that each of these layers provide services to various applications. A substantial part of the research effort dedicated to the Grid has concentrated on the computational and data services layers. However, growing interest in the recently established "Semantic Grid" working group at the Global Grid Forum (GGF) indicates the importance of services provided by the Semantic Grid.

3.5 SEMANTIC GRID ACTIVITIES

The Semantic Grid is a promising area of research. In the context of the Semantic Grid, apart from computational services, the Grid can also provide domain-specific problem-solving and knowledge-based services. A Grid application can be automatically composed from Grid services based on semantically matching the needs of an application. However, the Semantic Grid is still in its infancy.

3.5 SEMANTIC GRID ACTIVITIES

In this section, we present some of the Semantic Grid research that is currently being undertaken.

3.5.1 Ontology-based Grid resource matching

As we will discuss in Chapter 6, a Grid scheduling system performs resource description and selection when scheduling jobs to resources. However, as indicated in Tangmunarunkit *et al.* [37], existing resource description and selection mechanisms in the Grid are too restrictive. Traditional resource matching, as exemplified by the Condor Matchmaker or Portable Batch System (PBS) that will be described in Chapter 6, is based on symmetric, attribute-based matching. In these systems, the values of attributes advertised by resources are compared with those required by jobs or tasks. For a comparison to be meaningful and effective, the resource providers and consumers have to agree upon attribute names and values. The exact matching and coordination between providers and consumers make such systems inflexible and difficult to extend to new characteristics or concepts. Moreover, in a heterogeneous multi-institutional environment such as the Grid, it is difficult to enforce the syntax and semantics of resource descriptions.

Tangmunarunkit *et al.* [37] present a flexible and extensible approach for performing Grid resource selection using an RDFS ontology-based matchmaker which performs semantic matching using terms defined in those ontologies instead of exact syntax matching. The loose coupling between resource and request descriptions removes the tight coordination required between resource providers and consumers. Unlike traditional Grid resource selectors that describe resource/request properties based on symmetric and flat attributes (which might become unmanageable as the number of attributes grows), separate ontologies are created to declaratively describe resources and job requests using an expressive ontology language. Figure 3.10 shows the layout of the matchmaker.

The ontology-based matchmaker consists of three components:

- *Domain ontologies*: Provides the domain model and vocabulary for expressing resource advertisements and job requests.
- *Domain background knowledge*: Captures additional knowledge about the domain.
- *Matchmaking rules*: Defines when a resource matches a job description.

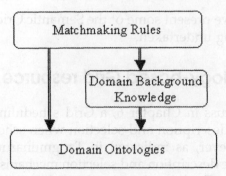

Figure 3.10 The layout of the ontology-based resource matchmaker

A matchmaker prototype has been implemented based on TRIPLE/XSB [38], a deductive database system that supports RDFS and TRIPLErule language. Protégé, an ontology editor that supports the RDFS, is used to build ontologies. This work is at an early stage of development, and the developers do intend to compare their efforts with the existing resource matchmakers, such as Condor Matchmaker.

3.5.2 Semantic workflow registration and discovery in myGrid

As we will discuss in Chapter 7, workflow systems provide users with the ability to build and manage composite applications. A Grid workflow system supports the construction of applications from Grid services. Once a workflow is constructed, the workflow system will generate a description using a flow language. It is desirable and necessary that the creation of a new workflow should reuse existing workflow descriptions instead of starting from scratch, which leads to the need for further workflow registration and discovery. To quickly and precisely locate a workflow, the workflow should be annotated with metadata to semantically describe itself. myGrid [39] supports this feature, and has a focus on semantic workflow registry and discovery for *in silico* experiments.

In myGrid, a UDDI-based registry has been implemented as a Web service combined with an information model for semantic workflow registration. The metadata associated with a workflow can be annotated with RDFS or OWL. The metadata could be a

3.5 SEMANTIC GRID ACTIVITIES

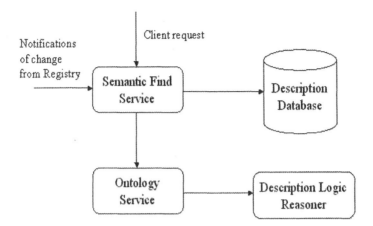

Figure 3.11 The semantic find service in myGrid

simple string recording, e.g. an estimate of the average time a workflow takes to execute. Alternatively, it can be the URI of a concept in the ontology.

The semantic find service provides discovery over specific descriptions by reference to domain ontologies. The find service makes use of several additional components as shown in Figure 3.11. The description database holds semantic descriptions gathered from resources published in registries and views, which act as proxies displaying a subset of services registered in a registry. The ontology service provides access to the domain ontologies and manages interaction with FaCT (Fast Classification of Terminologies), a description logic reasoner.

In summary, the find service is mainly responsible for:

- Gathering semantic descriptions from a view and maintaining a reference back to the entry in the view so that details for communicating with the services can later be retrieved.
- Using the ontology service and associated reasoner to index items in the descriptions database to ensure efficient retrieval of entries at the time of discovery.

3.5.3 Semantic workflow enactment in Geodise

In the Geodise project, efforts have been focused on the application of Semantic Web technologies to assist users in solving complex problems in Engineering Design Search and Optimization (EDSO) [40], in particular, allowing semantically enriched resource sharing and reuse. Geodise provides the following semantic support for the Grid.

3.5.3.1 EDSO ontologies

The acquisition of knowledge in the EDSO domain has been collected and modelled as either ontologies or rules in knowledge bases. DAML+OIL and OWL are used to build EDSO ontologies and DAML-S is used to specify the properties and functionality of EDSO services (tasks). An ontology service has been implemented as a Web service, which is independent of any specific domain, to facilitate the deployment of the EDSO ontology.

The ontology service consists of four components:

- An underlying data model that holds the ontology (the knowledge model) and allows the application to interact with it through a well-defined API;
- An ontology server that provides access to concepts in an underlying ontology data model and their relationships;
- The FaCT reasoner provides reasoning capabilities;
- A set of user APIs that interface user's applications and the ontology.

By using the service's APIs and the FaCT reasoner, common ontological operations – such as checking the relationship between two concepts, retrieving information, navigating concept hierarchies and retrieving lexical information – can be performed when required.

3.5.3.2 Semantic annotation of EDSO resources

The goal of semantic annotation is to add semantics to Web pages and documents as well as computational resources. In Geodise, OntoMat Annotizer [41] is used to perform semantic annotation. In addition, EDSO resources, such as tasks, have been semantically described to support automatic semantic enrichment, e.g. a workflow composed of semantic EDSO tasks can be automatically

3.5 SEMANTIC GRID ACTIVITIES

enriched with a semantic description. With semantic task descriptions, a semantically enriched task archive can be created based on previously performed tasks, which can be searched and reused.

3.5.3.3 Semantic workflow enactment

Geodise provides tools to support the graphical construction of workflows, which are composed from semantic EDSO tasks. Tools, such as the Ontology Concept Browser, the Workflow Editor, the Workflow Advisor, the Component Editor, the Ontological Reasoner and the State Monitor, have been implemented to assist the workflow construction process; as shown in Figure 3.12, the functionality of each component is described below.

- The *Ontology Concept Browser* presents the conceptual models of the EDSO tasks in a hierarchical structure. Every task is described with properties which specify the relations among conceptual task models.
- The *Component Editor* is used for task definition. It dynamically generates an ontology-driven form in which each slot of the form

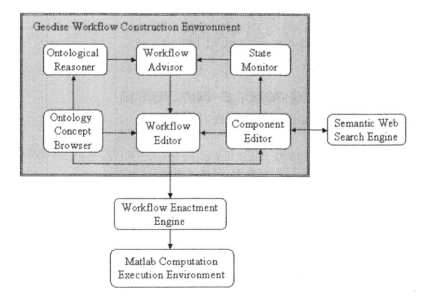

Figure 3.12 Workflow enactment in Geodise

represents a property of a task with an explicitly specified ontological concept type – the semantic link. A task can be defined by specifying every property following the ontological links or reused from an existing task.

- The *Semantic Web Search Engine* provides the ability to search for similar tasks in terms of algorithm performance, run time or accuracy of these tasks.
- The *Workflow Editor* provides editing functions such as modification and removal of functions as well as the graphical representation of tasks and workflow.
- The *State Monitor* holds state information about each task such as its inputs and output parameters.
- The *Ontological Reasoner* performs ontological reasoning based on a task's ontology and its state information.
- *The Workflow Advisor* gives advice on which task(s) should be undertaken next.
- The *Workflow Enactment Engine* resolves an abstract specification of a task in a workflow into a concrete task instance and establishes dynamic binding for service invocation. Matlab has been used as a computation environment; therefore, the workflow enactment engine will convert an ontology-represented workflow to a Matlab script file.
- The *Matlab Computation Execution Environment* provides the exact environment for the execution of EDSO tasks.

3.5.4 Semantic service annotation and adaptation in ICENI

ICENI provides the following support for semantic service annotation and adaptation based on RDF and OWL [42].

- *Metadata space*: ICENI introduces the concept of *metadata space* which is an environment with a standard metadata publication and discovery protocol to facilitate the processing of metadata and semantic interaction between Grid resources. The advantage of the metadata space is that it decouples Grid resources that have metadata from their implementations and hosting environments. Every participant in the metadata space is characterized

3.5 SEMANTIC GRID ACTIVITIES

as a metadata publisher. The published metadata falls into one of the three categories – requirement, implementation or domain.

- *Service implementation publisher*: The *service implementation publisher* behaves as a typical service provider within its hosting environment and projects its semantic representation into the metadata space by the publication of its semantic annotation.
- *Service requirement publisher*: A *service requirement publisher* is any Grid service consumer with the capability of publishing requirements with semantic annotations into the metadata space.
- *Ontology publisher*: The *ontology publisher* writes and publishes ontology information into the metadata space.
- *Semantic service annotation*: The annotation of services with semantics in ICENI has to be manually undertaken by a user. Different aspects of a service method's signature need to be described in RDF with a concentration on expressing the syntactic meaning of the service by annotating the semantics of the definition of the service method.
- *Semantic matching*: The requirements for a service interface method from a user are expressed through the semantic annotation of the interface, which will be semantically matched with a service implementation's annotation provided by a *service implementation publisher*. The match is made based on class and property inference in which class properties are represented by method signatures.
- *Service adaptation*: Upon receiving a list of conceptually equivalent implementations from the matching service, the *service adaptation* dynamically generates the adaptor proxy on demand.

The work on semantic service annotation and adaptation is currently being integrated into ICENI.

3.5.5 PortalLab – A Semantic Grid portal toolkit

As we will discuss in Chapter 8, in the development of portals, the use of portlets is gaining increasing attention from the Grid community for building portals. However, it is envisioned that more and more portlets will be developed by this community and it will be increasingly difficult for users to quickly find the right

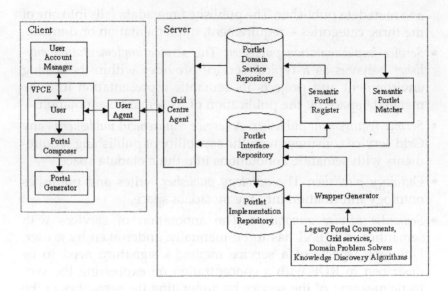

Figure 3.13 The software architecture of PortalLab

portlet they need. Portlets should be annotated with semantics to improve the efficiency of portlet discovery. PortalLab [43] is an ongoing project towards this goal. It is a GT3-based toolkit for building Semantic Grid portals with portlets. Figure 3.13 shows the software architecture of PortalLab. The components in PortalLab are described below.

3.5.5.1 Portlets in PortalLab

There are three kinds of portlets in PortalLab: Web-page portlets, Grid system portlets and Grid application portlets. Web-page portlets are used to deal with user inputs and outputs through Web pages. Grid system portlets provide Grid system level services such as job submission and monitoring, user authentication and authorization, and data transfer. Grid application portlets provide domain-specific services for problem-solving (solvers) and knowledge-related activities, such as extracting knowledge from a dataset using a data-mining algorithm. Portlets with semantics are called Semantic Grid Portlets (SGP). An OGSA compliant SGP is called OGSA-SGP, which is associated with an OGSA compliant Grid service.

3.5 SEMANTIC GRID ACTIVITIES

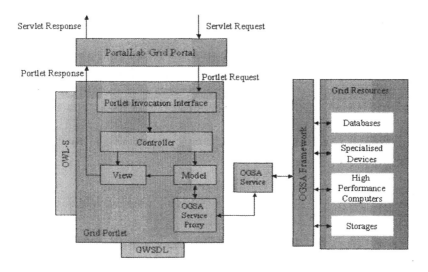

Figure 3.14 The OGSA compliant SGP infrastructure

As shown in Figure 3.14, an OGSA-SGP is associated with a Grid service managed by an OGSA framework such as the GT3. An OGSA-SGP can directly access an OGSA service through an OGSA service proxy. In addition, an OGSA-SGP has an interface defined in extended WSDL, which is called GWSDL in GT3. Each SGP in PortalLab can be annotated via OWL-S to specify its semantic meanings. When a portal receives a Java Servlet request, it generates and dispatches events to the portlet using parameters in the request and then invokes the portlet to be displayed through the portlet invocation interface (PII). The portlet internal design normally follows the MVC model [44], which separates the portlet functionality into a controller receiving incoming requests from the portlet PII, invoking commands operating on a model that encapsulates application data and logic to access backend resources. A portlet ontology built with OWL-S can be used to semantically match portlets.

3.5.5.2 The Visual Portal Composition Environment (VPCE)

A portal can be visually composed by plugging and playing with portlets by the toolkit's integrated VPCE using a Portal Composer (PC). The VPCE also incorporates a Portal Generator (PG) which will generate a portal once its composition is complete. The VPCE also supports portlet workflow enactment in which portlets can be visually connected and then published as a portlet. The PG will check the compatibility of the portlets used in a portal and will generate a task graph that can be described in a flow language. The task graph describes the dependency relationships between portlets. A portal template with some basic functionality can be easily customized to meet user-specific requirements to speed up overall portal composition.

3.5.5.3 Portlet repositories

There are three levels of portlet repositories in the toolkit, Portlet Interface Repository (PInR), Portlet Implementation Repository (PImR) and Portlet Domain Service Repository (PDSR).

- The PDSR is used to register a portlet with semantic capabilities. A Web-page portlet entry in the PDSR describes how data can be organized in a Web page. A Grid system portlet entry in the PDSR describes the functionality and the system requirements of the portlet. A Grid application portlet entry in the PDSR describes portlet ontology, portlet constraint and semantic data requirement. A service provider can also add quality of services to each portlet in the PDSR. For an application portlet, such as a domain problem solver, the quality of service describes the extent to which a domain problem can be solved. For a system portlet, such as a domain-related problem solver, the quality of service describes the job type that the portlet is most suitable for processing.
- The PInR is used to store interface-related information and an OGSA Grid Service Handle (GSH) of the Grid service associated with it. An interface describes how to use each portlet's input/output parameters. For example, the interface of a Web-page portlet describes that the input could be a plain data file marked up in HTML or XML, and the output could be an image or a table. Each portlet interface in the PInR has a unique entry in the PDSR.
- The PImR is used to store the implementation of each portlet through an OGSA Grid Service Reference (GSR) associated with a Grid service.

3.5.5.4 Semantic Portlet Matcher

The Semantic Portlet Matcher (SPM), a matching engine, is used to semantically match the closest portlet candidate. The matching engine will perform portlet matching of ontologies, constraints, quality of service, and semantic data requirement on each portlet registered with the PDSR.

3.5.5.5 Semantic Portlet Register

The Semantic Portlet Register (SPR) provides a GUI for portlet annotation. Once the annotation of a portlet is finished, the SPR registers the annotation with the PDSR for semantic portlet matching.

3.5.5.6 Agents in PortalLab

When using a Grid service via a portal, a service consumer, such as an end-user, needs to negotiate with a service provider about the terms under which services can be provided. The underlying complexity of the Semantic Grid infrastructure and the need for speedy interaction make software agents the most likely candidate to handle these negotiations. Software agents are autonomous components that have the following characteristics [45, 46]:

- *Autonomous*: Control over their internal states and behaviour;
- *Reactive*: Respond in a timely fashion to environmental changes;
- *Proactive*: Act in anticipation of future goals;
- *Interactive*: Communicate with other agents, perhaps including people too;
- *Deductive*: Deduce the user's interactive intention from their previous actions.

Figure 3.15 shows the data flow for semantically matching a portlet. When building a Grid portal, each user accesses portlets through a User Agent (UA). Each UA acts on behalf of a user as a service consumer and interacts with a Grid System Agent (GSA) which acts as a service provider. A UA can assist a user in formulating a portlet request, and submit the user request to a GSA. Upon receiving a user portlet request, the GSA will first perform

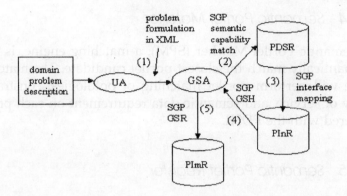

Figure 3.15 The data flow for semantically matching a SGP in PortalLab

a semantic match in the PDSR. If a portlet interface can be found, a service instance will be created and associated with a portlet by referring a GSH to its corresponding GSR.

3.5.5.7 A Portlet Wrapper Generator

Each Grid application portlet can be created from scratch or wrapped from legacy codes using the Portlet Wrapper Generator (PWG). Figure 3.16 shows the structure of the PWG. The PWG provides a GUI for portlet creation, of which there are two stages. First, the PWG will automatically wrap a legacy code as a Grid service factory. At the same time, the PWG also generates a Grid service client template which is used to request the service factory to create an instance. The wrapped Grid service will be automatically deployed in a Grid service container. In the second stage, the PWG will expose the wrapped Grid service as a portlet. To do this, it generates a JSP template for the presentation of the portlet, registers the portlet interface with the PInR and automatically generates an implementation of the portlet which will then be registered with the PImR. The portlet implementation will incorporate a call to the Grid service client through which to call the Grid service instance associated with it.

3.5.5.8 Peer-to-Peer (P2P) support in PortalLab

In PortalLab, each UA interacts with a GSA and multiple GSAs work in a P2P model [47]. When building a Grid portal, a UA will request that a GSA find the SGPs required. If a GSA does not have

3.5 SEMANTIC GRID ACTIVITIES

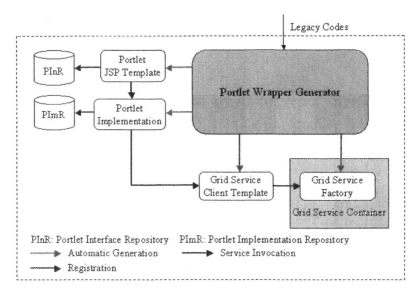

Figure 3.16 The structure of the Portlet Wrapper Generator

the required SGP, it may send a request to other GSAs to ask for the requested SGP. In this way, a Grid portal built from PortalLab may use SGPs provided by different Grid systems which form a Grid-enabled P2P system. (Figure 3.17). The benefits of the P2P

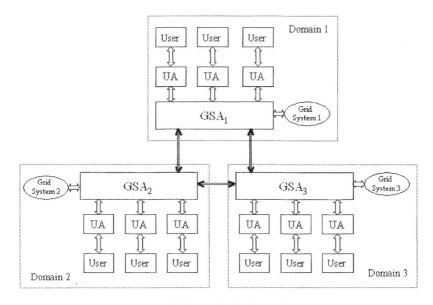

Figure 3.17 A Grid-enabled P2P model in PortalLab

paradigm are portlet interoperability and aggregation, as well as the simpler management of a large Grid environment which may involve several Virtual Organizations (VOs).

3.5.6 Data provenance on the Grid

Data provenance [48] is an annotation that explains how a particular result has been produced and what the result is about. This can be generalized to the Grid where any user of data needs to know the origin and history of data. Data provenance can be thought of as an audit trail that traces each step in sourcing, moving and processing data. It can apply to a single data item, a logical data record, a subset of a database or an entire database. Although clearly needed, the methods for recording provenance are not well set out, and the Semantic Grid offers an important technology for specifying such data lineage. Support for provenance is an essential requirement in an e-Science environment as data sharing is central to the basic concept of a VO. Provenance is the key area to help establish the quality, reliability and value of data in the discovery process.

Grid data will be annotated with its history. One way to ensure the integrity of a specific data set is to require that the generation of derived data sets be a reproducible process. In order to ensure that this is the case, all data products and transformations that go into generating a data set have to remain accessible. Provenance can include references to where data was first produced, in historical terms how it has been stored, curated and transferred, as well as the sequence of experimental processes applied and computations involved. Provenance can be represented as semantic metadata using RDF.

Data provenance is a research area that has not been thoroughly addressed by the Grid community. Currently only a few projects are working in the area of provenance, these are introduced below.

3.5.6.1 Provenance generation and discovery in myGrid

In the myGrid project, workflows representing *in silico* experiments describe the orchestration of bioinformatics data and analysis services that are used to derive output. The output of one workflow may form the input to another so that a complete *in silico* experiment includes a network of related workflow invocations. The myGrid workflow enactment engine ensures that any output can be associated with its corresponding workflow invocation record and the associated provenance data. myGrid supports two forms of provenance, derivation path and object annotations [49].

- A derivation path records the process by which results are generated from input data, e.g. a derivation path could be the logging of services invoked by a workflow or could be a record regarding alterations of the parameters of an activity while a workflow is being enacted.
- Object annotations specify when an object was created, last updated, who owns it and its format.

When a workflow is executed, the workflow enactment engine extracts needed resources, such as input/output data and parameters. The provenance logs are generated at the same time in the form of XML files by the workflow enactment engine, recording the start time, end time and service instances operated in this workflow. In addition, provenance logs can also be automatically annotated with the ontologies provided in myGrid for provenance document discovery.

3.5.6.2 PASOA

The Provenance-Aware Service-Oriented Architecture (PASOA) project [50] is an ongoing project with an aim to investigate the concept of provenance and its use for reasoning about the quality and accuracy of data and services in the context of UK e-Science programme.

3.5.7 A summary on the Semantic Grid

The development of the Semantic Grid is based on the application of technologies from the Semantic Web to the Grid. Metadata and ontologies play a crucial role in the evolution of the Semantic Grid. The Semantic Grid is still in its infancy, however, the results, as exemplified by current activities, have shown that it is narrowing the gap between users and the Grid. Grid systems enriched with semantics can automate the process of service and resource discovery. For example, autonomic Grid services can be built with metadata annotations plus ontologies so that these services can have the capability to self-configure, self-optimize, self-heal and self-protect. This leads us onto research about autonomic computing.

3.6 AUTONOMIC COMPUTING

In this section, we introduce autonomic computing, e.g. why we need it, what kinds of features an autonomic system has, how to apply autonomic computing to the Grid and what kinds of benefits it can bring to the Grid. Finally, we review some current works on autonomic computing.

3.6.1 What is autonomic computing?

Broadly speaking, autonomic computing refers to an infrastructure that automatically adapts to meet the demands of the applications that are running in it. Autonomic computing is a self-managing computing model named after, and patterned on, a human body's autonomic nervous system. An autonomic computing system is one that is resilient, and able to take both preemptive and *post facto* measures to ensure a high quality of service with a minimum of human intervention, in the same way that the autonomic nervous system regulates body systems without conscious input from the individual.

The goal of autonomic computing is to reduce the complexity in the management of large computing systems such as the Grid. The Grid needs autonomic computing for following reasons.

- *Complexity*: The Grid is complex in nature because it tries to couple large-scale disparate, distributed and heterogeneous resources – such as data, computers, operating systems, database systems, applications and special devices – which may run across multiple virtual organizations to provide a uniform computing platform.
- *Dynamic nature*: The Grid is a dynamic computing environment in that resources and services can join and leave at any time.

3.6.2 Features of autonomic computing systems

A system that is to be classified as an autonomic system should have the following major features.

Self-protection
A self-protecting system can detect and identify hostile behaviour and take autonomous actions to protect itself against intrusive behaviour. Self-protecting systems, as envisioned, could safeguard themselves against two types of behaviour: accidental human errors and malicious intentional actions.

To protect themselves against accidental human errors, e.g. self-protecting systems could provide a warning if the system administrators were to initiate a process that might interrupt services. To defend themselves against malicious intentional actions, self-protecting systems would scan for suspicious activities and react accordingly without users being aware that such protection is in process. Besides simply responding to component failure or running periodic checks for symptoms, an autonomic system will always remain on alert, anticipating threats and preparing to take necessary actions. Autonomic systems also aim to provide the right information to the right users at the right time through actions that grant access based on the users' roles and pre-established policies.

Self-optimizing
Self-optimizing components in an autonomic system are able to dynamically tune themselves to meet end-user or business needs with minimal human intervention. The tuning actions involve the reallocation of resources based on load balancing functions and system run-time state information to improve overall resource utilization and system performance.

Self-healing
Self-healing is the ability of a system to recover from faults that might cause some parts of it to malfunction. For a system to be self-healing, it must be able to recover from a failed component by first detecting and isolating the failed component, taking it off line, fixing and reintroducing the fixed or replacement component into service without any apparent overall disruption. A self-healing system will also need to predict problems and take actions to prevent the failure from having an impact on applications. The self-healing

objective must be to minimize all outages in order to keep the system up and available at all times.

Self-configuring
Installing, configuring and integrating large, complex systems is challenging, time consuming and error-prone even for experts. A self-configuring system can adapt automatically to dynamically changing environments in that system components including software components and hardware components can be dynamically added to the system with no disruption of system services and with minimum human intervention.

Open standards
An autonomic system should be based on open standards and provide a standard way to interoperate with other systems.

Self-learning
An autonomic system should be integrated with a machine-learning component that can build knowledge rules based on a certain time of the system running to improve system performance, robustness and resilience and anticipating foreseeable failures.

3.6.3 Autonomic computing projects

Autonomic computing is in its infancy. In this section, we review some projects that are being carried out in this area.

3.6.3.1 Industry efforts

IBM eLiza
eLiza [51] is IBM's initiative to add autonomic capabilities into existing products such as their servers. An autonomic server will be enhanced with capabilities such as the detection and isolation of bad memory chips, protection against hacker attacks, automatically configuring itself when new features are added, and optimizing CPU, storage and resources in order to handle different levels of internal traffic.

IBM OptimalGrid
OptimalGrid [52] is middleware that aims to simplify the creation and management of large scale, connected, parallel Grid

applications. It incorporates the core tenants of the features of autonomic computing such as self-configuring, self-healing and self-optimizing to create an environment in which it is possible for application developers to exploit these features, without the need either to build them or to code external APIs.

Intel Itanium 2
Intel built into its Itanium 2 [53] processor features of autonomic computing called the Machine Check Architecture (MCA). The MCA is an infrastructure that allows systems to continue executing transactions as it recovers from error conditions. It has the ability to detect and correct errors and to report these errors to the operating system. It also has the capability to analyse data and respond in a way that provides higher overall system reliability and availability.

Sun N1
The aim of the Sun N1 [54] is to manage N computers as 1 entity. The autonomic capability in N1 has been focused on the automation of software and hardware installation and configuration for new business service deployment.

3.6.3.2 Academic efforts

Autonomia [55], University of Arizona, USA
The objective of this project is to automate the deployment of mobile agents that have self-manageable attributes. Autonomia provides dynamically programmable control and management services to support the development and deployment of smart (intelligent) applications. The Autonomia environment provides the application developers with the tools required to specify the appropriate control and management schemes to maintain any quality of service requirement or application attribute/functionality, such as performance, fault or security. Autonomia also provides core middleware services to maintain the autonomic requirements of a wide range of network applications and services.

AutoMate [56], Rutgers University, USA
The overall objective of AutoMate is to investigate the technologies needed for the development of context aware Grid applications

with autonomic capabilities. Specifically, this project is investigating the definition of autonomic components, the development of autonomic applications as the dynamic composition of autonomic components, and the design of enhancements to existing Grid middleware and run-time services to support the execution of these applications. Based on AutoMate, Accord [57] provides a component-based programming framework for the development of autonomic self-managed applications. The Accord programming framework, which is currently based on CCA [58] and OGSA, introduces the following four concepts:

- An application context that defines a common semantic basis for components and the application.
- The definition of autonomic components as the basic building blocks for autonomic application.
- The definition of rules and mechanisms for the management and dynamic composition of autonomic components.
- Rule enforcement to enable autonomic application behaviours.

AMUSE [59], Imperial College, London
The Autonomic Management of Ubiquitous Systems for e-Health (AMUSE) project is investigating techniques to be used for building future e-Science and e-Health applications that can be self-managed.

Open Overlays [60], Oxford Brookes University and Lancaster University, UK
The Open Overlays, Component-based Communication Support for the Grid project is investigating the support for future Grid applications that can be self-managed.

ASA [61], University of St Andrews, UK
The Secure Location-Independent Autonomic Storage Architectures project is an autonomic system for a global storage infrastructure that approximates a "Utopian" set of ideal characteristics such as unbounded capacity, zero latency, zero cost, complete reliability, location independence; a simple interface for users; complete security; and provision of a complete historical archive.

3.6.4 A vision of autonomic Grid services

In this section, we present a vision of how to apply autonomic computing to the Grid, and thus develop autonomic Grid systems. As we have outlined in Chapter 2, OGSA is the *de facto* standard for building service-oriented Grid applications. Grid services, in this context are Web services with some extensions to copy dynamic behaviour and the need for state. In the context of the Semantic Grid, Grid services are annotated with metadata and have ontologies to specify their semantic functions and the domain they apply to. The research on the Semantic Grid has been focused on the automatic annotation, enactment, registration and discovery of semantic workflows that are composed from Grid services. At this stage, a Grid service can provide transient and stateful services, and it also has some knowledge about itself, e.g. what kind of input and output it semantically needs, and what kind of domain it applies to. However, it does not have any autonomic features. It is envisioned that a future Grid system will be composed from autonomic Grid services, as shown in Figure 3.18. An autonomic Grid service could have the following components:

- *Core*: This is the core component of the service to provide core functionality such as performing a computation.
- *Advertising*: The advertising component registers the service name, domain-related problems it could solve with a registry

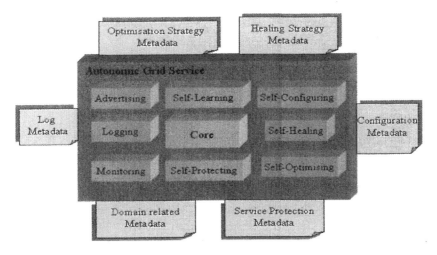

Figure 3.18 The concept of autonomic Grid services

server in a Grid system. Domain problems can be annotated with metadata.
- *Self-learning*: Based on a certain time of running of the service, this component can gain some knowledge on service component, e.g. how to tune the component, what kinds of fault it could have, how to fix faults.
- *Self-configuring*: This component automatically configures the environment for the service to execute, based on the configuration metadata. It may also install or remove additional software, if necessary.
- *Logging*: This component records events during the execution of the service for self-healing. The events will be annotated with log metadata.
- *Self-healing*: Once an error occurred to a service, this component uses the log metadata and healing strategy metadata to decide how to heal the service.
- *Monitoring*: This component monitors the execution of the service and periodically invokes the logging to record events. It also detects the correct execution of the service. If something incorrect happens, it will invoke the self-healing component to heal the service.
- *Self-protecting*: Based on the service protection metadata, this component can authenticate and authorize who can use the service.
- *Self-optimizing*: This component periodically checks the state information of the service. If necessary, it will optimize the service with the optimization strategy metadata.

3.7 CHAPTER SUMMARY

In this chapter, we have studied the Semantic Grid and autonomic computing. The development of the Semantic Grid is based on the application of Semantic Web technologies to the Grid. The goal of the Semantic Web is to convert unstructured information to structured information to improve efficiency in information discovery and use. Metadata and Ontologies play a crucial role in annotating the structured information on the Web. Metadata can help computers understand the meaning of a resource such as a Web page, a hyperlink, a data set, a software service or a service request. Ontologies specify the relationships between the terms used in metadata specific to one domain. With metadata and ontologies, computers can determine if a service can provide a service requested by a consumer based on semantic service matching. The RDF is the foundation of the Semantic Web. RDFS, DAML+OIL and OWL are ontology languages used to build domain ontologies. Among them, OWL is becoming the most widely use standard. Based on DAML+OIL, DAML-S can be used to annotate standard Web services (services that use WSDL for description, SOAP for message exchange and UDDI for service registration and discovery) with semantic meaning to specify the domain to which they apply. Evolved from DAML-S, OWL-S is becoming a standard way to annotate Web services with semantic capabilities. The Semantic Grid is still evolving; currently research has been focused on semantic workflow annotation, enactment, registration and discovery.

While the Semantic Grid pushes the Grid closer towards its goal of providing users with a uniform platform for solving domain-specific problems, autonomic computing can give help further by providing higher levels of availability and reliability. The goal of autonomic computing is to reduce the complexity of the management of large computing systems such as the Grid. The Grid needs autonomic computing because it is complex and dynamic in nature. We have provided a vision of autonomic Grid services which will extend current concepts of Grid and Semantic Grid services. Future Grid systems will be composed from autonomic services to enrich the Grid with autonomic features to provide self-configured, self-healing, self-optimizing, self-protecting, self-learning and self-advertising services to its users.

3.8 FURTHER READING AND TESTING

The research on metadata and ontologies belongs to the area of knowledge representation and artificial intelligence. Books such as [62–67] can be used for a background study.

To test work on the Semantic Grid, we suggest you start with RDF because it is the foundation of the Semantic Web and also the Semantic Grid. To test work on autonomic computing, we recommend that you start with software agent programming because agents are normally involved in the development of autonomic systems.

3.9 KEY POINTS

- The development of the Semantic Grid is based on the application of Semantic Web technologies to the Grid.
- The goal of the Semantic Web is to augment unstructured Web content to structured information to improve computer interpretability and the efficiency in Web information discovery and use.
- Metadata and ontologies play a critical role in the evolution of the Semantic Web and the Semantic Grid.
- RDF is the foundation of the Semantic Web.
- RDFS, DAML+OIL and OWL are ontology languages.
- DAML-S and OWL-S can be used to annotate Web services (WSDL, SOAP and UDDI) with semantics.
- Data provenance is an annotation that helps explain the origin and history of a data.
- Semantic portlets will play an important role in the development of the Semantic Grid portals.
- An autonomic system is a computer system that is capable of self-configuring, self-healing, self-optimizing, self-protecting and self-learning.
- The goal of autonomic computing is to reduce the complexity in the management of large computer-based systems such as the Grid.
- A future autonomic Grid system can be composed from autonomic Grid services.

3.10 REFERENCES

[1] The Semantic Grid, http://www.semanticgrid.org.
[2] Lee, T.B., Hendler, J. and Lassila, O. (2001). The Semantic Web. *Scientific American*, 284(5): 34–43.
[3] Gruber, T.R. (1993). A Translation Approach to Portable Ontology Specifications. *Knowledge Acquisition*, 5(2): 199–220. Academic Press.
[4] Kephart, J.O. and Chess, D.M. (2003). The Vision of Autonomic Computing. *IEEE Computer*, 36(1): 41–50.
[5] W3C, http://www.w3.org/.
[6] Cannataro, M. and Comito, C. (2003). *A Data Mining Ontology for Grid Programming*. Proceedings of the 1st Workshop on Semantics in Peer-to-Peer and Grid Computing, Budapest, Hungary. Lecture Notes in Computer Science, Springer-Verlag.

3.10 REFERENCES

[7] Miller, E. (1998). An Introduction to the Resource Description Framework. *D-Lib Magazine*, http://www.dlib.org/dlib/may98/miller/05miller.html.

[8] RDF Schema, http://www.w3.org/TR/rdf-schema/.

[9] DAML+OIL, http://www.daml.org/.

[10] Horrocks, I. (2002). *DAML+OIL: A Reason-able Web Ontology Language*. Proceedings of the VIII Conference on Extending Database Technology (EDBT 2002), Prague, Czech Republic. Lecture Notes in Computer Science, Springer-Verlag.

[11] OWL, http://www.w3.org/TR/2004/REC-owl-features-2004 0210/.

[12] PICS Specification, http://www.w3.org/PICS/.

[13] Layman, A., Paoli, J., Rose, S. and Thompson, H. (June 1997). XML-Data, http://www.microsoft.com/standards/xml/xmldata.htm.

[14] Guha, R. and Bray, T. (June 1997). Meta Content Framework Using XML, http://www.w3.org/TR/NOTE-MCF-XML/.

[15] Dublin Core, http://dublincore.org/.

[16] Lagoze, C., Lynch, C. and Daniel, R.J. (1996). The Warwick Framework: A Container Architecture for Aggregating Sets of Metadata. *Cornell Computer Science Technical Report TR96-1593*.

[17] Lassila, O. and McGuinness, D.L. (2001). The Role of Frame-Based Representation on the Semantic Web. *KSL Tech Report Number KSL-01-02*, http://www.ida.liu.se/ext/epa/ej/etai/2001/018/01018-etaibody.pdf.

[18] Horrocks, I. (2003). Implementation and Optimization Techniques. *Description Logic Handbook*, pp. 306–346. Cambridge University Press.

[19] Tsarkov, D. and Horrocks, I. (2003). *DL Reasoner vs. First-Order Prover*. Proceedings of the 2003 International Workshop on Description Logics (DL 2003), Rome, Italy. CEUR Publication.

[20] DAML-ONT, http://www.daml.org/2000/10/daml-ont.html.

[21] Fensel, D., Harmelen, F., Horrocks, I., McGuinness, D.L. and Patel-Schneider, P.F. (2001). OIL: An Ontology Infrastructure for the Semantic Web. *IEEE Intelligent Systems*, 16(2): 38–45.

[22] Horrocks, I. (1999). *FaCT and iFaCT*. Proceedings of the 1999 International Workshop on Description Logics (DL 1999), Linköping, Sweden. CEUR Publication.

[23] Bechhofer, S. and Horrocks, I. (2000). *Driving User Interfaces from FaCT*. Proceedings of the 1999 International Workshop on Description Logics (DL 2000), Aachen, Germany. CEUR Publication.

[24] Horrocks, I., Patel-Schneider, P.F. and Harmelen, F. (2002). *Reviewing the Design of DAML+OIL: An Ontology Language for the Semantic Web*. Proceedings of 18th National Conference on Artificial Intelligence, Edmonton, Alberta, Canada. American Association for Artificial Intelligence (AAAI) Publication.

[25] Web Ontology Working Group, http://www.w3.org/2001/sw/WebOnt/.

[26] OWL Use Case, http://www.w3.org/TR/webont-req/.

[27] Antoniou, G. and Harmelen, F. (2004). Web Ontology Language: OWL. *Handbook on Ontologies, International Handbooks on Information Systems*. Springer, pp. 67–92.

[28] Denny, M. (2002). Ontology Building: A Survey of Editing Tools, http://www.xml.com/pub/a/2002/11/06/ontologies.html.

[29] OntoEdit, http://www.ontoknowledge.org/tools/ontoedit.shtml.

[30] Sure, Y., Erdmann, M., Angele, J., Staab, S., Studer, R. and Wenke, D. (2002). *OntoEdit: Collaborative Ontology Engineering for the Semantic Web.* Proceedings of the International Semantic Web Conference 2002 (ISWC 2002), Sardinia, Italia. CS Press.

[31] Kifer, M., Lausen, G. and Wu, J. (1995). Logical foundations of object-oriented and frame based languages. *Journal of the ACM*, 42: 741–843.

[32] OilEd, http://oiled.man.ac.uk/.

[33] Eriksson, H., Fergerson, R.W., Shahar, Y. and Musen, M.A. (1999). *Automatic Generation of Ontology Editors.* Proceedings of the 12th Banff Knowledge Acquisition for Knowledge-based Systems Workshop (KAW-99), Banff, Alberta, Canada.

[34] Protégé, http://protege.stanford.edu/.

[35] Ankolekar, A., Burstein, M., Hobbs, J.R., Lassila, O., Martin, D.L., McDermott, D., McIlraith, S.A., Narayanan, S., Paolucci, M., Payne, T.R. and Sycara, K. (2002). *DAML-S: Web Service Description for the Semantic Web.* Proceedings of the 1st International Semantic Web Conference (ISWC), Sardinia, Italy. Lecture Notes in Computer Science, Springer-Verlag.

[36] OWL-S, http://www.daml.org/services/owl-s/1.0/owl-s.html.

[37] Tangmunarunkit, H., Decker, S. and Kesselman, C. (2003). *Ontology-Based Resource Matching in the Grid – The Grid Meets the Semantic Web.* Proceedings of the 2nd International Semantic Web Conference 2003 (ISWC 2003), Sanibel Island, Florida, USA. Lecture Notes in Computer Science, Springer-Verlag.

[38] Decker, S. and Sintek, M. (2002). *Triple – A Query, Inference, and Transformation Language for the Semantic Web.* Proceedings of the 1st International Semantic Web Conference (ISWC), Sardinia, Italy. Lecture Notes in Computer Science, Springer-Verlag.

[39] Lord, P., Wroe, C., Stevens, R., Goble, C., Miles, S., Moreau, L., Decker, K., Payne, T. and Papay, J. (2003). *Semantic and Personalised Service Discovery.* Proceedings of IEEE Workshop on Knowledge Grid and Grid Intelligence (KGGI '03), in Conjunction with 2003 IEEE/WIC International Conference on Web Intelligence/Intelligent Agent Technology, Halifax, Canada. CS Press.

[40] Chen, L., Shadbolt, N.R., Tao, F., Puleston, C., Goble, C. and Cox, S.J. (2003). *Exploiting Semantics for e-Science on the Semantic Grid.* Proceedings of IEEE Workshop on Knowledge Grid and Grid Intelligence (KGGI '03), in conjunction with 2003 IEEE/WIC International Conference on Web Intelligence/Intelligent Agent Technology, Halifax, Canada. CS Press.

[41] OntoMat Annotizer Annotation Tool, http://annotation.semanticweb.org/tools/ontomat.

[42] Hau, J., Lee, W. and Newhouse, S. (2003). *Autonomic Service Adaptation Using Ontological Annotation.* Proceedings of the 4th International Workshop on Grid Computing, Grid 2003, Phoenix, USA. CS Press.

[43] Li, M., Santen, P., Walker, D.W., Rana, O.F. and Baker, M.A. (2003). *PortalLab: A Web Services Oriented Toolkit for Semantic Grid Portals.* Proceedings of the 3rd IEEE/ACM International Symposium on Cluster Computing and the Grid (CCGrid 2003), Tokyo, Japan. CS Press.

[44] Krasner, G.E. and Pope, S.T. (1998). A Cookbook for Using the Model-View-Controller User Interface Paradigm in Smalltalk-80. *Journal of Object Oriented Programming*, 1(3): 26–49. ADT.

3.10 REFERENCES

[45] Jennings, N. (1999). *Agent-based Computing: Promises and Perils.* Proceedings of 16th International Joint Conference on Artificial Intelligence (IJCAI '99), Stockholm, Sweden.

[46] Moreau, L. (2002). *Agents for the Grid: A Comparison with Web Services.* Proceedings of the 2nd IEEE/ACM International Symposium on Cluster Computing and the Grid (CCGrid 2002), Berlin, Germany. CS Press.

[47] Milojicic, D.S., Kalogeraki, V., Lukose, R., Nagaraja, K., Pruyne, J., Richard, B., Rollins, S. and Xu, Z. (2002). Peer-to-Peer Computing. *HPL-2002-57*, HP Labs Technical Reports, http://www.hpl.hp.com/techreports/2002/HPL-2002-57.pdf.

[48] Szomszor, M. and Moreau, M. (2003). *Recording and Reasoning Over Data Provenance in Web and Grid Services.* Proceedings of International Conference on Ontologies, Databases and Applications of SEmantics (ODBASE '03), Catania, Sicily, Italy. Lecture Notes in Computer Science, Springer-Verlag.

[49] Zhao, J., Goble, C., Greenwood, M., Wroe, C. and Stevens, R. (2003). *Annotating, Linking and Browsing Provenance Logs for e-Science.* Proceedings of the 1st Workshop on Semantic Web Technologies for Searching and Retrieving Scientific Data, Sanibel Island, Florida, USA. CEUR Publication.

[50] PASOA, http://www.pasoa.org.

[51] IBM eLiza, http://www-1.ibm.com/servers/autonomic/.

[52] Deen, G., Lehman, T. and Kaufman, J. (2003). *The Almaden OptimalGrid Project.* Proceedings of the 5th Annual International Workshop on Active Middleware Services 2003 (AMS 2003), Seattle, USA. CS Press.

[53] Intel Itanium 2, http://www.intel.com/products/server/processors/server/itanium2/.

[54] Sun N1, http://www.sun.com/software/n1gridsystem/.

[55] Dong, X., Hariri, S., Xve, L., Chen, H., Zhang, M., Pavuluri, S. and Rao, S. (2003). *Autonomia: An Autonomic Computing Environment.* Proceedings of the 23rd IEEE International Performance Computing and Communications Conference (IPCCC-03). CS Press.

[56] AutoMate, http://automate.rutgers.edu/.

[57] Liu, H., Parashar M. and Hariri, S. (2004). *A Component-based Programming Framework for Autonomic Applications.* Proceedings of the 1st IEEE International Conference on Autonomic Computing (ICAC-04). New York, USA. May CS Press.

[58] CCA, http://www.cca-forum.org/.

[59] AMUSE, http://www.dcs.gla.ac.UK/amuse/.

[60] Coulson, G., Grace, P., Blair, G., Mathy, L., Duce, D., Cooper, C., Yeung, W. and Cai, W. (2004). *Towards a Component-based Middleware Framework for Configurable and Reconfigurable Grid Computing.* Proceedings of Workshop on Emerging Technologies for Next Generation Grid (ETNGRID-2004). CS Press.

[61] Dearle, A., Kirby, GNC., McCarthy, A., Diaz, Y. and Carballo, JC. (2004). *A Flexible and Secure Deployment Framework for Distributed Applications.* Proceedings of the 2nd International Working Conference on Component Deploymen. Lecture Notes in Computer Science, Springer-Verlag.

[62] Markman, A.B. (1998). *Knowledge Representation.* Mahwah, NJ. Erlbaum. ISBN 0805824413.

[63] Sowa, J.F. (1999). *Knowledge Representation: Logical, Philosophical, & Computational Foundations*. Pacific Grove, CA. Brooks/Cole. ISBN 0534949657.
[64] Russell, S.J. and Norvig, P. (2003). *Artificial Intelligence: A Modern Approach*. February 2003. 2nd edition. Upper Saddle River, NJ. Prentice Hall. ISBN 0137903952.
[65] Fayyad, U.M., Shapiro, G.P., Smyth, P. and Uthurusamy, R. (1996). *Advances in Knowledge Discovery & Data Mining*. Menlo Park, CA. MIT Press. ISBN 0262560976.
[66] Staab, S. and Studer, R. (2004). *Handbook on Ontologies (International Handbooks on Information Systems)*. Springer-Verlag. ISBN 3540408347.
[67] Luger, G.F. (2002). *Artificial Intelligence: Structures & Strategies for Complex Problem Solving*. Harlow. Addison-Wesley. ISBN 0201648660.

Part Two
Basic Services

Part Two

Basic Services

4
Grid Security

4.1 INTRODUCTION

In general, IT security is concerned with ensuring that critical information and the associated infrastructures are not compromised or put at risk by external agents. Here, the external agent might be anyone that is not authorized to access the aforementioned critical information or infrastructure. The critical infrastructure we are referring to is that which supports banking and financial institutions, information and communication systems, energy, transportation and other vital human services. The Grid is increasingly being taken up and used by all sectors of business, industry, academia and the government as the middleware infrastructure of choice. This means that Grid security is a vital aspect of its overall architecture if it is to be used for critical infrastructures.

A number of observations have been made on critical infrastructures [1]. It is clear that in today's world they are highly interdependent, both physically and in their reliance on national information infrastructure. Most critical infrastructures are largely owned by the private sector, where there tends to be a reluctance to invest in long-term and high-risk security-related technologies. Ongoing changes to business patterns are reducing the level of tolerance to errors in these infrastructures. However, there is insufficient awareness of critical infrastructure issues. The growth of

---

The Grid: Core Technologies Maozhen Li and Mark Baker
© 2005 John Wiley & Sons, Ltd

IT and the Internet can therefore have major implications for the economic and the military security of the world.

IT infrastructures are changing at a staggering rate. Their scale and complexity are becoming ever greater in scope and functional sophistication. Boundaries between computer systems are becoming indistinct; increasingly every device is networked, so the infrastructure is becoming a heterogeneous sea of components with a blurred human/device boundary. There is continuous and incremental development and deployment; systems evolve by adding new features and greater functionality at an unremitting pace. These systems are becoming capable of dynamic self-configuration and adaptation, where systems respond to changing circumstances and conditions of their environment [2]. Increasingly there are multiple innovative types of networked architectures and strategies for sharing resources. This obviously leaves gaps for a multiplicity of fault types and openings for malicious faults, as well as attacks from internal and external parties.

The actors that may want to compromise critical information or infrastructures are many and varied. They include those that pose national security threats, such as information warriors or agents involved in national intelligence. Alternatively the actors could be terrorists involved in industrial espionage or organized crime and who pose a shared threat to a country. Or the threats could just be local and come from institutional or recreational hackers intent on thrill, challenge or prestige.

4.2 A BRIEF SECURITY PRIMER

The goals of security are threefold [2]: first, prevention – prevent attackers from violating security policy; secondly detection – detect attackers' violation of security policy; finally recovery – stop an attack, assess and repair damage, and continue to function correctly even if the attack succeeds.

Obviously prevention is the ideal scenario. In this case there would be no successful attacks. Detection occurs only after someone violates the security policy. It is important when a violation has occurred or is underway that it is reported swiftly. The system must then respond appropriately. Recovery means that the system continues to function correctly, possibly after a period of degraded operation. Such systems are said to be intrusion tolerant.

4.2 A BRIEF SECURITY PRIMER

This is very difficult to do correctly. Usually, recovery means that the attack is stopped and the system fixed. This may involve shutting down the system for some time, or making it unavailable to all users except those fixing the problem, and then the system resumes correct operation.

The three classic security concerns of information security deal principally with data, and are:

- *Confidentiality*: Data is only available to those who are authorized;
- *Integrity*: Data is not changed except by controlled processes;
- *Availability*: Data is available when required.

Confidentiality is aimed at different issues. The content of a packet during a communication has to be secure to prevent malicious users from stealing the data. In order to prevent unauthorized users retrieving secret information, a common approach is to encrypt data from the sender before sending to the receiver. On the receiving end, the receiver can extract the original information by decrypting the encrypted text. Hence, confidentiality of data transmission is closely related to application of different encryption algorithms.

Integrity is the protection of data from modification by unauthorized users. This is not the same as data confidentiality. Data integrity requires that no unauthorized users can change or modify the data concerned. For example, you want to broadcast a message to the public, which is definitely not confidential to anyone. You have to ensure the data integrity of your message from modification by unauthorized people. In this instance, you may have to stamp or add your signature to certify the message.

The term "availability" addresses the degree to which a system, sub-system or equipment is operable and in a usable state.

Additional concerns deal more with people and their actions:

- *Authentication*: Ensuring that users are who they say they are;
- *Authorization*: Making a decision about who may access data or a service;
- *Assurance*: Being confident that the security system functions correctly;

- *Non-repudiation*: Ensuring that a user cannot deny an action;
- *Auditability*: Tracking what a user did to data or a service.

Authentication implies ensuring the right user executes the granted services or the origin of the data was from the real sender. There are a large number of techniques that may be used to authenticate a user – passwords, biometric techniques, smart cards or certificates. Before starting services between a server and client, there should be a mechanism to identify the privilege of the user. You may consider a user name and password for a logon scheme to be the most common authentication scheme. For example, you have to input your Visa Card password at an ATM terminal or enter a password to gain access to a Web portal. Hence, authenticity is mainly related to the identification of authorized users.

Authorization is often a first step towards providing the service of authentication. Authorization enables the decision to allow a particular operation when a request to perform the operation is received. Authorization in existing systems is usually based on information local to the server. This information may be present in Access Control Lists (ACL) associated with files or directories. ACLs are files listing individuals authorized to login to an account (e.g. the UNIX .rhosts file), configuration files naming authorized users of a node and sometimes files read over the network. When applied to distributed systems, authorization mechanisms are required to determine whether a particular task should be run on the current node when requested by a particular principal. Many applications, and in particular applications using distributed systems, can benefit from an authorization mechanism that supports delegation. Delegation is a means by which a user or process authorized to perform an operation can grant the authority to perform that operation to another process. Delegation can be used to implement distributed authorization where, for example, a resource manager might allocate a node to a job and delegate authority to use that node to the job's initiator.

Assurance is the counterpart to authorization. Authorization mechanisms allow the provider of a service to decide whether to perform an operation on behalf of the requester of the service. Assurance mechanisms allow the requester of a service to decide whether a candidate service provider meets the requesters' requirements for security, trustworthiness, reliability or other characteristics. Assurance mechanisms can be implemented through certificates (see Section 4.3.5 for a discussion of certificates) signed by a third party trusted to endorse, license or insure a service provider; certificates are checked as a client selects the providers to contact for particular operations.

Non-repudiation is the concept of ensuring that a contract, especially one agreed to via the Internet, cannot later be denied by one of the parties involved. With regard to digital security, non-repudiation means that it can be verified that the sender and the recipient were, in fact, the parties who claimed to send or receive the message, respectively.

Auditability is about keeping track of what is happening on a system. The idea is that if there is an intrusion, then the system operator can find out exactly what has been done and in whose name.

Other security concerns relate to:

- *Trust*: People can justifiably rely on computer-based systems to perform critical functions securely, and on systems to process, store and communicate sensitive information securely;
- *Reliability*: The system does what you want, when you want it to;
- *Privacy*: Within certain limits, no one should know who you are or what you do.

4.3 CRYPTOGRAPHY

4.3.1 Introduction

Cryptography is the most commonly used means of providing security; it can be used to address four goals:

- *Message confidentiality*: Only an authorized recipient is able to extract the contents of a message from its encrypted form;
- *Message integrity*: The recipient should be able to determine if the message has been altered during transmission;
- *Sender authentication*: The recipient can identify the sender, and verify that the purported sender did send the message;
- *Sender non-repudiation*: The sender cannot deny sending the message.

Obviously, not all cryptographic systems (or algorithms) realize, nor intend to, achieve all of these goals.

4.3.2 Symmetric cryptosystems

Using symmetric (conventional) cryptosystems, data is transformed (encrypted) using an encrypted key and scrambled in such a way that it can only be unscrambled (decrypted) by a symmetric transformation using the same encryption key. Besides protecting the confidentiality of data, encryption also protects data integrity. Knowledge of the encryption key is required to produce cipher text that will yield a predictable value when decrypted. Therefore modification of the data by someone who does not know the key can be detected by attaching a checksum before encryption, and verified after decryption. A sketch of symmetric key cryptography is shown in Figure 4.1.

4.3.2.1 Example: Data Encryption Standard (DES)

DES consists of two components – an algorithm and a key. The DES algorithm involves a number of iterations of a simple transformation which uses both transposition and substitution techniques applied alternately. DES is a so-called private-key cipher; here data is encrypted and decrypted with the same key. Both sender and receiver must keep the key a secret from others. The DES algorithm is publicly known, learning the encryption key would allow an encrypted message to be read by anyone.

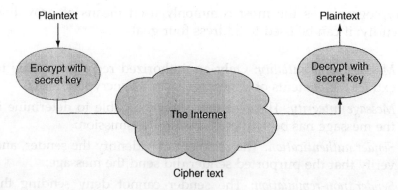

Figure 4.1 Symmetric key cryptography

4.3.3 Asymmetric cryptosystems

In asymmetric cryptography, encryption and decryption are performed using a pair of keys such that knowledge of one key does not provide knowledge of the other key in the pair. One key, called the public key, is published, and the other key, called the private key, is kept private. The main advantage of asymmetric cryptography is that secrecy is not needed for the public key. The public key can be published rather like a telephone number. However only someone in possession of the private key can perform decryption. A sketch of asymmetric key cryptography is shown in Figure 4.2.

4.3.3.1 Example: RSA

An example of a public-key cryptosystem is RSA, named after its developers, Rivest, Shamir and Adleman (RSA), who invented the algorithm at MIT in 1978. RSA provides authentication, as well as encryption, and uses two keys: a private key and a public key. With RSA, there is no distinction between the function of a user's public and private keys. A key can be used as either the public or the private key. The keys for the RSA algorithm are generated mathematically, in part, by combining prime numbers. The security of the RSA algorithm, and others similar to it, depends on the use of very large numbers (RSA uses 256 or 512 bit keys).

With both symmetric and asymmetric systems there is a need to secure the private key. The private key must be kept private. It should not be sent to others, and it should not be stored on a system where others may be able to find and use it. The stored keys should always be password protected. Another issue, with

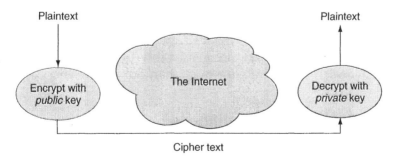

Figure 4.2 Asymmetric key cryptography

key-based systems, is that the algorithms that are used are public. This means that algorithms could be coded and used to decrypt a message via a brute force method of trying all the possible keys. Such a program would fortunately need a significant amount of computational power to accomplish such a process. With keys of sufficient length, the time to decode a message would be unreasonable. Currently, keys are typically 1024–2048 bits in length. However, the availability of cheap computational power, in the form of clusters of PCs, and their doubling in speed every eighteen months, means that the brute force approach may be viable in the future.

4.3.4 Digital signatures

Integrity is guaranteed in public-key systems by using digital signatures, which are a method of authenticating digital information, in the same manner that an individual would sign a paper document to authenticate it. A digital signature is itself a sequence of bits conforming to one of a number of standards.

Most digital signatures rely on public-key cryptography to work. Consider a scenario, where someone wants to send a message to another and prove its originator, but does not care whether anybody else reads it. In this case, they send an encrypted copy of the message, along with a copy of the message encrypted with their private (not public) key. A recipient can then check whether the message really came from originator by unscrambling the scrambled message with the sender's public key and comparing it with the unscrambled version. If they match, the message was really from the originator, because the private key was needed to create the encrypted copy and no one but the originator has it. Often, a cryptographically strong hash function [4] is applied to the message, and the resulting message digest is encrypted instead of the entire message, which makes the signature significantly shorter than the message and saves considerable time since hashing is generally much faster, byte for byte, than public-key encryption.

4.3.5 Public-key certificate

A public-key certificate is a file that contains a public key, together with identity information, such as a person's name, all of which is

4.3 CRYPTOGRAPHY

signed by a Certification Authority (CA). The CA is a guarantor who verifies that the public key belongs to the named entity.

Certificates are required for the large-scale use of public-key cryptography, since anybody can create a public–private-key pair. So in principle, if the originator is sending private information encrypted with the recipient's public key, a malicious user can fool the originator into using their public key, and so get access to the information, since it knows its corresponding private key. But if the originator only trusts public keys that have been signed ("certified") by an authority, then this type of attack can be prevented. In large-scale deployments one user may not be familiar with another's certification authority (perhaps they each have a different company CA), so a certificate may also include a CA's public key signed by a higher level CA, which is more widely recognized. This process can lead to a hierarchy of certificates and complex graphs representing trust relations.

Public Key Infrastructure (PKI) refers to the software that manages certificates in a large-scale setting. In X.509 PKI systems, the hierarchy of certificates is always a top-down tree, with a root certificate at the top, representing a CA that is so well known it does not need to be authenticated. A certificate may be revoked if it is known that the related private key has been exposed. In this circumstance, one needs to look up the Certificate Revocation List, which is often stored remotely, and updated frequently. A certificate typically includes:

- The public key being signed;
- A name, which can refer to a person, a computer or an organization;
- A validity period;
- The location (URL) of a revocation list.

The most common certificate standard is the ITU-T X.509 [5]. An X.509 certificate is generally a plaintext file that includes information in a specific syntax:

- *Subject*: This is the name of the user;
- *Subject's public key*: This includes the key itself, and other information such as the algorithm used to generate the public key;
- *Issuer's subject*: CA's distinguished name (Table 4.1);

Table 4.1

Distinguished Names (DN)
Names in X.509 certificates are not encoded simply as common names, such as "Mark Baker" or "Certificate Authority XYZ" or "System Administrator". Names are encoded as distinguished names, which are name–value pairs. An example of typical distinguished name attributes is shown below.

OU = Portsmouth, L = DSG, CN = Mark Baker
A DN can have several different attributes, and the most common are the following:

OU: Organizational Unit

L: Location

CN: Common Name (usually the user's name).

- *Digital signature*: The certificate includes a digital signature of all the information in the certificate. This digital signature is generated using the CA's private key. To verify the digital signature, we need the CA's public key, which is found in the CA's certificate.

4.3.6 Certification Authority (CA)

The CA exists to provide entities with a trustable digital identity which they can use to access resources in a secure way. Its role is to issue (create and sign) certificates, make the valid certificates publicly accessible, revoke certificates when necessary and regularly issue revocation lists. The CA must also keep records of all its transactions.

A CA can issue personal certificates to users; the purpose of the certificate is to allow users to identify themselves to remote entities. A personal certificate can also be used for digital signatures. A CA can also issue host (server) and service certificates. Each host and service connected to a network must be able to identify itself.

Some CAs can issue certificates which validate the identity of subordinate CAs. Some CAs are therefore subordinate CAs and some simply appoint themselves. In either case the CA publishes a document called the Certificate Policy Statement (CPS). This is a document which details the conditions under which it issues certificates, and the level of assurance which a relying party can place in certificates issued by the CA.

A CA issues public-key certificates that state that the CA trusts the owner of the certificate, and that they are who they purport to be. A CA should check an applicant's identity to ensure it matches the credentials on the certificate. A party relying on the certificate trusts the CA to verify identity so that the relying party can trust that the user of the certificate is not an imposter.

4.3.7 Firewalls

A firewall is a hardware or software component added to a network to prevent communication forbidden by an organization's administrative policy. Two types of firewalls are generally found, traditional and personal. A traditional firewall is typically a dedicated network device or computer positioned on the boundary of two or more networks. This type of firewall filters all traffic entering or leaving the connected networks. On the other hand, a personal firewall is a software application that can filter traffic entering or leaving a single computer.

Traditional firewalls come in several categories and subcategories. They all have the basic task of preventing intrusion in a connected network, but accomplish this in different ways, by working at the network and/or transport layer of the network.

A network layer firewall operates at the network level of the TCP/IP protocol stack; it undertakes IP-packet filtering, not allowing packets to pass the firewall unless they meet the rules defined by the firewall administrator. A more liberal set up could allow any packet to pass the filter as long as it does not match one or more negative-rules or deny rules.

Application layer firewalls operate at the application level of the TCP/IP protocol stack; intercepting, for example, all Web, telnet or ftp traffic. They will intercept all packets travelling to or from an application. These firewalls will block packets; usually dropping them without acknowledgement to the sender.

These firewalls can, in principle, prevent all unwanted traffic from reaching protected machines. The inspection of all packets for improper content means that firewalls can even prevent the spread of such things as viruses or Trojans. However, in practice, this becomes complex and difficult to attempt, in the light of the variety of applications and the diversity of content, so an all-inclusive firewall design will not attempt this approach.

Sometimes, a proxy device, which can again be implemented in hardware or software, can act as a firewall by responding to input packets (e.g. connection requests) in the manner of an application, whilst blocking other packets. Proxies help make tampering with the internal infrastructure from an external system more difficult, and misuse of one of its internal systems would not necessarily cause a security breach that would be exploitable from outside, assuming that the proxy itself remains intact. The use of internal address spaces enhances security; an intruder may still employ methods such as IP spoofing to attempt to pass packets to the target internal network.

Correctly configuring a firewall demands skill. It requires a good understanding of network protocols and of computer security in general. Errors or small mistakes in the configuration of a firewall can render it valueless as a security tool.

4.4 GRID SECURITY

4.4.1 The Grid Security Infrastructure (GSI)

Grid security is based on what is known as the Grid Security Infrastructure (GSI) (Figure 4.3), which is now a Global Grid Forum (GGF) standard [6, 7]. GSI is a set of tools, libraries, and protocols used in Globus (see Chapter 5), and other grid middleware, to allow users and applications to access resources securely. GSI is

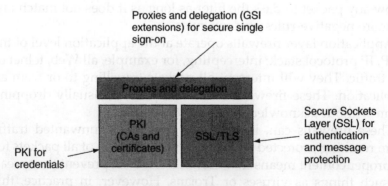

Figure 4.3 The Grid Security Infrastructure

4.4 GRID SECURITY

based on a Public Key Infrastructure, with certificate authorities and X.509 certificates. GSI provides:

- A public-key system;
- Mutual authentication through digital certificates;
- Credential delegation and single sign-on.

4.4.1.1 Introduction

GSI is a collection of well-known and trusted technologies. It can be configured to provide privacy, integrity and authentication; in addition, strong authentication is provided with the help of certificates. Not all of these features are always needed during communication. Typically, a GSI-based secure conversation must at least be authenticated. Integrity is desirable, but can be disabled and encryption can also be activated, when needed, to ensure privacy.

4.4.1.2 Mutual authentication through digital certificates

GSI uses X.509 certificates to guarantee strong authentication. Mutual authentication means that both parties in a secure conversation authenticate the other. So, when an originator wants to communicate with a remote party, the originator must establish trust in the remote party and vice versa. Here trust means that the each party must trust the certificate of the CA that signed the other's certificate. Otherwise, no trust can exist between the parties.

4.4.1.3 Credential delegation and single sign-on

GSI lets a user create and delegate proxy credentials to processes running on remote resources; this allows remote processes and resources to act on a user's behalf. This feature is important for complex applications that need to use a variety of remote resources. Proxy credentials are short-lived credentials created by a user; in essence they are a short-term binding of the user's identity to an alternate private key. It allows users to be authenticated once, and then perform multiple actions. The proxy credential is stored unencrypted for ease of repeated access, and has a short lifetime in case of theft. Single sign-on is an important feature as it simplifies the coordination of multiple resources; a user authenticates once, and can then perform multiple actions without further authentication; in addition, the system allows processes to act on the user's behalf without further authentication.

4.4.2 Authorization modes in GSI

GSI supports three authorization modes on both the server and client.

Server-side authorization
Depending on the authorization mode chosen, the server will decide if it accepts or declines an incoming security request.

- *None*: This is the simplest type of authorization. No authorization will be performed.
- *Self*: A client will be allowed to use a grid service if the client's identity is the same as the service's identity.
- *Gridmap*: The gridmap file contains a list of authorized users; this is similar to an ACL. Here only the users listed in the service's gridmap file may invoke it.

Client-side authorization
A client in GSI can chose to use a remote service or not based on it having the appropriate security credentials.

- *None*: No authorization will be performed.
- *Self*: The client will authorize an invocation if the service's identity is the same as the client. If both client- and server-side use self-authorization, a service can be invoked if its identity matches the client's.
- *Host*: The client will authorize a security request if the host returns an identity containing the hostname. This is done using host certificates.

4.4.2.1 Requesting a certificate

To request a certificate a user starts by generating a key pair. The private key is stored in an encrypted form secured with a pass phrase. The public key is put into a certificate request that is sent to the CA. The CA usually includes a number of Registration Authorities (RA). RAs verify the request by ensuring facts such as the name is unique with respect to the CA, and it is the real name of the user. If the RA is happy then the CA signs the certificate request and issues a certificate for the user.

4.4.2.2 Mutual authentication

The GSI uses the Secure Sockets Layer (SSL) [8] for its mutual authentication protocol. Before mutual authentication can occur, the entities involved must first trust the CAs that signed each other's certificates. Each entity will have a copy of the other CA's certificate, which contains its public keys. The authentication process is illustrated in Figure 4.4.

To mutually authenticate:

1. The first entity (A) establishes a connection with the second entity (B). To start the authentication process, A gives B its certificate. The certificate tells B who A is claiming to be (the identity), A's public key, and which CA is being used to certify the certificate.

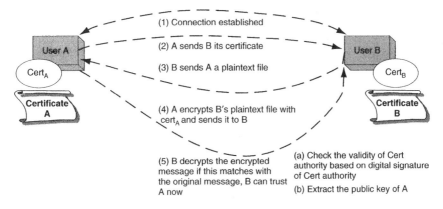

Figure 4.4 Mutual authentication

2. First, B makes sure that the certificate is valid by checking the CA's digital signature to ensure that the CA actually signed the certificate and that the certificate has not been tampered with. This is the point where B must trust the CA that signed A's certificate.
3. Once B has checked out A's certificate, B makes sure that A really is the entity identified in the certificate. To do this, B generates a random plaintext message and sends it to A, asking A to encrypt it.
4. A encrypts the message using its private key, and sends it back to B. B decrypts the message using A's public key. If the result is the same as the original random message, then B knows that A is who they say they are and B trusts A's identity.
5. The same operation now happens in reverse. B sends A its certificate. A validates the certificate and sends a challenge message to be encrypted. B encrypts the message and sends it back to A. A decrypts it and compares it with the original. If it matches, then A knows that B is who they say they are.
6. At this point, A and B have established a connection to each other and are certain that they know each others' identities, i.e. they trust each other.

4.4.2.3 Confidential communication

The default behaviour of GSI is not to establish encrypted communication between entities. In GSI, once mutual authentication has occurred, communications occur without the overhead of encryption. Confidential communication can be established again when it is needed. By default, GSI provides communication integrity.

4.4.2.4 Securing private keys

GSI software expects a user's private key to be stored in a file, which is encrypted via a password (also known as a pass phrase), in a safe location on a computer's file system. The user needs to enter the required pass phrase to decrypt the file containing the private key.

4.4 GRID SECURITY

4.4.2.5 Delegation and single sign-on

The GSI delegation capability is the means to reduce the number of times a user must enter the pass phrase. If an activity requires that several resources be used, or a broker or agent is acting upon a user's behalf – each requiring mutual authentication, a proxy can be created (as shown in Figure 4.5) that avoids the need to enter the pass phrase repeatedly.

A proxy consists of a new certificate (with a new public key) and a new private key. The new certificate contains the owner's identity, modified slightly to indicate that it is a proxy. The owner, rather than a CA, signs the new certificate. The certificate also includes a time notation, after which others should no longer accept the proxy. Proxies have limited lifetimes and cannot live past the expiry of the original certificate.

The proxy's private key must be kept secure. As the proxy's key is not valid for very long, it is typically kept on the local storage system without being encrypted, but with file permissions that prevent it being examined easily. Once a proxy is created and stored, the user can use the proxy certificate and private key for mutual authentication without entering a password.

Mutual authentication differs when proxies are used. The remote party receives the proxy's certificate (signed by the owner), and

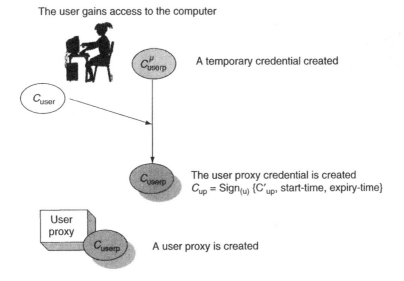

The user gains access to the computer

C^{μ}_{userp} A temporary credential created

C_{user}

C_{userp} The user proxy credential is created
$C_{up} = \text{Sign}_{(u)} \{C'_{up}, \text{start-time, expiry-time}\}$

User proxy C_{userp} A user proxy is created

Figure 4.5 User proxy creation

also the owner's certificate. During mutual authentication, the owner's public key (obtained from their certificate) is used to validate the signature on the proxy certificate. The CA's public key is then used to validate the signature on the owner's certificate. This establishes a chain of trust from the CA to the proxy through the owner, where the proxy process impersonates the owner for the lifetime of the proxy certificate.

4.5 PUTTING IT ALL TOGETHER

4.5.1 Getting an e-Science certificate

One of the first steps that a Grid user needs to undertake is to get the appropriate certificate so that they can interact and use Grid-based resources. In the UK, the e-Science programme has set up a CA to support Grid projects and users. The UK e-Science CA provides a comprehensive service that ensures that users have all the help necessary to use certificates. To get a UK e-Science certificate, the first action that a user must take is to go to the CA's support Web site (Figure 4.6) and read the associated documentation about how to proceed. For the purpose of a demonstration we will run through the Java-based Certificate Request process. It should be noted there is also support for using a Web browser to create a request for a certificate.

Step 0: Read the appropriate documents.

Step 1: Go to http://ca.grid-support.ac.uk/.

Step 2: Click on "Get the CA Root Certificate" – this loads the CA Root Certificate.

Step 3: Go to http://ca.grid-support.ac.uk/jcr (Figure 4.7).

Step 4: Read the JCR documentation.

As the Web page states, the following components must be installed and correctly configured on the client system (the download locations are specified in the user documentation):

- Java Run Time Environment (JRE) 1.4.2;
- Sun's unrestricted policy files for JCE – (see Table 4.2);

4.5 PUTTING IT ALL TOGETHER

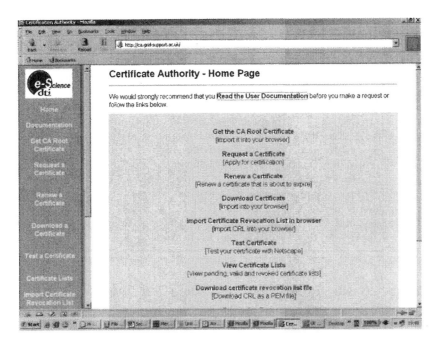

Figure 4.6 The UK e-Science Certificate Authority Home Page

Figure 4.7 e-Science Java Certificate Request (JCR) Page

Table 4.2

JCE
The Java Cryptography Extension (JCE) is a package that provides a framework and implementations for encryption, key generation and agreement and message authentication code (MAC) algorithms. The support for encryption includes symmetric, asymmetric, block and stream ciphers. JCE has the ability to enforce restrictions regarding the cryptographic algorithms and maximum cryptographic strengths available to applets/applications in different (jurisdiction contexts) locations. Sun's JCE provides support for the likes of DES, Blowfish, PBEWithMD5AndDES, Diffie-Hellman key agreement among multiple parties, HmacMD5 and HmacSHA1. Unfortunately, RSA is not included. Another third party, Bouncy Castle JCE, provides a larger collection of cryptographic algorithms, including Blowfish, DES, IDEA, RC2-6, AES, RSA, MD2, MD5, SHA-1 and DSA. Bouncy Castle JCE is an open-source Apache style project that provides the most complete, freely available JCE.

- Bouncy Castle cryptographic provider – (see Table 4.2);
- The CA root certificate loaded into your browser in Step 2.

Step 5: Install and set up your system as specified.

Step 6: Launch the JCR applet (Figure 4.8).

Step 7: Click User Certificate Request (Figure 4.9), complete the form and submit.

Note that the pass phrase is used to secure your private key. When you receive your certificate, the same pass phrase will be used to protect the certificate.

Step 8: If you are successful you will get the JCR message shown in Figure 4.10. Check that your certificate request details are correct.

Step 9: You will be prompted to save a file to your hard disk (Figure 4.11). JCR uses the filename to store a copy of the certificate request. It has the extension **.csr**. It also saves a copy of the private key, which is generated along with the request. The private-key file is encrypted and has the extension **.enc**.

Step 10: If you are successful you will get the JCR message shown in Figure 4.12.

Step 11: If you successfully get to Step 10, then you should take your photo ID and the PIN returned in Figure 4.12 to your local RA.

4.5 PUTTING IT ALL TOGETHER

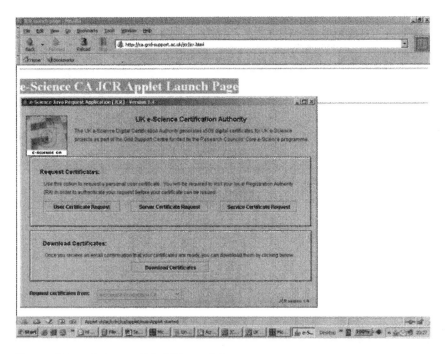

Figure 4.8 e-Science CA JCR applet

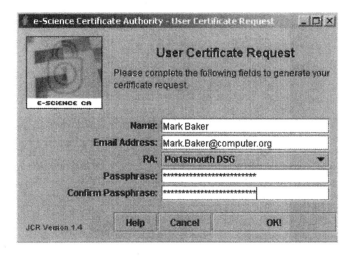

Figure 4.9 User Certificate Request

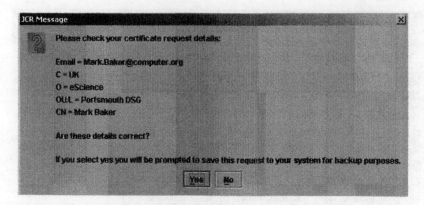

Figure 4.10 JCR check message

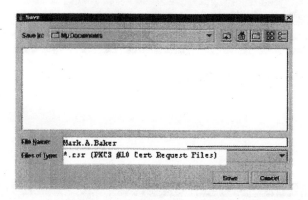

Figure 4.11 Saving the Cert Request File

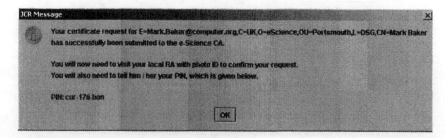

Figure 4.12 JCR successful request message

4.5 PUTTING IT ALL TOGETHER

The RA will check your ID and send a request to the CA on your behalf for an e-Science certificate.

Step 12: At some stage soon (between 24 and 48 hours later) the CA will email you. The email will contain the serial number of your certificate.

Step 13: Start-up the JCR applet again and click on the "Download Certificates" button (Figure 4.8). You will be presented with a window shown in Figure 4.13.

You should do the following:

- Enter the certificate serial number;
- Use the "Browse" button to locate the **.enc** file that was saved when you made your request;
- Enter the pass phrase you used to protect the private key;
- Click on "Go!"

The applet will then do the following:

- Retrieve your certificate and write it to a **.cer** file;
- Un-encrypt your private key;
- Write a **.p12** file, which is protected with the pass phrase;
- Delete the **.enc, .csr** and **.cer** files.

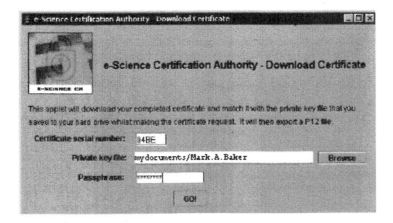

Figure 4.13 JCR certificate download

The .p12 file contains your certificate with its private key. You should take care of this file. It is in a form which is suitable for importing into other applications such as a Web browser. It is not suitable in this form for Globus. It is necessary to use OpenSSL [9] to convert the .p12 file to the format used by Globus [10].

4.5.2 Managing credentials in Globus

Once a user has been authenticated to a system and proven their GSI identity, they must be authorized before they gain any access. Globus comes with an authorization mechanism, which uses the gridmap file. The gridmap file is the main point of control for the local system administrator to control who can access their system using Globus.

The gridmap file is a plaintext file containing a list of authorized GSI identities and mappings from those identities to local user identities (e.g. UNIX account names). Thus it is both an ACL and a mapping mechanism. The local system administrator, when requested by a user who presents their GSI identity to the administrator, adds entries to it. The administrator then determines what the local account name is for the user and adds this mapping to the gridmap file.

The gridmap file is used by GSI after authenticating a user. A gridmap file at each site specifies grid-id to local-id mapping. GSI first checks to see if the user's grid identity is listed in the gridmap file. If the user is not listed, they are denied access to the resource.

An example of a gridmap file entry is:

```
"/C=UK/O=eScience/OU=Portsmouth/L=DSG/ CN=Mark Baker" mab
```

The Globus command `grid_mapfile_add_user` can be used by an administrator to add entries to the gridmap file. For example, the following command would add a user to the gridmap file:

```
grid_mapfile_add_user -dn
"/C=UK/O=eScience/O=Portsmouth/L=DSG/CN=Mark Baker"
```

The distinguished name must be exactly as it appears in the user's certificate. Not doing so will result in an authentication failure.

4.5 PUTTING IT ALL TOGETHER

You can get this information directly from the user's certificate by using the following command:

```
grid-cert-info --subject --file <certificate file>
```

For example, the following command prints the certificate subject for the X.509 certificate in the file $HOME/.globus/usercert.pem. This is the default location for a user's certificate file (Table 4.3):

```
grid-cert-info --subject --file $HOME/.globus/usercert.pem
```

Table 4.3 Digital certificate file formats

Digital certificates
An X.509 digital certificate is based upon what is known as Abstract Syntax Notation (ASN.1). The X.509 data can be encoded in either a binary or ASCII form. These files are known as DER distinguished encoding rules – or (Privacy Enhanced Mail) PEM-based files.
DER, a binary representation, is specified in the ISO Standard X.690. DER files are usually characterized by a ".der" extension.
A Certificate Signing Request (CSR) consists of a distinguished name, a public key and an optional set of attributes. The CSR is signed by the entity requesting certification. The CSR is sent to a CA that transfers the request into an X.509 public-key certificate. CSR files are usually characterized by a ".csr" extension.
A PEM (RFC 1421 RFC 1424) encoded format is essentially the same X.509 digital certificate but in an ASCII form. PEM is a Base64 version of the DER file wrapped with the "----BEGIN CERTIFICATE----" and "----END CERTIFICATE---" lines. This format is used for ease of transport and to facilitate cutting and pasting of the data.
PKCS#12 is a standard for storing private keys and certificates in a portable format. The specification is actually called Personal Information Exchange Syntax. A PKCS#12 file contains a user's private key and digital certificate, along with other "secret" information about the user. PKCS#12 files are usually characterized by a ".p12" extension.
Digital certificate files can have a ".cer" extension.
During the JCR process (see Step 9) the private key is encrypted and saved in a file with the extension ".enc".

4.5.3 Generate a client proxy

Before running a client, it is necessary to generate a proxy for the client. Proxies are intended for short-term use, when the user is submitting many jobs and does not want to repeatedly enter their password for every job. The subject of a proxy certificate is the same as the subject of the certificate that signed it, with /CN=proxy added to the distinguished name. A gatekeeper accepts any job requests submitted by the user, as well as any proxies he has created. Proxies may be a convenient alternative to constantly entering passwords, but they are also less secure than the user's normal security credential. Therefore, they should always be user-readable only, and should be deleted after they are no longer needed (or after they expire). Globus uses MyProxy [11] as a credential repository.

The command to create a proxy in Globus is:

```
grid_proxy_init
```

Generating a proxy uses the $HOME/.globus/usercert.pem and $HOME/.globus/userkey.pem files by default to generate a proxy. You will have to enter the pass phrase you provided when your public/private-key pair was initially generated.

```
Your identity: /C=UK/O=eScience/OU=Portsmouth/L=DSG/CN=Mark Baker
Enter GRID pass phrase for this identity:
Creating proxy ..............................
Done
Your proxy is valid until: Sat March 27 05:19:30 2004
```

This will create a proxy file in /tmp/x509up_u$UID. For example with an UID of 1004 the proxy file will be /tmp/x509up_u1004. This proxy has a default expiry time of 12 hours.

4.5.4 Firewall traversal

The Globus Toolkit (see Chapter 5) provides a way to restrict the ports which client and server programs can use. This is done by the use of the GLOBUS_TCP_PORT_RANGE environment variable. Setting this variable to a range of ports will cause Globus tools to only use ports in this range, see Table 4.4 for further details.

Table 4.4 TCP ports

Ephemeral	A non-deterministic port assigned by the system in the untrusted port range (>1024)
Controllable ephemeral	An ephemeral port selected by the Globus Toolkit libraries that can be constrained by use of the GLOBUS_TCP_PORT_RANGE environment variable to a given port range
Grid service	Static ports for well-known Grid services. These ports are, for example: • 22/TCP (GSI-enabled OpenSSH) • 2119/TCP (Globus Gatekeeper) • 2135/TCP (MDS) • 2811/TCP (GridFTP server).

4.6 POSSIBLE VULNERABILITIES

Security mechanisms are not fool proof. There always tends to be tension between security and usability. So, there will always be vulnerabilities, which range from technical to physical ones. This section deals with Grid security flaws.

4.6.1 Authentication

On the Grid, both sides achieve authentication by using certified public keys. An authorized CA decides who should be certified. An obvious concern here is the amount of checking done to confirm the identity of someone before issuing the certificate. A CA should publish a policy statement (CPS) about how they operate.

We recommend that you read Mike Surridge's e-Science report, *A Rough Guide to Grid Security* [12], especially note the social engineering techniques that can be used to obtain user credentials.

4.6.2 Proxies

The proxy mechanism is useful as it provides local authorization without the need to contact the user site for every remote action or service access. There are some drawbacks that compromise security:

- The proxy's private key resides on a remote system outside the user's direct control; however, this is used to sign messages that the Grid infrastructure will trust as coming from the user.
- When creating a proxy certificate the user acts like a CA for the proxy; however, the user does not have to publish a certification policy or a revocation process.

If the remote system were compromised, the user's proxy private key would no longer be private, and even if the user were aware of this there is no revocation process. Globus attempts to minimize the risk by recommending that proxies are short-lived, so the stolen credential can only be misused for a short time. Here active detection of stolen credentials is very important.

4.6.3 Authorization

Authorization is the step of deciding what access rights a user has to services. In the Grid, most systems assert that authorization to a service should be granted by the service owner. Most Grid systems do this by mapping a remote Grid user's identity onto a local account. The local system administrator can then define the services the local account is able to access. The greatest problem with this methodology is that for every user, service or resource, there need to be entries in the gridmap file. The number of such entities to be managed by the gridmap file can be huge, and system administrators will be swamped and start giving all users default rather than particular access rights, which may mean that vulnerabilities are introduced. This model also means that it is likely that it takes time to add or remove user access rights, which can cause problems also.

In general, it is clear that the greater the access rights that a user is granted, the more likely that a service or resource will be exposed to security compromise. If a user's credentials are stolen and if they are not revoked quickly, services and resources that the user had access right to are at risk. This area has been addressed by the GridPP project [13], who have created a patch which provides a modification for Globus Toolkit v2+, which enables the dynamic allocation of local UNIX usernames to Grid users. The system administrator using the normal account creation method can create a pool of local accounts, and these are leased to incoming Grid users. This is similar to the way DHCP allocates temporary leases on IP numbers from a pool. A UNIX cron job can be run at regular intervals to revoke the leases after a specified time (e.g. hourly) and make them available to other users.

4.7 SUMMARY

Security can be seen as both sociological and technical in nature, and there are many challenges that need to be addressed to ensure that users and resources are secure. Some of the technical challenges are being met; there are many, such as those relating to trust, which are being considered; however, there are others which have still to be addressed.

As the Grid is increasingly being taken up and used by all sectors of business, industry, academia and the government as the middleware infrastructure of choice, it is crucial that Grid security is watertight.

In this chapter we first provided a security primer that described potential risks, then we moved on to outline the basic principles of cryptography, explaining briefly the typical terms and technologies as will be seen when security is mentioned. The chapter then discussed the GSI, and the various authorization modes used in GSI. We then put together all the security components previously mentioned and described how a user would get a certificate; manage credentials and transverse a firewall. Finally we briefly highlighted possible vulnerabilities that possibly still exist in Grid security.

4.8 ACKNOWLEDGEMENTS

/C=UK/O=eScience/OU=CLRC/L=RAL/CN=Alistair Mills
/C=UK/O=eScience/OU=Portsmouth/L=DSG/CN=Hong Ong

4.9 FURTHER READING

Azzedin, F. and Maheswaran, M., Towards Trust-Aware Resource Management in Grid Computing Systems, *Cluster Computing and the Grid 2nd IEEE/ACM International Symposium CCGrid 2002*, pp. 452–457.

Butler, R., Welch, V., Engert, D., Foster, I., Tuecke, S., Volmer, I. and Kesselman, C., A National-Scale Authentication Infrastructure, *Computer*, 33 (12), December 2000, pp. 60–66.

Butt, A., Adabala, S., Kapadia, N., Figueiredo, R. and Fortes, J., Fine-grain Access Control for Securing Shared Resources in Computational Grids, *Parallel and Distributed Processing Symposium*, Proceedings International, IPDPS 2002, Abstracts and CD-ROM, 2002, pp. 206–213.

Humphrey, M. and Thompson, M., *Security Implications of Typical Grid Computing Usage Scenarios*, Security Working Group GRIP forum draft, October 2000.

Internet Security Glossary, ftp://ftp.isi.edu/in-notes/rfc2828.txt.

4.10 REFERENCES

[1] Critical Infrastructure Information Security Act, Bob, Bennett, http://bennett.senate.gov/bennettinthesenate/speeches/2001Sep25_Crit_Infrast_Inf_Sec.htm.
[2] eLiza Project, http://www.ibm.com/servers/autonomic/.
[3] Bosworth, S. and Kabay, M.E. (eds), *Computer Security Handbook*, Wiley, US, 4th edition (5 April 2002), ISBN: 0471412589.
[4] Hash Function, http://www.nist.gov/dads/HTML/hash.html.
[5] ITU-T, http://www.itu.int/rec/recommendation.asp?type=folders&lang=e&parent=T-REC-X.509.
[6] GSI Working Group, http://forge.gridforum.org/projects/gsi-wg.
[7] Foster, I., Kesselman, C., Tsudik, G. and Tuecke, S., *A Security Architecture for Computational Grids*. Proceedings of 5th ACM Conference on Computer and Communications Security Conference, 1998, pp. 83–92.
[8] SSL, http://docs.sun.com/source/816-6156-10/contents.htm.
[9] OpenSSL, http://www.openssl.org/.
[10] Key Conversion, http://www.grid-support.ac.uk/ca/user-documentation/Globus.html.
[11] Novotny, J., Tuecke, S. and Welch, V., *An Online Credential Repository for the Grid: MyProxy*, High Performance Distributed Computing, Proceedings of 10th IEEE International Symposium on, Los Alamitos, CA, USA, IEEE Computer Society Press, 2001, pp. 104–111, http://csdl.computer.org/comp/proceedings/hpdc/2001/1296/00/12960104abs.htm.
[12] Surridge, M. *A Rough Guide to Grid Security*, http://www.nesc.ac.uk/technical papers, e-Science Technical Report 2002.
[13] `gridmapdir` patch for Globus, http://www.gridpp.ac.uk/gridmapdir/.

5
Grid Monitoring

5.1 INTRODUCTION

A Grid environment is potentially a complex globally distributed system that involves large sets of diverse, geographically distributed components used for a number of applications. The components discussed here include all the software and hardware services and resources needed by applications.

The diversity of these components and their large number of users render them vulnerable to faults, failure and excessive loads. Suitable mechanisms are needed to monitor the components, and their use, hopefully detecting conditions that may lead to bottlenecks, faults or failures. Grid monitoring is a critical facet for providing a robust, reliable and efficient environment.

The goal of Grid monitoring is to measure and publish the state of resources at a particular point in time. To be effective, monitoring must be "end-to-end", meaning that all components in an environment must be monitored. This includes software (e.g. applications, services, processes and operating systems), host hardware (e.g. CPUs, disks, memory and sensors) and networks (e.g. routers, switches, bandwidth and latency). Monitoring data is needed to understand performance, identify problems and to tune a system for better overall performance. Fault detection and recovery mechanisms need the monitoring data to help determine if parts of an environment are not functioning correctly, and whether to restart

---

The Grid: Core Technologies Maozhen Li and Mark Baker
© 2005 John Wiley & Sons, Ltd

a component or redirect service requests elsewhere. A service that can forecast performance might use monitoring data as input for a prediction model, which could in turn be used by a scheduler to determine which components to use.

In this chapter, we will study Grid monitoring related techniques. In Section 5.2, we introduce the Grid Monitoring Architecture (GMA), an open architecture proposed by the GGF's [1] Grid Monitoring Architecture Working Group (GMA-WG). In Section 5.3, we define the criteria we use to review the systems discussed in this chapter. This is followed by an overview of representative monitoring systems and we provide a comparison of them in terms of openness, scalability, resources to be monitored, performance forecasting, analysis and visualization in Section 5.4. In Section 5.5, we outline six alternative systems that are not strictly Grid resource monitoring systems. In Section 5.6, we discuss some issues that need to be taken into account when using or implementing a Grid monitoring system. Section 5.7 summarizes the chapter.

5.2 GRID MONITORING ARCHITECTURE (GMA)

The GMA [2] consists of three types of components (see Figure 5.1):

- A Directory Service which supports the publication and discovery of producers, consumers and monitoring data (events);
- Producers that are the sensors that produce performance data;
- Consumers that access and use performance data.

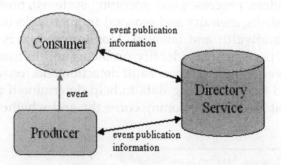

Figure 5.1 The Grid Monitoring Architecture

5.2.1 Consumer

Any program that receives monitoring data (events) from a producer can be a consumer. The steps supported by consumers are listed in Table 5.1. An event-naming schema is normally used to describe the meaning of an event type. All producers that handle new event types should dynamically provide a naming schema for event description. Consumers that initiate the flow of events should support steps 2–5; consumers that allow a producer to initiate the flow of events should support steps 6–8.

It is possible to have a number of different types of consumers:

- The archiving consumer aggregates and stores monitoring data (events) for later retrieval and/or analysis. An archiving consumer subscribes to producers, receives event data and places it in long-term storage. A monitoring system should provide this component, as it is important to archive event data in order to provide the ability to undertake historical analysis of system performance, and determine when/where changes occurred.

Table 5.1 Consumer steps

1. *Locate events*: Consumers search a schema repository for a new event type. The schema repository can be a part of the GMA Directory Service.
2. *Locate producers*: Consumers search the Directory Service to find a suitable producer.
3. *Initiate a query*: Consumers request event(s) from a producer, which are delivered as part of the reply.
4. *Initiate a subscription*: Consumers can subscribe to a producer for certain kinds of events they are interested in. Consumers request event(s) from a producer.
5. *Initiate an unsubscribe*: Consumers terminate a subscription to a producer.
6. *Register*: Consumers can add/remove/update one or more entries in the Directory Service that describe events that the consumer will accept from producers.
7. *Accept query*: Consumers can also accept a query request from a producer. The "query" will also contain the response.
8. *Accept subscribe*: Consumers accept a subscribe request from a producer. The producer will be notified automatically once there are requests from the consumers.
9. *Accept unsubscribe*: Consumers accept an unsubscribe request from a producer. If this succeeds, no more events will be accepted for this subscription.

While it may not be a good idea to archive all monitoring data, it is desirable to archive a reasonable sample of both "normal" and "abnormal" system operations, so that when problems arise it is possible to compare the current system to a previously working system. Archive consumers may also act as GMA producers to make the data available to other consumers.

- As the name implies, real-time consumers collect monitoring data in real time. A real-time consumer potentially subscribes to multiple events of interest, and receives one or more streams of event data. In this way, data from many sources can be aggregated for real-time performance analysis.
- Overview consumers collect events from several sources, and use the combined information to make some decision that could not be made on the basis of data from only one producer.
- Job monitoring consumers can be used to trigger an action based on an event from a job, e.g. to restart the job.

5.2.2 The Directory Service

The GMA Directory Service provides information about producers or consumers that accept requests. When producers and consumers publish their existence in a directory service they typically specify the event types they produce or consume. In addition, they may publish static values for some event data elements, further restricting the range of data that they will produce or consume. This publication information allows other producers and consumers to discover the types of events that are currently available, the characteristics of that data, and the sources or sinks that will produce or accept each type of data. The Directory Service is not responsible for the storage of event data; only information about which event instances can be provided. The event-naming schema may, optionally, be made available by the Directory Service.

The functions supported by the Directory Service can be summarized as:

- *Authorise a search*: Establish the identity (via authentication) of a consumer that wants to undertake a search.
- *Authorise a modification*: Establish the identity of a consumer that wishes to modify entries.

5.2 GRID MONITORING ARCHITECTURE (GMA)

- *Add*: Add a record to the directory.
- *Update*: Change the state of a record in the directory.
- *Remove*: Remove a record from the directory.
- *Search*: Perform a search for a producer or consumer of a particular type, possibly with fixed values for some of the event elements. A consumer can indicate whether only one result, or more if available, should be returned. An optional extension would allow a consumer to get multiple results, one element at a time using a "get next" query in subsequent searches.

In a Grid monitoring system, there can be one central Directory Service or multiple services managed by a Directory Service Gateway. Figure 5.2 shows an extended Grid Monitoring Architecture with multiple Directory Services.

5.2.3 Producers

A producer is a software component that sends monitoring data (events) to a consumer. The steps supported by a producer are listed in Table 5.2. Producers that wish to handle new event types dynamically should support the first step. Producers that allow

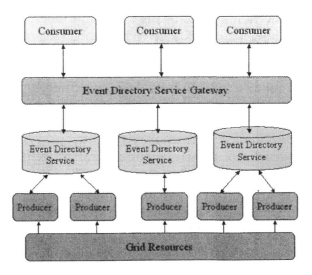

Figure 5.2 Grid Monitoring Architecture

Table 5.2 Producer steps

---

1. *Locate event*: Search the Event Directory Service for the description of an event.
2. *Locate consumer*: Search the Event Directory Service for a consumer.
3. *Register*: Add/remove/update one or more entries in the Event Directory Service describing events that the producer will accept from the consumer.
4. *Accept query*: Accept a query request from a consumer. One or more event(s) are returned in the reply.
5. *Accept subscribe*: Accept a subscribe request from a consumer. Further details about the event stream are returned in the reply.
6. *Accept unsubscribe*: Accept an unsubscribe request from the consumer. If this succeeds, no more events will be sent for this subscription.
7. *Initiate query*: Send a single set of event(s) to a consumer as part of a query "request".
8. *Initiate subscribe*: Request to send events to consumers, which are delivered in a stream. Further details about the event stream are returned in the reply.
9. *Initiate unsubscribe*: Terminate a subscription to a consumer. If this succeeds, no more data will be sent for this subscription.

---

consumers to initiate the flow of events should support steps 2–6. Producers that initiate the flow of events should support steps 7–9.

Producers can deliver events in a stream or as a single response per request. In streaming mode, a virtual connection is established between the producer and consumer and events can be delivered along this connection until an explicit action is taken to terminate it. In query mode, the event is delivered as part of the reply to a consumer-initiated query, or as part of the request in a producer-initiated query.

Producers are also used to provide access control to the event, allowing dissimilar access to different classes of users. Since a Grid can consist of multiple organizations that control the components being monitored, there may be different access policies, varying frequencies of measurement and ranges of performance detail for consumers "inside" or "outside" the organization owning a component. Some sites may allow internal access to real-time event streams, while providing only summary data outside a site. The producers would potentially enforce these policy decisions. This mechanism is important for monitoring clusters or computer farms, where there may be extensive internal monitoring, but only limited monitoring data accessible to the Grid.

5.2.3.1 Optional producer tasks

There are many other services that producers might provide, such as event filtering and caching. For example, producers could optionally perform any intermediate processing of the data the consumer might require. A consumer might request that a prediction algorithm be applied to historical data from a particular sensor. On the other hand, a producer may filter the data for the consumer and deliver it according to a predetermined consumer schedule. Another example is where a consumer requests that an event be sent only if its value crosses a certain threshold; such as CPU utilization becomes greater than 50%, or changes by more than 20%. The producer might also be configured to calculate summary data; such as 1, 10 and 60-minute averages of CPU use, and make this information available to consumers. Information on the services a producer provides would be published in the directory service, along with associated event information.

5.2.4 Monitoring data

The data used for monitoring purposes needs to have timing, flow and content information associated with it.

5.2.4.1 Time-related data

- Time-stamped dynamic data comes within a flow with several regular messages and temporal information that may be provided by a counter related to the sampling rate (frequency). This data includes performance events and status monitoring.
- Time-stamped asynchronous data used to indicate when an event happens. This data is used for alerts and checkpoint notifications.
- Non-time-related data includes static information such as OS type and version, hardware characteristics or the update time of monitoring information. The term "static" here refers to fact that the data remains almost constant, and is generally operator-updated. Whereas "dynamic" refers to information, like status or performance, that change over time.

5.2.4.2 Information flow data

- Direct producer–consumer flow does not need a central component involved in data transfer. A monitor may be active or passive depending on whether the communication is producer or consumer initiated. Three interactions are described by the GMA document:
  - Publish/subscribe,
  - Query/response,
  - Notification.
- Indirect data distribution via a centralized repository. This may be useful for static information, where there is a relatively small amount of data that is seldom updated, and where the cost of the publication/discovery process is comparable to that of information gathering. In this case interaction is via the initial notification of the producers to the directory service, and consumers can pick up data from this source too.
- Following a workflow's path, where monitoring information is produced and stored locally. The data is tagged so that it can be associated with a particular part of a workflow. At the end of the job the monitoring information and tag, together with the workflow output, may be returned to a consumer or discarded. A consumer can gather tags and monitoring data by following the job's path, which may be combined to provide a summarized view, or sent independently to the consumer.

5.2.4.3 Monitoring categories

- Static monitoring is where the cost of information gathering, in terms of time and used bandwidth, is less or comparable to the cost of resource discovery, for example like a query to a central Directory Service to find the information provider. The information changes rarely and a central repository can directly provide the needed data. Information in this category could include system configuration and descriptions.
- Dynamic monitoring is where the cost of information gathering is generally greater and usually involves time series, like when a continuous data flow is provided or a large amount of data is needed. Classical examples of this category are network and system performance monitoring.

- Workflow monitoring is where a variable amount of data is produced as the processing of a job/task takes place and all or part of it may be of some interest for a consumer. Examples are job/task processing status information, error reporting and job/task tracing.

5.3 REVIEW CRITERIA

The Grid monitoring systems reviewed here were categorized and classified using the following criteria.

5.3.1 Scalable wide-area monitoring

To operate in a Grid context a system must be capable of supporting concurrent interaction of potentially thousands of clients and millions of resources. System architectures should support the features desired of distributed systems, which include:

- *Scalability*: A system's ability to maintain or increase levels of performance or quality of service under an increased system load, by adding resources.
- *Fault tolerance*: Systems that are capable of operating successfully even when a number of their components are unavailable or experiencing errors, by avoiding a single point of failure for critical components.

5.3.2 Resource monitoring

The systems reviewed in this chapter primarily focus on monitoring computer-based resources and services. While network and application monitoring are important, they are not considered our main interest, which is the health and performance of the core grid infrastructure.

5.3.3 Cross-API monitoring

An important aspect of a system is the integration of monitoring data collected by legacy and specialized software. Given the existing investment in time and money for administrating resources across an organization, we feel it is important to utilize the existing infrastructure as much as possible. This implies that monitoring systems should not dictate that their own custom agents or sensors be installed across the resources to be monitored.

5.3.4 Homogeneous data presentation

In order to efficiently use heterogeneous resources, it is important that retrieved information is meaningful, clear and presented in a standard way to clients, regardless of its source. For example, when comparing resource memory capacities, heterogeneous resources may report in bits, bytes or megabytes. Clients should not be exposed to inconsistencies between the ways different resources report their configuration or status.

5.3.5 Information searching

Clients must be capable of locating appropriate resources, in a timely manner, in order to efficiently perform their work. This implies it must be possible to locate resources based on the functionality or services they provide. Standard definitions of resource categories are required to achieve this and resources should be capable of belonging to more than one category as their functionality dictates. Furthermore, it should be possible to select only those resources within a given category that meet certain criteria; for example, a CPU load lower than a specified threshold.

5.3.6 Run-time extensibility

Many resources within a Grid will reflect the transient nature of virtual organizations; as project collaborations are created to meet a short-term need and then torn down afterwards, so resources will join and leave. Monitoring systems must expect and support rapid transitions in the number and types of available resources.

5.3.7 Filtering/fusing of data

Mechanisms should be supported to reduce network traffic, as well as host and client loads, by providing the ability to filter and fuse data from potentially multiplexed data streams.

5.3.8 Open and standard protocols

Open and standard protocols are necessary to provide a robust infrastructure that is capable of interoperating with existing and emerging middleware tools and utilities. Open standards allow developers to implement systems that can interoperate with standards compliant systems from different organizations. Therefore, open and standard protocols will avoid organizations becoming tied to a single platform and promote acceptance for a system within the community.

5.3.9 Security

Standard security mechanisms are required to promote interoperability with third-party middleware providers. Examples include GSI [3] and SSL [4].

5.3.10 Software availability and dependencies

State-of-the-art projects can be classified as those that have released substantial software at the time of this review. Determining whether monitoring software can be installed on demand, independent of other components, is important to ascertain the utility of the system and the potential overhead required for installation, configuration and management. Ideally, a monitoring package will not require the installation of third-party software components.

5.3.11 Projects that are active and supported; plus licensing

It should established whether a project is actively supported or in a dormant state. Also it is important to determine what type of license the software produced by a project will be released under, as this will determine how the software can be used, developed and released downstream.

5.4 AN OVERVIEW OF GRID MONITORING SYSTEMS

In this section, we will review some of the most popular monitoring systems that can be deployed in a Grid environment. Section 5.5 briefly mentions other monitoring systems that are being used or developed.

5.4.1 Autopilot

5.4.1.1 Overview

Autopilot [5, 6] is an infrastructure for real-time adaptive control of parallel and distributed computing resources. The objective of Autopilot is to create an environment which provides distributed applications with real-time adaptive control so that they can automatically select and configure resource management features based on request patterns and observed system performance. To achieve this, Autopilot provides components to facilitate the collection and distribution of host, service and network performance information. Autopilot was developed by the Pablo Research Group, University of Illinois at Urbana Champaign, and is used in a number of projects including the Grid Application Development Software Project (GrADS) [7, 8].

5.4.1.2 Architecture: General

Autopilot's infrastructure is based on the GMA and uses the Globus Toolkit to perform wide-area communication between its components. Figure 5.3 shown a general architecture of Autopilot. The Pablo Self-Defining Data Format (SDDF) [9] is used for describing resource information. Autopilot monitoring components include:

- The Sensor, which corresponds to a GMA producer; sensors are installed on monitored hosts to capture application and system

5.4 AN OVERVIEW OF GRID MONITORING SYSTEMS

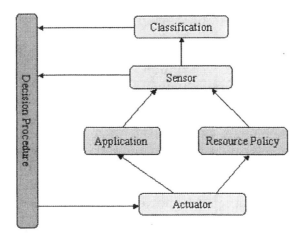

Figure 5.3 The architecture of Autopilot

performance information. Sensors can be configured to perform data buffering, local data reduction before transmission, and to change the frequency at which information is communicated to remote clients. Upon start-up, sensors register with the Autopilot Manager (AM).

- Actuators, which correspond to GMA producers and provide mechanisms for steering remote application behaviour and controlling sensor operation. Upon start-up, actuators register with the AM.
- The AM, which performs the duties of a GMA registry; it supports registration requests by remote sensors and actuators, and provides a mechanism for clients to locate resource information.

An Autopilot client corresponds to a GMA consumer; it locates sensors and actuators by searching the AM for registered keywords. For each producer found, a Globus URI is returned so that the consumer can connect to producers directly. Once connected, the client can instruct sensors to extract performance information, or actuators to modify the behaviour of remote applications.

The Autopilot Performance Daemon (APD) provides mechanisms to retrieve and record system performance information from remote hosts. The APD consists of collectors and recorders. Collectors execute on the machines being monitored and retrieve local resource information. Recorders receive resource information from collectors for output or storage.

5.4.1.3 Architecture: Scalability and fault tolerance

The AM binds together multiple concurrent clients and producers, and provides a seamless mechanism for locating and retrieving resource information from remote sensors. Therefore the AM is a key component for ensuring fault tolerance and scalability of the system. However, while multiple AMs can exist within the monitored environment, there is currently no support for communication between multiple AMs, therefore if an AM fails, the sensor registrations that it holds will be unavailable. Sensors can potentially register with multiple AMs, and clients can query those AMs; however, mechanisms are not provided to locate available AMs. Due to the lack of communication between managers, it is not possible to create hierarchies of managers; each manager contains information from sensors that report directly to it.

5.4.1.4 Monitoring and extensibility

The APD periodically captures network and operating system information from the computers on which they execute. For consistency in heterogeneous networks, only a common subset of host monitoring information, from the range of operating systems supported, is available. Typical host information includes processor utilization, disk activity, context switches, system interrupts, memory utilization, paging activity and network latencies.

Developers could extend the scope of monitoring information by inserting sensors into existing source code that is used to perform local monitoring functions. These sensors can be configured to return specified resource information. Autopilot does not provide a query interface for sensors; clients retrieve information that has been previously configured for collection.

5.4.1.5 Data request and presentation

Sensors periodically gather information and cache it locally regardless of client interest. Client requests are fulfilled from the sensor's cache. Historical data is collected by the APM's collectors and made available to clients. Sensors are capable of filtering and reducing the information returned to clients by using customized functions in the sensors. Aggregated records can also be used to combine information relating to a given host. Mechanisms to support a homogeneous view of data from heterogeneous resources are not provided by Autopilot.

The Autodriver Java graphical user interface allows sensor information to be viewed and actuators to be controlled. Virtue [10], an immersive environment that accepts real-time performance data from Autopilot, interacts with sensors and actuators using SDDF and provides graphical features to view and control Autopilot components.

5.4.1.6 Searching and standards

Clients locate sensors based on the attributes they register with the AM. Given a match, clients connect to the sensors and retrieve the available information. Clients need to be aware of the relationship between the attribute name registered by a sensor and the information it produces. Autopilot uses the Globus Toolkit 2 [11] to perform wide-area communications and follows the GMA. The data format used is SDDF, which although self-describing is non-standard. Additional tools can be utilized to convert SDDF into XML.

5.4.1.7 Security

Autopilot does not provide security support, but instead assumes that applications will utilize Globus security mechanisms.

5.4.1.8 Software implementation

Autopilot is available for download; it is actively supported and released under the Pablo Project Software License [12] Agreement. Autopilot is freely available without fee for education, research and non-profit purposes.

Software portability is limited to UNIX-based platforms. System dependencies include the Globus Toolkit 2.2 and the Pablo SDDF library.

5.4.2 Control and Observation in Distributed Environments (CODE)

5.4.2.1 Overview

CODE [13, 14] is a GMA-like system that attempts to provide an extensible approach for monitoring and managing the Grid. CODE allows administrators to monitor distributed resources, services or applications and react to changes in their status by remotely performing predefined system tasks to the remote hosts. CODE was developed at the NASA Ames Research Center [15] and is used in the NASA Information Power Grid (IPG) [16] to ensure that resources are operating correctly.

5.4.2.2 Architecture: General

The CODE framework is designed to provide the functionality for performing monitoring and management tasks (Figure 5.4). Users extend this framework by adding customized monitoring modules. Monitoring information is propagated through CODE as

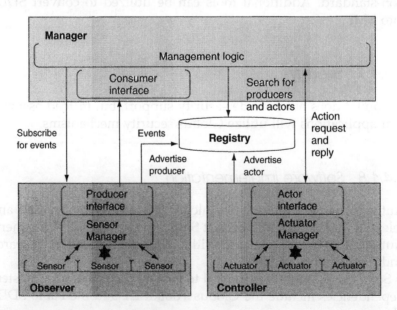

Figure 5.4 The CODE architecture

5.4 AN OVERVIEW OF GRID MONITORING SYSTEMS

events that contain a type followed by name–value pairs. The core framework is made up of Observers, Controllers, Managers and Registries:

- Sensors are installed on monitored hosts and gather monitoring data. Each sensor generates one or more monitoring events that contain monitoring information described in terms of the sensor's naming schema. Sensors can be queried to determine the type of information they produce. Sensors gather resource information only in response to a direct request from a Sensor Manager (SM).
- The SM supervises a set of local sensors and determines which should be executed in order to fulfil client requests. The SM receives query requests and subscriptions from the Observer. In response to a specific query, the SM sends a request to an appropriate sensor and returns the results through the Observer's producer interface to the requesting consumer.
- The Observer encapsulates the SM and sensor mechanisms on a monitored host and provides a Producer Interface (PI) that consumers query to receive monitoring information. The PI supports both query-response and subscription-based requests. Observers implement access control mechanisms based on user identity, client location and information type.
- The Controller resides on a monitored host and provides mechanisms that allow consumers to execute actions on that host. The controller consists of an AM that interacts with a number of locally installed actuator components, which are used to perform a specific function, for example to start an operating system daemon. Like sensors, actuators are passive components that only perform an action when requested by their manager.
- The Manager (consumer) connects to an observer to query for the monitoring data it provides, to subscribe for events and to modify event subscriptions. The Manager connects to Controllers to modify the execution of daemons or applications on a remote host. Users can implement management logic within the Manager in order to automatically respond to changes in the monitored environment by controlling remote hosts. For example, the Manager might detect that a remote job manager is failing to respond and so automatically instruct a remote controller to kill all associated job processes and start a new instance.

Management logic can be implemented using Java code or through an expert system using appropriate management rules.
- The Registry stores the locations of Observers and Controllers, and describes the sensors and actuators they provide. The Manager uses the Registry to locate these remote components.

5.4.2.3 Architecture: Scalability and fault tolerance

Multiple Managers can concurrently monitor data from multiple remote hosts. The Registry provided as part of CODE version 1.0 beta is a Java application providing in-memory registration of producers. This is a temporary measure to allow other developers to download and experiment with CODE. The CODE developers have previously reported the use of an LDAP server to provide registry functionality. The use of multiple LDAP-based registries that potentially perform LDAP referrals (hierarchies of servers) and provide LDAP replication mechanisms could be used to increase scalability and guard against a single point of failure.

Event subscription mechanisms can potentially reduce the amount of traffic generated by the system, as clients are not required to continuously poll for resource information. Subscription requests include details of how frequently the SM should query the sensor and an event filter that the SM uses to determine which results should be streamed back to the consumer. The SM queries sensors in accordance with a specified frequency. The SM uses the event filter to determine if the current results match consumer requirements and should therefore be transmitted. For example, a consumer may require notification only if CPU load is greater than 50%.

5.4.2.4 Monitoring and extensibility

Sensors are installed on all hosts that are to be monitored. New sensors can be registered with a SM, which advertises an Observer's current monitoring capability with the Registry. Sensors are registered by a keyword that describes their function. Clients can locate monitoring functionality based on a keyword search of the Registry. A small set of sensors is provided, administrators are expected to create their own or employ sensors created by third

5.4 AN OVERVIEW OF GRID MONITORING SYSTEMS

parties, to meet their own requirements. Currently there are sensors that report CPU and disk utilization, process status, network interface statistics, file details, Portable Batch System (PBS) queue status and the contents of a Globus GT2 `grid_mapfile`.

The intrusiveness of monitoring can potentially be controlled in the SM by caching results from a sensor query. Cached results from one request could then be used to fulfil requests from further clients that arrive within a suitable time frame. Sensors execute only when directed by the SM, therefore constant polling of resource status can potentially be avoided. However, this is subject to the update frequency rate requested in client subscription requests.

5.4.2.5 Data request and presentation

CODE provides near real-time access to resource data, using either query-response or subscription-based requests. Event notification can be based on a client's needs, so that only a subset of available events is transmitted from the SM to the consumer for a given sensor.

A homogeneous view of heterogeneous data can be provided by CODE. Event-naming schemas are used to describe the data returned by sensors. Sensors are required to individually format their output in order to meet the naming schema they support.

5.4.2.6 Searching and standards

Clients locate Observers in the Registry and then connect directly to suitable Observers. To ascertain the sensors an Observer supports, the Registry can either be searched, or a given Observer can be queried directly. If a consumer executes a subscription query to an Observer, then it is possible for the SM to return only those results that match a consumer-specified criteria, e.g. CPU load greater than 50%.

CODE consumers and producers communicate using XML encoded data over UDP, TCP or GSI SSL/TLS.

5.4.2.7 Security

CODE supports authentication and authorization based on host name and X.509 certificates. CODE supports the Grid Security

Infrastructure (GSI) so that clients can delegate their identity in order for tasks to be performed on their behalf by a server.

5.4.2.8 Software implementation

Version 1.0 beta of CODE is free and available for download under the NASA Open Source Agreement [17]. The project is active and supported. CODE is implemented in Java and has been tested on Linux, Solaris, Irix and MacOS X. CODE's requirements include Java 1.3 or greater, the Xerces Java XML Parser version 2.x and Globus Java CoG kit version 1.1.a. The Controller, Actuator Manager and Actuator components are not implemented in the current software release; therefore control mechanisms are not available.

5.4.3 GridICE

5.4.3.1 Overview

GridICE [18–20] is targeted at monitoring Grid resources in order to analyse their use, behaviour and performance. The project aims to provide client reporting mechanisms for fault detection, service-level agreement violations and user-defined events. GridICE is intended for integration with Grid Information Services (GIS) and currently uses the Globus MDS2 [21, 22] to discover new resources. GridICE queries EDG Lemon [23] agents installed on resources for GLUE [78] information, which is then published into the MDS2. A Web-based interface provides resource views based on virtual organization, grid site and user requirements. GridICE has been developed from work within the INFN-Grid [24] and European DataTAG [25] projects and is used by the LHC Computing Grid (LCG) [26] and INFN Production Grid [27].

5.4.3.2 Architecture: General

GridICE, shown in Figure 5.5, consists of the following layers:

- The Measurement Service (MS) uses the EDG Lemon monitoring infrastructure [23] to query resources and cache information in an internal, centralized repository. Lemon requires agents to be

5.4 AN OVERVIEW OF GRID MONITORING SYSTEMS 173

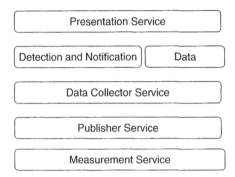

Figure 5.5 The layers of GridICE

installed on each monitored resource to control the operation of individual sensor components. Sensors execute local scripts or applications in order to retrieve resource information, which they are then required to output in an extended version of GLUE. The extended version of GLUE uses roles to describe the services a computer provides, for example job submission or brokering services. Sensors must be configured individually to advertise, gather and format the resource information generated by a host. The Publisher Service (Pub) classifies resources for users based on resource roles.

- The Publisher Service provides the captured resource information to consumers by inserting the latest resource values into a GIS. The GIS is additionally required to publish definitions of the GLUE naming schema to clients. The use of a GIS is intended to provide clients with a common interface to GridICE monitoring information. Currently GridICE uses the Globus MDS2.

- The Data Collector Service (DCS) gathers and persistently stores historical monitoring data. A resource detection component periodically scans a local MDS2, in order to automatically detect new resources suitable for monitoring. The contact information for new resources is passed to a scheduler component that periodically queries resources to discover the information they provide. Resource information is gathered and persistently stored by the GridICE server.

- The Detection and Notification Service (DNS) provides a configurable mechanism for event detection and notification using the event mechanisms provided by the Nagios [28] service and host-monitoring programme. The DNS is designed to allow

a pre-defined set of events to be checked and for sending notifications to clients.
- The Data Analyser (DA) is designed to provide performance and usage analysis, and generate statistical output.
- The Presentation Service (PS) provides a Web interface for role-based views of resources intended to meet the needs of different classes of user. For example, for a virtual organization's manager, it presents a view of all the resources available and jobs that are executing. For a Grid site manager the view may show the status of local resources, and the user view may include details such as accessible processor levels.

5.4.3.3 Architecture: Scalability and fault tolerance

Multiple users can concurrently use the GridICE Web interface to view the status information of resources. Alternatively, clients may interact directly with the MDS2. A seamless view of resources is achieved through the MDS2 query interface and GLUE, which allows resources to be uniformly described.

Architecturally, although GridICE only provides a centralized point for gathering information, fault-tolerance and scalability can be achieved through the introduction of multiple GridICE servers monitoring different parts of a site and each reporting data into different MDS2s. Given that the MDS2 within a site can be federated into a hierarchy, with possibly multiple root MDS2s, fault tolerance can be achieved. The root MDS2 from individual Grid sites can then be incorporated into the virtual organization MDS2 federation.

While MDS2 provides a distributed query engine and standard interface, the authors report that a pull model is required that involves the continual polling of resource data to populate the MDS2 with current values. This introduces a scalability issue for the GridICE server and resource layer. The DCS may be of use to provide caching functionality in an attempt to reduce overhead. However, this service is still required to periodically query resources regardless of interest by users interacting via the MDS2.

5.4.3.4 Monitoring and extensibility

The DCS's "resource detection component" periodically scans the MDS2 for new resources. GridICE does not have an event

5.4 AN OVERVIEW OF GRID MONITORING SYSTEMS

mechanism to provide notification of new resources arriving at the MDS2, therefore, a balance must be achieved between the frequency of probes, the rate at which resources are added to a Grid and the timeliness with which new resources are visualized by users.

The type of information (or role) provided by a known resource may change over time. For example, a computer host may be upgraded to provide new job submission features. The DCS periodically queries resources to discover the new classes of information on offer. While this approach provides updated information on resource roles and capabilities, it is expected that an event-based mechanism would be more scalable.

GridICE utilizes EDG Lemon as the local data collector within the Publisher Service. Lemon sensors provide GLUE-formatted resource information. To include new information, multiple EDG sensors must be configured on monitored hosts.

The recommended approach to provide information from existing cluster monitoring systems, for example Ganglia [29], appears to be as follows: A proxy must be created that periodically queries a ganglia daemon, converts the output into GLUE and inserts the results into the local MDS2. While this follows the standard MDS2-provider approach, it implies that historical information will not be incorporated within GridICE, as the MDS2 typically stores latest state information only, and the GridICE internal repository is not utilized.

5.4.3.5 Data request and presentation

The DCS provides access to historical information. Event subscription and notification is under development; however, it is not clear how this functionality will operate with an existing GIS, like MDS2, to provide real-time asynchronous events to users. The GridICE data access model allows client pull queries from the MDS2 and portal interfaces. Homogeneous views of information are achieved at the resource level with sensors required to support GLUE.

5.4.3.6 Searching and standards

Data searching mechanisms are currently implemented using MDS2; clients will be required to understand the LDAP syntax, semantics and ordering of information within the server.

GridICE acts as an information provider to MDS2, and due to this relationship, monitored information can be utilized within existing Globus testbeds. Interaction with later versions of the MDS, for example MDS3 and MDS4, has not been reported.

5.4.3.7 Security

Currently there are no security mechanisms employed within GridICE; all information is open to anonymous client requests from the MDS2 and the GridICE Web interface. However, X.509-based authentication for the Web interface is planned.

5.4.3.8 Software implementation

GridICE is an open-source software released under the INFN license [30] and is free and available for download. The project is active and provides mailing list support. The software is packaged in Linux RPM format and access to the source code is provided via anonymous CVS.

GridICE requires network access to an external information service to operate. The reference implementation requires access to MDS2. In addition, EDG Lemon and Nagios are required for monitoring resources. Currently the Data Analyser is not available.

5.4.4 Grid Portals Information Repository (GPIR)

5.4.4.1 Overview

The aim of GPIR [31] is to pre-fetch, aggregate and cache information from Grid resources into a central location in order to support the development of Grid portals. In particular, the work focuses on reducing the frequency of queried to access resource information and minimize complexity for portal developers by removing the need to interact with different classes of resource. Information is "ingested" into GPIR from a range of resources that use custom information providers. GPIR is packaged as part of the GridPort Grid portal toolkit [32] from the Texas Advanced Computing Center and is used in the NPACI Hotpage project [33].

5.4.4.2 Architecture: General

The GPIR database is a centralized relational database used for caching resource information from producers. The GPIR architecture is shown in Figure 5.6. Web services interfaces are responsible for receiving information from resources and providing query mechanisms for clients. GPIR provides a number of XML naming schemas that describe how producers should present information to the database for specific aspects of the monitored environment. Currently GPIR defines nine naming schemas that describe:

- A GPIR Information Provider (GIP) executes on monitored resources, gathers local information and outputs an XML document that adheres to one of the naming schemas. The client presents the XML to the GPIRIngester; if the XML document adheres to a registered naming schema it is stored in the GPIR database. Sample clients are supplied to automatically perform these steps.
- The GPIRQuery service provides an interface for clients to query the information cached in the databases. Resources can

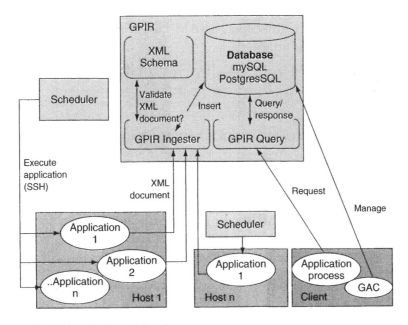

Figure 5.6 The GPIR architecture

be queried by resource or by virtual organization name. Query requests are formulated using the name of a predefined naming schema, for example "load" represents machine load and "services" represents the availability of services executing on a system. The GPIRQuery retrieves information from the database and returns it to the client as an XML document.

- The GPIR Admin Client (GAC) is a Web-based administrative client for defining virtual organizations and managing resource information. The GAC can be used to manually provide additional information about resources, for example a resource's physical location, and system administrator contact details.

1. Static host details, including machine name, its architecture, location and administrator contact details.
2. Host load details including CPU and memory averages.
3. Host status: up, down and unreachable.
4. The downtime for a resource.
5. Job information including queue, job status and constraints.
6. Host MOTD messages.
7. Node status information for a cluster's compute node.
8. The status (pass, fail, timeout) for a number of predefined Grid services: Globus Gatekeeper, GRAM, GIIS, GRIS, GridFTP, NWS and Batch Job Submission.
9. Network Weather Service (NWS) bandwidth and latency measurements.

5.4.4.3 Architecture: Scalability and fault tolerance

Information from multiple resources is held in the database and can be concurrently queried by multiple clients. GPIR providers reside at each resource and feed the database with information formatted to one of the predefined XML naming schemas. Client interaction with the GPIRQuery interface and the use of predefined XML naming schemas provide a seamless view of available resources.

The centralized approach to the database is not fault-tolerant and scalability concerns may become an issue as the number of users and resources increases. These issues could be addressed using distributed or replicated databases.

5.4 AN OVERVIEW OF GRID MONITORING SYSTEMS

Monitoring intrusiveness is determined by the frequency with which a producer is called to populate the database and the overhead associated with starting the producer and obtaining, parsing and outputting information in the required XML format. All user requests for information are fulfilled by the database, thereby shielding resources from direct user interaction.

5.4.4.4 Monitoring and extensibility

Providers must be installed on monitored resources. New resources that need monitoring can be added at run time by installing and executing a suitable provider on the given resource and notifying the GPIRIngester of the resource's host IP address.

The categories of resource information are described by XML naming schemas. The GPIRIngester only accepts XML documents that correspond to predefined and valid XML naming schemas; thereby preventing "unauthorised" XML documents entering the system. To extend the scope of monitored resources, new naming schemas would need to be defined, the GPIR's relational database tables updated and the GPIRIngester configured to accept the new document type. Finally, custom producer code would need to be developed to execute on monitored resources and format output in accordance with the new XML naming schema.

5.4.4.5 Data request and presentation

All client requests are fulfilled by the data cached in the GPIR database. Depending on the frequency with which producers are queried, the database can contain near real time as well as historical data.

Client access to resource information is coordinated by the GPIR Query service, with results returned in GPIR XML format. Event subscription and notification are not supported. Producers push data into the database, while clients pull cached data from the database. GPIR performs filtering and fusing of data by providing summary information for a VO and statistics for a range of resources.

Normalization of complex resource data values to a standard format is not enforced in XML schema; individual resource producers are free to return data values in the format they wish.

Therefore, any normalization would have to be implemented in the producers on an individual basis, using agreed upon methods, to which all should adhere.

5.4.4.6 Searching and standards

In GPIR a client can search for information by resource category, class, functionality, capability and value. The use of Web services provides an open approach for clients to access the GPIR database. However, standards like GLUE are not used to describe resource data, so clients may need to perform extra processing to transform data into a format suitable for their needs.

5.4.4.7 Security

The GPIRQuery service has no security and provides open access to clients. The GPIRIngester service is protected by a list of valid producer IP addresses. The GPIR administrative client is secured by JBoss security controls [34] that are implemented using standard username/password access control lists. More advanced Grid security mechanisms are not provided.

5.4.4.8 Software implementation

The GPIR project is active and the software is available for free download. The software is licensed using the UT TACC Public License version 1. Linux is currently the only supported platform and user support is via a mailing list. Software dependencies include PostgresSQL [35], the PostgresSQL JDBC driver, Java 1.2 and the JBoss application server [34].

5.4.5 GridRM

5.4.5.1 Overview

GridRM [36, 37] is a generic open-source Grid resource-monitoring framework designed to harvest resource data from a range of networked devices and services and provide information to a variety of clients in a form that is useful for their needs. GridRM is not intended to interact with instrumented wide-area applications; rather it is designed to monitor the resources that an application may use. The main objective of the project is to take a standards-based approach, independent of any particular Grid middleware to provide a homogeneous view of heterogeneous resources, using a seamless and extensible architecture.

GridRM provides a consistent SQL-based query language for clients to interact with legacy and emerging resource-monitoring technologies through a number of drivers. GridRM drivers accept SQL queries and remotely query the underlying resources using appropriate native protocols. Native resource data is gathered and transformed into a standard, normalized format according to a user-selected naming schema. The GLUE naming schema is used extensively and default drivers provide access to resources that include Ganglia [29], the Simple Network Management Protocol (SNMP), Network Weather Service (NWS) [38] and Linux/proc.

GridRM was developed by the Distributed Systems Group (DSG) [39] at the University of Portsmouth and is being used on a testbed to monitor resources at institutions that include the Grid Computing and Distributed Systems (GRIDS) Lab, University of Melbourne, Australia; University of Veszprem, Hungary; Institute of Information Technology, National University of Sciences and Technology (NUST), Pakistan; High-Performance Computing and Networking Center, Kasetsart University, Thailand; CCLRC e-Science Centre, Rutherford Appleton Laboratory, UK.

5.4.5.2 Architecture: General

GridRM has a hierarchical architecture with a global, and potentially multiple local layers, each of which has gateways that provide access to a site's local resource information (Figure 5.7):

- A Naming Schema (NS) defines the semantics by which resources are defined. By default GridRM uses GLUE to define the attributes and values of computer-based resources. Drivers use naming schemas to translate raw data from heterogeneous resources into a standard form.
- A Driver is a modular plug-in that is used to retrieve selected information from native monitoring agents.

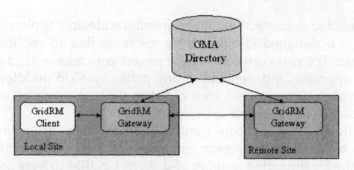

Figure 5.7 The architecture of GridRM

- The Local Layer provides access to real-time and historical information gathered from local resources. Administrators interact with the Local Layer to configure drivers, naming schema and resource interaction.
- The Global Layer provides inter-grid site or VO interaction between GridRM gateways, using a lightweight implementation of the GMA, known as jGMA [40, 41]. GridRM gateways are GMA producers that register with the GMA directory, and respond to consumer requests for resource information. Requests are received in an SQL form and passed to the Local Layer for processing. Results received from the Local Layer are processed into XML and returned to the consumer. The Global Layer provides consumer access control mechanisms and is responsible for controlling the release of information. A large Grid site or VO, may have multiple gateways, in which case a hierarchy of gateways can be constructed to provide resource information.
- Consumers interact with gateways at the Global Layer. Gateways are located using the GMA registry and their resources queried using SQL syntax. Consumers can use the registry to discover the resources currently available at a Grid site and then directly query those resources.

5.4.5.3 Architecture: Scalability and fault tolerance

Multiple clients can concurrently interact with each gateway to retrieve resource information. A gateway can be configured to monitor any number of resources. An SQL query syntax and database abstraction is used to provide a seamless interface to heterogeneous resources. The Local Layer performs caching and provides consumers with the ability to determine cache policy. For example, with each request consumers can indicate the age of cached information they are willing to tolerate. If the cached entries are too old, drivers are called to retrieve new information. This approach allows the results from one request to be shared by multiple consumers, and allows a consumer to dictate the maximum age of information that is useful to them.

Fault tolerance is achieved through the use of multiple GridRM gateways and the replicated registry provided by jGMA. Scalability is addressed through the partitioning of monitored resources across multiple, potentially hierarchical gateways with caching used at each. Consumers locate resources and their controlling gateways using the registry and then directly interact with the appropriate gateways.

5.4.5.4 Monitoring and extensibility

GridRM is designed to monitor computer-based resources by utilizing existing native monitoring agents. GridRM drivers reside on the gateway and interact with remote resources using appropriate native protocols. Default drivers are provided for a range of systems, including Ganglia, SNMP, NWS and Linux/proc.

By default, drivers use GLUE to normalize native data. However, drivers can support multiple naming schemas, which clients select dynamically when submitting a resource query.

Monitoring intrusiveness can be controlled though the use of caching at the Local Layer. Resources are queried either in direct response to a client request for specific information, or periodically if the gateway is explicitly instructed to automatically gather specified resource data. In addition, driver developers have the choice of implementing their own caching and data request strategies within the drivers, as appropriate for a particular native agent.

5.4.5.5 Data request and presentation

GridRM provides access to real-time and historical information. Consumers primarily interact with gateways using a request–response protocol; however, gateways are designed to provide event notification, which is currently being implemented.

A homogeneous view of resource information is achieved through the use of naming schemas, standard SQL syntax and drivers that abstract resources, making them appear as relational databases. The SQL syntax provides fine-grained access to information, allowing the consumer to specify exactly what values should be returned.

When processing queries, drivers filter the requested data at the resource's native protocol level and fuse different values in order to generate information appropriate to the SQL request and current naming schema. Further filtering and fusing of information occurs at the Local Layer when multiple resources are queried and at the consumer when multiple gateways are used to retrieve the available resources within a VO.

GridRM has a Web interface that provides a graphical representation of gateways and resources. Users can either interact with resources using custom forms, or through an SQL command-line interface.

5.4.5.6 Searching and standards

Resources can be searched by the naming schemas their drivers provide and by standard SQL queries. For example, the GridRM Web interface allows users to search for computer hosts based on operating system, CPU load and installed memory size.

GridRM uses open standards including Servlets, JSPs, JDBC, SQL and relational databases.

5.4.5.7 Security

The GridRM gateways provide access control lists, which map client's identities to native resource-level usernames and passwords. jGMA provides encrypted wide-area communications. In the future the GSI will be supported.

5.4.5.8 Software implementation

GridRM is actively being developed and supported, and the beta version is free and available for download under the open-source GNU Public License [42]. The software is written in Java and portable across operating systems. Software dependencies include Java 1.4, Apache Tomcat and mySQL.

5.4.6 Hawkeye

5.4.6.1 Overview

Hawkeye [43] is a monitoring tool, from the University of Wisconsin, that provides mechanisms for monitoring distributed collections of computers by gathering computer-based resource information. Hawkeye's design goals include retrieval of host resource information in a consistent and extensible manner and the ability to automatically execute tasks in response to observed conditions on the monitored hosts. Although Hawkeye is based on technology from the Condor project [44] and utilizes Classified Advertisements (ClassAds) [45] for collecting and publishing resource information, it is packaged as a self-contained system. Data collected by Hawkeye is made available for applications and users to manage monitored resources. Hawkeye is deployed primarily for monitoring of Condor workstation pools at the University of Wisconsin.

5.4.6.2 Architecture: General

Agents gather local resource information and report it to a centralized manager. Hawkeye requires that an agent be installed on each monitored host. Figure 5.8 shows the architecture of Hawkeye.

- The Hawkeye Monitoring Agent (HMA) periodically executes monitoring modules that independently collect resource information and return it in the form of Condor ClassAd attribute–value pairs. The monitoring agent is responsible for consolidating information from multiple modules into a single ClassAd that describes the combined status of a local resource. Periodically, the HMA pushes this combined information to a remote Hawkeye Manager (HM).
- The Hawkeye Manager (HM) caches information submitted to it, in order to provide clients with potentially low latency access to recent resource information. All client requests are submitted

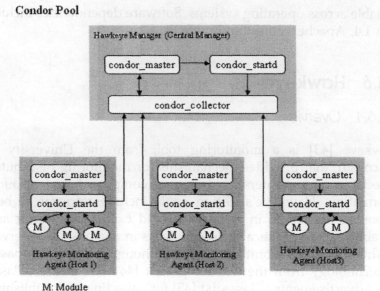

Figure 5.8 The architecture of Hawkeye

to the HM and fulfilled from the HM's cache. The HM passively receives updates from HMA. Updates also serve as a heartbeat mechanism and are used to notify clients of HMA that are no longer responding.

5.4.6.3 Architecture: Scalability and fault tolerance

Multiple users can connect to the HM and retrieve cached resource information. The combination of resource data and use of ClassAds can provide a seamless view of remote resource information to clients. Although the HM performs caching to reduce the overheads of client requests, the use of a single centralized HM is a potential bottleneck and a single point of failure.

HMAs push resource information to the HM using address information from their local configuration file. Currently it is not possible to specify multiple HMs to transmit information to. Furthermore, the HM removes cached resource information after 15 minutes, requiring HMA to continually send updates, potentially creating high levels of overhead even when clients do not require information.

5.4.6.4 Monitoring and extensibility

Hawkeye agents must be installed on each monitored host. Monitoring functionality is provided through a set of default modules that provide access to host resource information, typically via local scripts. Example modules report the following:

- Free disk space, memory use, network interface status, CPU load, process monitoring, open files, logged-in users;
- Individual Condor node and Condor pool status;
- Portable Batch System queue status.

Customized modules can be implemented to extend the range of information retrieved. When executed, all modules are required to produce ClassAd attribute–value pairs, which are then combined into a host-wide ClassAd that describes the combined state of local resources. Although ClassAds provide a standard framework for presenting attribute–value pairs, a common naming schema is not prescribed to define attribute names or to format data for a given class of resource.

Each Hawkeye module can use a number of different local commands to get the appropriate data to construct resource information. For example, the core modules are tailored for Solaris or Linux. In this instance the `df` command is used to report free disk capacity, the process reporting uses `top` and system load is reported by `w`. Scripts implemented in this manner will have limited portability to other operating systems. Due to the reliance on the local operating system commands, and the lack of standard naming schemas, there is a risk that implementing a given module across different operating system platforms will result in inconsistent results when monitoring heterogeneous environments.

5.4.6.5 Data request and presentation

Near real-time information is provided by agents that periodically push resource information to the manager. Hawkeye does not currently provide historical information; however, there are plans to provide this in the future.

Clients interact with the HM using simple pull requests for the cached information. The manager provides filtering and fusing mechanisms that allow clients to select a subset of information from selected hosts. Furthermore the manager is capable of executing tasks in response to the periodic updates it receives from agents, for example, to notify a user by email if a host is running low on disk space.

Given that no uniform naming schema is employed and that modules are free to use native operating system tools to retrieve resource information, support for a homogeneous view of heterogeneous resources is not provided. Examples of Web-based user interfaces for monitoring the status of the Condor pool at the University of Wisconsin are given in [43].

5.4.6.6 Searching and standards

The HM is designed to support client queries for subsets of resource information. For example, clients are able to return a list of all hosts that have a load average greater than some threshold. Clients interact with the HM using HTTP. Agents publish information to the manager in ClassAd format using XML. While the XML provides self-describing data, the ClassAd format is non-standard.

5.4.6.7 Security

Secured network connections using X.509 or Kerberos are provided for sending resource information from a Hawkeye monitor to its manager. Other security measures are not present.

5.4.6.8 Software implementation

Hawkeye is actively supported, available for free download, and released under the Condor Public License. Portability is currently limited to Solaris and Linux platforms. Hawkeye can be installed on demand independently of any other middleware, including Condor. Software dependencies are limited to the availability of Perl on the supported operating systems.

5.4.7 Java Agents for Monitoring and Management (JAMM)

5.4.7.1 Overview

JAMM [46, 47] is a wide-area GMA-like system that uses sensors to collect and publish computer host monitoring data. Clients can control the execution of remote sensors and receive monitoring data in the form of time-stamped events. JAMM has been used to provide distributed control and transport mechanisms for projects using the NetLogger Toolkit [48] and has been used in the DARPA MATISSE [49] project. JAMM was developed at the Lawrence Berkeley National Laboratory by the Data Intensive Distributed Computing Group [50]. The project is no longer active although some JAMM components have been incorporated into the NetLogger Activation Service [51].

5.4.7.2 Architecture: General

JAMM consists of a set of distributed components that collect and publish data about monitored resources (Figure 5.9):

- Sensors (producers) execute on host systems and collect monitoring data from locally executing processes. Data may be collected

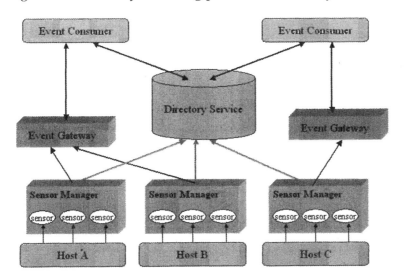

Figure 5.9 The JAMM architecture

from processes that execute once and then exit, or from continuously executing processes. On start-up, sensors register with a Directory Service. A sensor is any application that generates time-stamped monitoring events, which are used throughout the system as the means of propagating data.

- The Sensor Manager (SM) controls the execution of sensors and their registration with the sensor directory. The SM provides a GUI where users can configure sensors' execution. Sensors can be instructed to execute continuously or dynamically, in response to client requests from the SM GUI or Port Manager Agent (PMA).
- The PMA determines which sensors are to be executed based on the applications currently executing on a host. The PMA assumes that applications are started remotely by an SM GUI request on a well-known port. For example, the PMA can be used to start CPU monitoring sensors when processor intensive applications are started by remote client request.
- The Event Gateway (EG) provides mechanisms for clients to control and subscribe to sensors. A single gateway can serve as a point-of-contact for multiple hosts and provide request handling and event filtering duties. An EG caches events from monitored resources and provides consumers with the API that they can query for specific monitoring data. EGs support query-response and streaming requests from consumers. The consumer may request all event data or only events of a certain type.
- The Sensor Directory (SD) is used to publish the location of sensors and their associated gateways. Consumers use the directory to determine which sensors are available and the EG they must subscribe to in order to consume a sensor's output.
- An Event Consumer (EC) locates sensors using the SD and subscribes to receive events from an appropriate EG. Consumers process event data and transform it into custom resource information that can be visualized or further processed.

5.4.7.3 Architecture: Scalability and fault tolerance

Multiple clients can retrieve monitoring data concurrently from EGs. Sensor location information is registered with the SD. JAMM uses LDAP servers for the SD. The JAMM developers acknowledge the importance of avoiding a single physical SD and employ

LDAP referral mechanisms (aggregation of LDAP servers into a hierarchy) for scalability, and the replication of SDs for fault tolerance.

5.4.7.4 Monitoring and extensibility

JAMM requires that sensors and SMs execute on monitored hosts. Sensors can be updated and new types of sensor deployed into the system at run time. Example sensors include host, network, process and application sensors that can be used to monitor CPU, memory, network usage and application error conditions. The monitoring performed by sensors can be configured to execute continuously or in response to specified conditions, for example when network activity is detected on a given port.

EGs cache data received from their sensors and use the cache to respond to client requests. EGs can perform filtering operations to reduce the amount of data returned to a client. Multiple sensors configured for continuous execution can potentially place high loads on the EG's host, which is independent of client interaction. Therefore, to reduce the intrusiveness on monitored hosts, the developers recommend that the EG be located on a remote, but nearby, host.

5.4.7.5 Data request and presentation

Sensors gather monitoring data and submit it to an EG as the data is generated at the monitored resource. Clients can request to receive all data cached by an EG, or subscribe to receive data as the EG receives it from sensors. The EG performs filtering functions whereby clients can select particular types of data. The EG is designed to return summary data to clients: for example, by taking 30-second CPU load readings from sensors and computing 1-, 10- and 15-minute CPU load averages. Default JAMM consumers are provided to perform data archiving services.

A homogeneous view of data from heterogeneous resources is not provided by JAMM. However, mechanisms exist for a sensor to execute a parsing module to extract and format the output from an application into NetLogger format. In cases where a parsing module is not used, JAMM assumes that the application's output is already formatted and the data is passed directly to consumers without modification.

Example consumers include an event collector used to collect data for use by real-time analysis tools, an archiver for persistent event storage, a process monitor that notifies system administrators by e-mail when a process terminates abnormally and an overview monitor that constructs resource status information from event data generated by multiple hosts.

5.4.7.6 Searching and standards

Consumers can request the EG to stream data to them only if a value meets some threshold; for example, if CPU load for a given sensor changes by more than 10%. Resource data is reported in either XML or Universal Logger Message (ULM) [52] formats.

5.4.7.7 Security

An EG is designed to provide access control to sensors based on X.509 certificates and SSL connections. It is not clear if these security measures have been implemented.

5.4.7.8 Software implementation

JAMM is free, available for download and has an open-source license. However, the project is no longer active or supported. JAMM is written in Java and uses Remote Method Invocation (RMI) for communications.

5.4.8 MapCenter

5.4.8.1 Overview

MapCenter [53–56] is used to monitor and display the availability of services across a Grid using graphical maps, logical views and trees of computing resources, in a client's Web browser. MapCenter uses an extensible model to visualize different levels of

5.4 AN OVERVIEW OF GRID MONITORING SYSTEMS

resource based on department, organization and virtual organization views. Appropriate native protocols are used to determine service availability and a presentation layer is provided for Globus MDS2 LDAP servers.

MapCenter was developed by the Network Work Package of the European DataGrid, and has been deployed in a number of environments including DataGrid [57], DataTAG [58], CrossGrid [59], PlanetLab [60], L-Bone [61] and AtlasGrid [62].

5.4.8.2 Architecture: General

The MapCenter architecture (Figure 5.10) is composed of a centralized monitoring server that consists of a data store, as well as a monitoring and presentation layer:

- The Data Store (DS) provides configuration and status information for MapCenter elements (objects, symbols, maps, link, services and URLs). A system administrator can modify the resources referenced in the DS, or automatically using data obtained from remote information services (e.g. Globus MDS2). The DS data model provides a hierarchical description of

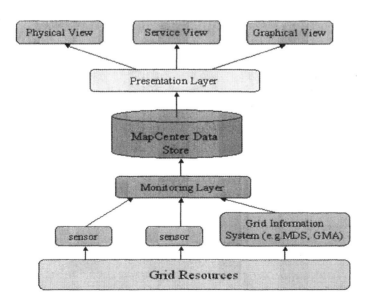

Figure 5.10 The architecture of MapCenter

resource status for generating logical or graphical resource views. The data model is based on four entities:
- *Object*: The basic element that models a computer system, or set of services on a host that are polled at a fixed frequency;
- *Symbol*: A visual representation of an object, symbols are grouped into maps;
- *Map*: Contains symbols, links and sub-maps;
- *Link*: An abstraction of a logical interconnection between maps.

- The Monitoring Layer (ML) uses sensors to poll resources. Sensors implement the native protocol for the class of resource they are designed to poll. For example, sensors provide: ping requests, to check the availability of a machine; TCP connection requests to test for services running on given ports; and HTTP get requests. Sensors are invoked in a uniform manner and output native data representing poll status. MapCenter objects are used to define the number of ports or URLs, and poll rate for a given resource in the DS. Status determination rules are used to assign a standard "poll severity" value (such as "responding" or "unavailable") based on an object's configuration and native protocol response.
- The Presentation Layer (PL) generates and visualizes different views of the monitored resources. Symbols are animated with corresponding resource status. The "best" and "worst" resource status are propagated through multi-level map views, so that a snapshot of overall resource availability can be visualized at a high level.

5.4.8.3 Architecture: Scalability and fault tolerance

Multiple resources at different sites can be polled at various frequencies and the results cached within the server's data store. Multiple users can connect to the MapCenter Web server to view HTML-based status pages. A homogeneous view of resource status is obtained from hierarchical resource maps, which detail the availability of services running on hosts. Computer hosts are displayed either logically or geographically, with the services they support listed and colour coded to indicate their status.

In addition to automatic polling, MapCenter permits users to probe services on demand. The native information is displayed temporarily and MapCenter processes the response in order to derive the standard "poll severity". Furthermore, invoking a MDS2 URL results in the presentation of a LDAP Web interface, where the user must use native LDAP semantics and syntax. In both cases, the user is presented with service- and protocol-specific native output that may prove confusing when interacting with large numbers of diverse resources.

The MapCenter architecture provides a centralized server. All monitoring features are performed from this central point; it is recommended [53] that servers should be replicated for fault tolerance. However, replication of servers will result in an increase of requests to monitored resources as each MapCenter server attempts to poll the same resources as its peers for status.

5.4.8.4 Monitoring and extensibility

MapCenter connects to monitored resources through sensors installed on the MapCenter server. Resources can be located automatically by querying an information service and incorporated into the DS at run time.

The Web user interface displays the services exported by hosts, for example MDS2, GSIFTP, mySQL, Apache Tomcat or a particular URL. Essentially any service or daemon that can be accessed over the network can be probed and assigned a colour code to indicate availability. Services are polled using sensors that plug into the MapCenter architecture, implement an appropriate protocol to connect to a service and compute a poll severity by parsing the returned response. Poll severities represent the logical state a service can be in and provide a simple value that can be used to compare the availability of diverse services. The querying of an information service is achieved using customized interfaces that expose underlying information. Example interfaces exist for LDAP, SQL databases and R-GMA.

5.4.8.5 Data request and presentation

MapCenter provides real-time and historical data. The MapCenter server initiates all polling in a server pull fashion. Event subscription and notification is not supported. The returned results are fused and filtered to provide a number of map views for the user to navigate.

A homogeneous view of heterogeneous data sources is achieved for service polling only. The mechanism to poll a service and the "poll severity" result presented to the DS are uniform across different resource types. However, when manually polling a service, the user is presented with information detailing the native protocol results from that poll.

5.4.8.6 Searching and standards

Information searching by resource category, functionality, capability or load is not supported natively within MapCenter. Interfaces to third-party information systems may provide this functionality independently; for example, the LDAP interface provides searching by job queue or CPU load.

MapCenter does not appear to support any open, standard protocols suitable for interoperating with other monitoring systems. However, it is reported that support for the Open Grid Services Architecture (OGSA) is planned [53].

5.4.8.7 Security

No security mechanisms are provided.

5.4.8.8 Software implementation

MapCenter software is open source, written in C, and is available in a Linux RPM and tgz format. Mailing list support is not provided on the project Web site, but contributor contact names are given. Software dependencies are limited to the Apache Web server.

5.4.9 Monitoring and Discovery Service (MDS3)

5.4.9.1 Overview

MDS3 [63] is the information service for the Globus Toolkit 3.x (GT3) developed by the Globus Alliance [63]. MDS3 information

5.4 AN OVERVIEW OF GRID MONITORING SYSTEMS

services are intended to provide scalable, uniform and efficient access to distributed information sources in order to support the discovery, selection and optimization of resources for users and applications within a Globus environment. The approach is intended to mask underlying resource heterogeneity through standardized reporting of static and dynamic resource information.

The GT3 is based on the OGSA [64], with components implemented to the Open Grid Services Infrastructure (OGSI) specification [65]. This service-oriented approach moves away from the LDAP-based architecture found in earlier MDS versions [66]. As a result, components of MDS3 are represented as Grid services. Each Grid service instance has associated Service Data (SD) [65] that reveals resource information. The SD is represented in a standardized manner; various operations are provided for a Grid service to register new information and for clients to retrieve that information. By default, each Grid service exposes computer host-based performance information. A pluggable information provider framework permits new types of information to be made available.

5.4.9.2 Architecture: General

The SD associated with each Grid service instance represents status information for that service, for example host processor and memory utilization. Figure 5.11 shows MDS3 architecture. The SD mechanisms can also be used to expose additional information that is gathered, queried or probed as a result of performing the service's task. Resource information is presented according to defined SD descriptions and advertised using the service's service type WSDL definition. A generic interface for mapping queries and subscriptions for resource information to service implementation mechanisms is provided by OGSI. This approach ensures that a client can utilize standard operations when retrieving information from a service instance.

The MDS3 is a distributed information system consisting of a resource and collective layer of Grid services:

- The Resource Layer, of MDS3 consists of one or more service instances that produce SD. Typically these services will monitor and provide access to resources, such as job submission mechanisms or replica catalogues. Grid services may be persistent or

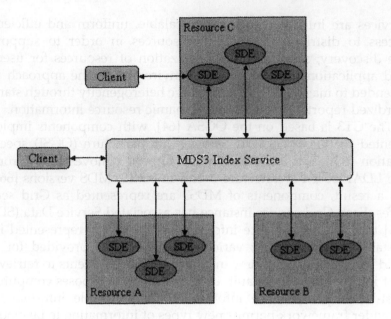

Figure 5.11 The MDS3 architecture

transient. A transient service may, for example, be a data-mining operation, network bandwidth allocation or a reporting mechanism for an experiment. A persistent service might be a queue for a batch processing system.

- The Collective Layer, of MDS3 consists aggregates information from multiple "Resource Layer" services. The Index Service (IS) is an example of a Collective Layer service that supports operations for accessing, aggregating, generating and querying SD from remote services. Potentially, information can also be obtained from applications running locally on the IS host. In this case a provider mechanism is utilized to dynamically generate Service Data Elements (SDE) from the application so that it becomes available via the SD interface. The IS's indexing and caching functionality is utilized to reduce client overhead when locating appropriate information from a set of remote resources. The indexing provides service lookup mechanisms and caching provides clients with access to recent information from Resource Layer services.
- Clients, e.g. user applications, interact with the IS or resource-level services directly, using subscription and query requests. MDS3 provides dynamic service registration via a registration

5.4 AN OVERVIEW OF GRID MONITORING SYSTEMS

protocol. Client interactions with services are based on a resource discovery and an information protocol.

5.4.9.3 Architecture: Scalability and fault tolerance

There may be more than one instance of a Grid service with multiple clients querying those services concurrently. The OGSI SD interface provides a common means of retrieving information, regardless of the underlying platform.

Fault tolerance and scalability are achieved through the replication of resource and collective layer services. Furthermore, soft-state registration allows resources to dynamically enter and leave the IS, while stale references are identified and removed. ISs aggregate information from any other service, including remote ISs. Therefore hierarchies of ISs can be established to aggregate and cache resource information. Cached resource information is refreshed through OGSI subscription to registered services.

5.4.9.4 Monitoring and extensibility

SD providers and SDEs are used to extend the information services provided by MDS3. An SD Provider collects data from a local application and creates an XML document by marking up the native data against a naming schema to create a description of resource information. The resulting XML document is injected into the Grid service's SD interface and made available to clients as an SDE.

As well as creating custom SD providers to extend monitoring capability, a number of default providers can be utilized:

- `SimpleSystemInformationProvider`: A Java-based host information provider of CPU count, memory statistics, OS type and logical disk volumes.
- `HostScriptProvider`: Shell scripts for UNIX systems that output various details about host resource information.
- `AsyncDocumentProvider`: A provider that periodically reads an XML document from disk.
- `ScriptExecutionProvider`: Provides a wrapper to execute applications that generate an XML document via their standard output stream.

Resources are queried periodically by SD managers; monitoring is independent of a client's interest. Polling frequency is defined separately for each SDE. Administrators have the option to modify the rate a resource is polled in order to reflect the frequency with which underlying information is expected to change, or to meet client demand for up-to-date information.

5.4.9.5 Cross-API monitoring

The source of resource information is generally expected to be the host on which a Grid service is executed. However, SD providers are not restricted in their operation and can connect to remote resources, such as Ganglia. SD providers must interact with remote resources using their native API and format resulting data according to naming schema requirements. This implies that SD providers are capable of monitoring cross-API resources, without the need to install MDS components on remote monitored resources.

5.4.9.6 Data request and presentation

All client queries retrieve information from a cache. This data may be near real time, based on the poll frequency of SD providers. Historical data can be provided by services that perform archival duties.

MDS3 supports client-initiated query/response and subscription/notification protocols. Clients register their interest in services, which in return send notification messages to the client, when specified information is updated.

The host information provided by default GT3 SD providers is represented in GLUE. Customized data providers can use GLUE or any other naming schema that meets their needs. However, the use of standard naming schemas is desirable to promote interoperability between Grid services.

GUIs are available for interacting with the IS. For example, the Service Data Browser is a Java application that provides a framework to plug in "visualisors" that are used to format XML resource information for specific user requirements. Another GUI is the Web Service Data browser, which uses XSLT style sheets to convert SD XML into HTML in a manner that allows the addition of new types of SD to be incorporated for display.

5.4.9.7 Searching and standards

MDS3 supports both simple and complex queries. Query-by-name is performed by specifying a service and one or more SDE names. Optionally clients may also provide an XPath expression to refine the information returned from a query. The approaches used allow information to be returned by resource type and by the comparison of resource values. All responses are returned in XML. The use of XML and GLUE ensures that information retrieved from the system has wide utility for non-GT3 components.

5.4.9.8 Security

GT3 services, including MDS3, can use the GSI [3] to provide certificate-based authentication and authorization. By default, anonymous interaction to MDS3 is provided. MDS3 can be configured to provide access control for both registrations and queries based on a list of authorized users.

5.4.9.9 Software implementation

MDS3 is available within the Globus Toolkit release 3.x. The project is active and supported. GT3 is released under an open-source licence and is free to download and use. While the Grid services components of the toolkit are written in Java, the underlying core of the Globus code is currently targeted at UNIX platforms. MDS3 is integrated with the Globus Toolkit, which must be installed and configured before the MDS3 can be utilized. Software dependencies include Java 1.3 and Jakarta Ant. Optional MDS3 IS dependencies include the Apache Xindice, an XML database for transparent SD persistence.

5.4.10 Mercury

5.4.10.1 Overview

Mercury [67–69] is a monitoring system that aims to support application steering and self-tuning, performance analysis and prediction. Mercury provides a general Grid monitoring infrastructure that extends the GMA, with actuators and actuator controllers in order to influence the operation of the monitoring system. These extensions are intended to offer intelligent and adaptive control mechanisms to react to changes in the monitored environment. Mercury can perform application and resource monitoring as well as adaptive management functions. Mercury is developed by the EU GridLab [70] project and utilized in a number of Grid testbeds, including the European DataGrid [57].

5.4.10.2 Architecture: General

Mercury is based on the GGF's GMA and concepts from the Autopilot [5] and OMIS [71] projects, for providing adaptive control within a distributed environment. Figure 5.12 shows Mercury's distributed architecture. Mercury consists of local monitors, main monitors, and the monitoring service. "P" refers to a process executing on the same host as the Local Monitor (LM).

- The Local Monitor resides on each monitored host, gathers raw host performance and application information from a locally executing sensor process (P), which it transmits to a MM. The LM

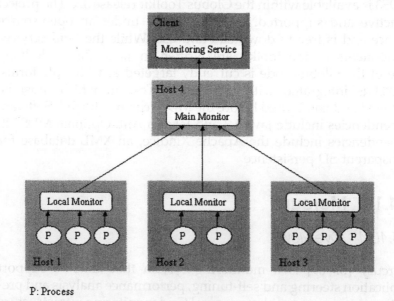

Figure 5.12 The architecture of Mercury

is responsible for controlling a sensor's execution in response to client requests.

- The Main Monitor (MM) aggregates raw resource information from multiple LMs and provides a central location on each resource (host or cluster) for clients to access information. In response to a client request, the MM instructs appropriate LMs to take measurements and pass back their results. The LM and MM constitute the Local Monitoring Architecture (LMA). As well as providing information from LMs, the LMA can potentially aggregate information from other monitoring systems.

- The Monitoring Service (MS) provides a gateway for remote clients to access a Mercury resource's monitoring information. The MS gathers raw information recorded by the LMA and converts local measurements into a standardized format for use by clients. The MS provides client authentication and authorization, and also coordinates the monitoring of resources in line with client requirements. Information retrieved by the MS can potentially be stored in persistent remote storage service instead of being delivered directly to the requesting client.

- Actuators extend the GMA by providing mechanisms for adaptive control of monitored resources. Actuators are similar to GMA producers, but instead of performing monitoring duties they provide a mechanism for clients to exert control over a part of the monitored system. Actuators are managed by actuator controllers. The LM contains both producer and actuator functionalities.

5.4.10.3 Architecture: Scalability and fault tolerance

Mercury's architecture is based on the GMA. The MS equates to GMA producers and clients to consumers. Mercury uses multiple MMs on large clusters; cluster nodes can be logically grouped to report to different MMs. Multiple clients can monitor multiple resources concurrently. The MS coordinates client requests across remote resources and provides a single point of contact for retrieving information and controlling sensor behaviour.

5.4.10.4 Monitoring and extensibility

Mercury provides mechanisms to monitor host and application information using sensors that must be installed on the monitored hosts. By default, Mercury defines a range of host information that can potentially be collected from hosts, including processor, memory, disk, network and operating system details. The host sensors provided by Mercury are operating-system specific; and are only capable of providing a subset of the defined range of host information. For example, some sensors do not provide CPU load average. In Mercury version 2.2, host sensors exist for versions of Linux, Mac OS X, UNIX and Mach.

Other monitoring information is made available to Mercury using loadable modules that link sensors into the LM. Sensors are user-defined codes that can retrieve information using local system commands and applications or remote services. Clients have the ability to control the operation of LM and so potentially have some influence over the monitoring intrusiveness of sensors.

5.4.10.5 Data request and presentation

Clients can request real-time information by request–response queries or subscription. Simple consumers that capture information persistently for off-line historical analysis have been provided. Furthermore, the MS can be instructed to redirect streams of information to remote storage facilities for archiving, instead of returning results directly to the requesting client.

Mercury supports the provision of a homogeneous view of heterogeneous data by utilizing the MS to transform raw information produced by sensors into normalized form. The normalized information is resource independent and adheres to a predefined format. Normalization is performed at the MS level in order to allow resource administrators to work with detailed raw information within a site, while external clients receive aggregated or potentially less detailed information.

The protocol used between client (consumer) and MS (producer) consists of commands, responses and information values. Channels define a logical connection between a producer and consumer and may be initiated in either direction in order to support simple request–response queries and subscription.

5.4.10.6 Searching and standards

The MS allows the selection of information based on unique template names. Clients need to provide a hostname when querying a MS for resource information. It is assumed that the MS provides mechanisms to find registered hosts with sensors. Searching can be performed by value comparison, e.g. (cpuLoad > specifiedvalue).

Mercury uses the External Data Representation (XDR) [72] format, to encode information. XML is not utilized due to concerns over parsing overhead and message size.

5.4.10.7 Security

Mercury uses Generic Security Services Application Programming Interface (GSSAPI) [4] authentication and encryption, and basic SSL [4] support is provided. Access Control Lists (ACL) are used to determine which services clients are permitted to use. ACL policies can allow or deny access to certain functions for specified users.

5.4.10.8 Software implementation

Mercury version 2.2.0 is available for download free of charge under the GNU Lesser General Public License. The project is active and provides support by a mailing list.

Mercury does not require the installation of Grid middleware. A small number of external libraries are required, depending on configuration and functionality required. Installation is limited to UNIX platforms.

5.4.11 Network Weather Service

5.4.11.1 Overview

The goal of the Network Weather Service (NWS) [38, 73] is to provide recent historical information and short-term forecasts of computer and network performance within distributed systems. NWS was developed for use by dynamic schedulers and to provide quality-of-service readings for distributed computational environments. NWS sensors provide periodic network measurements in the form of end-to-end TCP/IP bandwidth and latency, and host CPU usage and memory availability. The NWS has a distributed, GMA-like architecture, and is used within a number of projects including the NSF Middleware Initiative (NMI) Grid [74] and NPACI Grid [75].

5.4.11.2 Architecture: General

The NWS architecture (Figure 5.13) consists of a number of distributed components, including a name server, memories, forecasters and sensors.

- The Name Server (NS) is a centralized registry that contains address bindings for NWS components. The NS is the only part of the system that requires components to have prior knowledge of its address. All other NWS components are located by and periodically register with the NS.
- Sensors are installed on monitored hosts and periodically transmit time-stamped information of local host and network performance to memory components.
- Memories, also known as Persistent State, provide measurement information storage and retrieval services for sensors and clients. Memories store time-stamped resource observations from individual sensors. The information is stored to local disk, using circular buffer techniques so that only recent resource information is retained. If a memory component fails, the history of recent information is available to clients when the memory is re-started. A simple request mechanism is provided for clients to read information.
- Forecasters process resource observations from memories and perform forecasts over a given time frame.

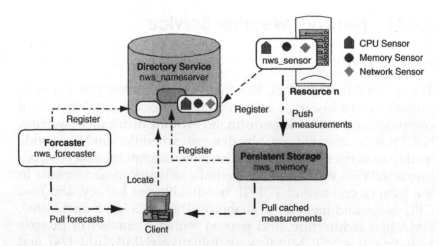

Figure 5.13 The architecture of NWS

5.4.11.3 Architecture: Scalability and fault tolerance

Numerous resources can be monitored concurrently, with multiple clients accessing resource measurements. Network and host performance measurements are accessed in a seamless manner using the memory API that provides uniform access to information.

The developers acknowledge that the NS is a potentially weak link in the system and provide replication mechanisms to synchronize registration information across multiple NSs. Sensors include a fail-over mechanism that allows them to switch over to a different NSs in the event of their default NS failing. The use of multiple NSs and sensor fail-over techniques provides fault tolerance. Scalability can be addressed by adding additional NSs and memories to the system as the number of sensors increases.

All other components in NWS can fail individually without affecting the operation of the remainder of the system. Periodic component rebinding with the NS ensures stale bindings are found and removed.

All sensor information is cached in NWS memory components in a circular buffer ordered by a timestamp. Clients query memory components for resource information. It is expected that information is periodically read from memory components and placed in a permanent store as required by the client. Clients never query sensor components directly. Providing that memories are located on a remote host, the sensor host's performance will not be affected by client requests.

5.4.11.4 Monitoring and extensibility

An NWS sensor must be installed on each host and configured to start a number of monitoring activities that periodically gather local information. A client can start and stop monitoring activities using the sensor control API. The set of default information available in NWS is:

- *TCP bandwidth/latency/connect-time*: The speed with which data can be sent to a target sensor in MBs, the amount of time in milliseconds required to transmit a message to a target sensor, and the amount of time required to establish a connection to a target sensor.

- *CPU*: The fraction of CPU that is available to:
  - a newly started process;
  - an existing process.
- *Storage/memory*: The amount of space in megabytes of unused disk/memory.

Further monitoring capabilities could be added by creating additional sensor functionality, for example, to SNMP; however, generally this appears to have not been the case. NWS does, however, provide an alternative approach, the nws_insert command, can be used to store information from external programs into NWS memory. Once inserted, the new type of information can be extracted, viewed and forecasts generated, using NWS' standard mechanisms.

NWS forms network sensors into cliques to reduce network-monitoring intrusiveness. Members of a clique coordinate their timing to avoid collecting network data at the same time.

5.4.11.5 Data request and presentation

Near real time and recent historical performance information is provided to clients. Short-term forecasts can be generated to provide near future quality of service predictions. Client event notification is not supported; clients have to initiate requests to retrieve data from NWS memory and forecaster components. Sensors push data into memory components.

NWS does not provide any mechanisms for normalizing data from the heterogeneous resources it monitors. The memory component combines recent measurements from a single sensor activity on a particular host. When clients query the memory component for a host's particular performance activity, a series of up to twenty, time-stamped measurements are returned in delimited text format. The request granularity is focused on retrieving specific information from a named host. User interaction with components (sensors, memories, name server) is via command-line applications or the C API.

5.4.11.6 Searching and standards

The NS is utilized to locate performance data and a search can be made on hostname, sensor, monitoring activity or clique. Queries are to an activity on a given host. The NS does not contain details of performance information so, for example, searching for a host by CPU load is not possible at this level.

The NWS data format is non-standard, consisting of a combination of name–value pairs and space-delimited text. The query language is based in part on an LDAP-like syntax. Standard Grid protocols are not used in NWS. However, NWS has been integrated with the Globus LDAP MDS through the NWSlapd portal. The NWSSlapd portal allows NWS to act as a Globus information provider, by allowing Globus users to query the NWS name server and obtain performance data.

5.4.11.7 Security

No security mechanisms are provided by NWS.

5.4.11.8 Software implementation

NWS is open source, free and available for general download. The project is active and supported. Portability is based around C code for UNIX platforms. No further software dependencies are required.

5.4.12 The Relational Grid Monitoring Architecture (R-GMA)

5.4.12.1 Overview

R-GMA [76–79] was developed within the European DataGrid project [57] as Grid information and monitoring system. R-GMA provides an implementation of the GMA that utilizes a relational model. R-GMA collects information about the state of the Grid and makes it available to other components in a way that provides a global view of information. R-GMA operates over the wide area and provides an information transport layer for host, network, service and application monitoring data. A key abstraction behind R-GMA is to promote the perception that data is being stored or streamed through one large data warehouse for each VO. R-GMA is used in the DataGrid test platform.

5.4.12.2 Architecture: General

R-GMA has a distributed architecture consisting of agents, producers, consumers, producer–consumers, registry (mediator) and naming schema (Figure 5.14):

- R-GMA clients (user code) are written by users to provide or consume information within R-GMA. User code interacts with agents through the agent APIs. In Figure 5.14 the boxes labelled "Producer" and "Consumer" represent user code, while the inner boxes represent the agent API. The agent API interacts with generic R-GMA agents that abstract the user from the underlying infrastructure.
- An Agent is the implementation of a producer or consumer component that performs R-GMA operations on behalf of the user code. An agent shields the user code from R-GMA implementation details and provides support for tasks such as registration, remote agent and data location, and schema management. In Figure 5.14, agents are represented by the producer and consumer servlets.
- A Producer is a component that makes available resource information for consumers to query. Producers register the type of

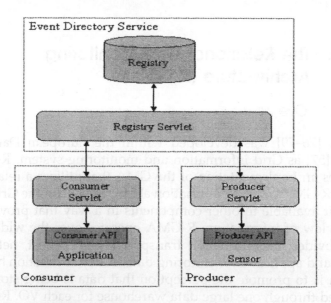

Figure 5.14 The R-GMA architecture

information they provide with the R-GMA registry, using SQL CREATE TABLE statements. Producers publish instance information using SQL INSERT statements.

- A Consumer provides the means by which user code accesses information made available by various producers. A consumer handles a single query, expressed as an SQL SELECT statement. There are two approaches for the transfer of information from a producer to a consumer:
  - *The pull model*: The consumer can perform a one-time query to a producer. This process may be repeated but if there is no new information available, the same results will be returned to the user code.
  - *The push model*: The consumer can perform a streaming query, so that whenever new data that matches the query is available, the data is transmitted back to the user code, via the consumer. While this follows a push model, it is the consumer that initiates the streaming of data from producer to consumer.
- A Producer–consumer is a component that contains both producer and consumer functionality. For example, Archivers are used to combine, and possibly filter, information streams from multiple producers and archive them in a database.
- The Mediator provides support to a consumer for searching the registry for an appropriate producer to obtain information from, in cases where there are potentially multiple producers capable of fulfilling the consumer query.
- The Registry records all producers and consumers that currently exist. The registry supports the abstraction of a single physical database of information that is physically provided by remote producers. The registry achieves this abstraction by mapping producers to the logical database table they produce information for. The registry uses a SQL WHERE predicate to define which partition of the logical table a producer provides data for.
- The Schema contains details about the relational tables that create the illusion of a single database spanning remote producers and consumers. That is, the Schema provides metadata that describes the semantics and scope of the information that producers will provide to consumers. In order to interact, producers and consumers must share a common naming schema for their information, which the Schema allows them to do. As well as

providing a global set of core naming schemas, defined using GLUE [78], producers are free to submit their own naming schemas (relations) that describe the data they produce.

5.4.12.3 Architecture: Scalability and fault tolerance

Multiple clients can utilize the R-GMA infrastructure concurrently. A virtual relational view of resources is achieved by publishing information from producers that use standard and custom naming schemas. Providing a client understands the naming schema utilized by a provider, information can be accessed in a seamless manner. The client's view is of a single virtual database. This abstraction hides the potentially complex underlying interactions between distributed R-GMA components.

The virtual database R-GMA provides is partitioned across independent and distributed producers, with the description of system partitioning held in the registry. If a producer or consumer fails, then the remainder of the system will continue functioning. Work is underway to avoid a single point of failure in the registry and schema, using dynamic replication techniques.

Soft state registration, using a heartbeat mechanism, of producers and consumers with the registry ensures that the system is self-healing and promotes scalability. The automatic removal of stale references to end points, when a heartbeat is not received, reduces the load on the registry, network and agents interacting with components that are not available.

5.4.12.4 Monitoring and extensibility

Five different types of producer exist in R-GMA. While all appear to be the same from the client's point of view, they have differing characteristics:

- `DataBaseProducer` – supports historical queries, by writing each record to an RDBMS.
- `StreamProducer` – supports continuous queries and stores information in-memory for consumer retrieval.
- `ResilientProducer` – similar to the `StreamProducer`, but information is also written to disk to avoid loss in the event of a system failure.

- `LatestProducer` – supports queries for latest information only, by overwriting earlier records in an RDBMS;
- `CanonicalProducer` – executes user code to retrieve information in response to a client query.

Monitoring intrusiveness can be controlled using hierarchies of Republishers that combine and cache information, thereby reducing the overhead on the original producer agent. In addition, Republishers could be located in strategic positions within the Grid, in an attempt to reduce client request latency by moving a copy of the data closer to consumers.

5.4.12.5 Data request and presentation

The consumer API refines SQL queries by performing continuous, latest state and historical queries on producers. Communication patterns are client pull for latest state and historical queries, and server push for continuous queries. R-GMA provides event notification by utilizing "continuous" producers to stream information as it is generated, back to consumers that previously indicated their interest.

Producers utilize a naming schema to publish data. This implies that if two producers wish to publish data using the same naming schema, but from different sources, they should implement functionality into their code to format the native data to meet the naming schema requirements, before submitting the data to their producer agent. Producers of similar types of information can register different naming schemas if they do not wish to manipulate native data into a different form.

5.4.12.6 Searching and standards

Data in R-GMA can be searched for by resource class, such as functionality or capability, or by resource values, such as cpuLoad >4.0. R-GMA uses the GMA as its basis, and goes on to define a data model, a query language and the functionality of the directory service. Currently R-GMA implements an XML protocol over HTTP(s).

5.4.12.7 Security

Security in R-GMA is optional; however, if selected, authentication using Grid certificates can be provided, while client-to-servlet and inter-servlet communications are protected by Secure Sockets Layer (SSL). Communication between the R-GMA components is via HTTP(s). The ability for an R-GMA client to present a Grid certificate signed by an accepted Certification Authority (CA), or to generate a proxy certificate for use when interacting with the R-GMA agent API follows the GSI approach.

5.4.12.8 Software implementation

R-GMA is available for free download and is being supported by the EGEE [80].

The core of R-GMA is based on Java servlet technologies. The current installation mechanism is based on Linux (Redhat) RPMs. Configuration mechanisms also appear to be particular to Linux, so providing a portable installation of R-GMA requires more than simply extracting the contents of RPMs into a cross platform package format.

R-GMA is released under an open-source licence. R-GMA is dependent on a number of EDG-mandated RPMs which must be installed (14 for standard and 30 for developer environments) before the core set R-GMA RPMs can be installed. The agent APIs for interacting with the R-GMA servlets are provided in Java, C++, C, Python and Perl.

5.4.13 visPerf

5.4.13.1 Overview

visPerf [81] is a Grid monitoring and visualization system that utilizes remote sensors to extract information from log files, as well as interacting with existing Grid middleware in order to remotely observe performance. visPerf interfaces with the NetSolve [82] system and provides information on client users, host resource performance and patterns of internal NetSolve function calls. visPerf was developed by the Innovative Computing Laboratory at the University of Tennessee, Knoxville, and is used within the GridSolve project [83].

5.4 AN OVERVIEW OF GRID MONITORING SYSTEMS

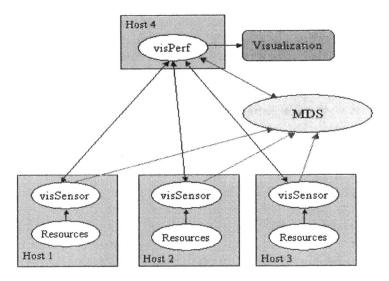

Figure 5.15 The architecture of visPerf

5.4.13.2 Architecture: General

The visPerf NetSolve Monitor (architecture shown in Figure 5.15) is a distributed application made up of a Java applet monitor and remote sensors:

- The visPerf Sensor is a monitoring component that resides locally on each host to be monitored and periodically gathers local host performance information, which is cached within the sensor. The monitoring of local Grid middleware is achieved through the parsing of application log files, or directly from the application if it provides an API for performance observations at run time. A direct connection between the sensor and its clients is used to return information. Sensors operate either in slave mode, where preconfigured monitoring duties are performed, or in a proxy mode, where the sensor aggregates information from slave sensors and makes this available to clients.

- The visPerf Main component is used for browsing and controlling the status of remote sensors. This component periodically transmits synchronization messages to registered sensors in order to determine if any are unreachable or experiencing operational problems. Sensors can be started, stopped and their status viewed. visPerf Main also provides visual components for

displaying graphical representations of a resource's status. For example, graphs display CPU load, while resource maps display the pattern of function calls executed internally by monitored middleware.
- A Directory Service is used to register sensors and provides clients with a mechanism to locate sensors.

5.4.13.3 Architecture: Scalability and fault tolerance

visPerf provides a GMA-like architecture that supports multiple concurrent clients and sensors. However, the directory service implementation appears to be centralized, which raises scalability and fault tolerance concerns. To reduce the amount of data transmitted to the client across the system, proxy sensors can be used to provide aggregation and reduction services for multiple remote slave sensors.

5.4.13.4 Monitoring and extensibility

Sensors are installed on all resources that require monitoring and provide access to local computer resource information, for example CPU workload and disk I/O, by periodically executing local operating system commands. Furthermore, the monitoring of third-party applications and middleware can be provided by using either log-based monitoring techniques or by connecting to exposed profiling and monitoring APIs, on an application-by-application basis.

5.4.13.5 Data request and presentation

Sensors support simple request/response (client pull) mechanisms to respond on demand to client requests. Furthermore, sensors can be configured to stream real-time information to a predefined client. Mechanisms to support a homogeneous view of heterogeneous resources are not provided, although a pre-processing filter is used to format raw application logs of a specific system into a compact semantic format. This approach is intended to reduce network overhead and provide clients with a single format for retrieving information obtained from different applications.

5.4.13.6 Searching and standards

Sensors can be located using the directory service and queried directly to provide a list of the monitoring services they provide. Monitoring information is grouped according to a predefined name by the sensor. The grouping and naming of available information is performed statically when the sensor is configured for operation.

5.4.13.7 Open standards

visPerf uses its own raw sensor monitor protocol, or an XML-RPC protocol can be used to communicate between clients and sensors.

5.4.13.8 Security

Sensors provide MD5 authentication and connections between clients and sensors are encrypted. Standard Grid security mechanisms are not used.

5.4.13.9 Software implementation

visPerf is under active development and currently in beta release. At the time of writing, the source code was only available by contacting the developers directly. Software requirements include Netsolve and Python 2.1. Sensors are supported on Linux, FreeBSD and Solaris. `sysstat` [84] is required for providing `iostat` and `vmstat` monitoring commands for Linux-based sensors.

5.5 OTHER MONITORING SYSTEMS

5.5.1 Ganglia

Ganglia [29] is a distributed monitoring system for high-performance computing systems such as clusters and the Grid. It is based on a hierarchical design targeted at federations of clusters. Ganglia relies on a multicast-based listen/announce protocol to

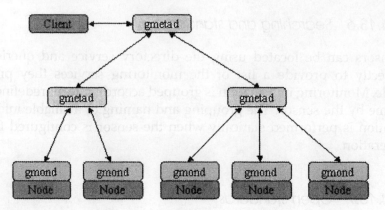

Figure 5.16 The architecture of Ganglia

monitor state within clusters and uses a tree of point-to-point connections amongst representative cluster nodes to federate clusters and aggregate their state. Ganglia uses XML for data representation, XDR [72] for data transport, and RRDtool [85] for data storage and visualization. Ganglia attempts to minimize per-node system overheads by using specialized data structures and algorithms. Figure 5.16 shows the architecture of Ganglia.

The Ganglia Monitoring Daemon (gmond) is a multi-threaded daemon, which runs on each cluster node to be monitored; it has four main responsibilities:

- monitor changes in host state;
- multicast relevant changes;
- listen to the state of all other Ganglia nodes via a multicast channel; and
- answer requests for an XML description of the cluster state.

Each daemon transmits information in two different ways: multicasting host state in XDR format or sending XML over a TCP connection.

Ganglia Meta Daemons (gmetad) are used to provide a federated view. At each node in the tree, a *gmetad* periodically polls a collection of child data sources, parses the collected XML, saves all numeric, volatile metrics to round-robin databases, and exports the aggregated XML over a TCP socket to clients. Data sources may be either *gmond* daemons, representing specific clusters or other *gmetad* daemons, representing sets of clusters. Data sources use source IP addresses for access control and can be specified using multiple IP addresses for failure. The latter capability is natural for aggregating data from clusters since each *gmond* daemon contains the entire state of its cluster.

Ganglia PHP Web User Interface provides a view of the gathered information via real-time dynamic Web pages. Most importantly, it displays Ganglia data in a meaningful way for system administrators and computer users. Although the Web UI to Ganglia started as a simple HTML view of the XML tree, it has evolved into a system that keeps a colourful history of all collected data.

The UI depends on the existence of the *gmetad*, which provides it with data from several Ganglia sources. Specifically, the Web UI will open the local port 8651 (by default) and expects to receive a Ganglia XML tree. The Web pages themselves are dynamic; any change to the Ganglia data appears immediately on the site. This behaviour leads to a very responsive site, but requires that the full XML tree be parsed on every page access. Therefore, the Ganglia UI should run on a dedicated machine if it presents a large amount of data.

The UI is written in the PHP scripting language, and uses graphs generated by *gmetad* to display historical information.

5.5.2 GridMon

GridMon [86] is a network performance monitoring toolkit to identify faults and inefficiencies. The toolkit is composed of a set of tools that are able to provide measures concerning different aspects related to network performance: connectivity, inter-packet jitter, packet loss, Round Trip Time (RTP) and TCP and UDP throughput. The main components of GridMon are shown in Figure 5.17.

- PingER is a collection of Perl scripts, which use the ICMP ping utility to send 10 ping requests to remote hosts. The results are recorded for later analysis.
- IperfER is based on NCSA's Iperf [87] utility, used for measuring the network's view of TCP or UDP throughput between two hosts. Iperf consists of client and server executables, which sit at either end of a TCP/UDP connection, streaming data

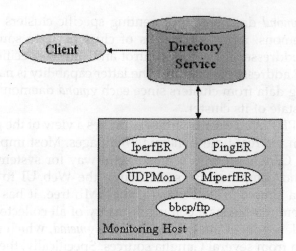

Figure 5.17 The architecture of GridMon

between each other. UDPmon is essentially a UDP equivalent of IperfER.

- bbcp and bbftp are basic tools for copying files between sites, albeit using multiple TCP streams and large TCP window sizes.

5.5.3 GRM/PROVE

GRM [88] is an online monitoring tool for performance monitoring of message passing parallel applications running on the Grid. PROVE [89] is a performance visualization tool for GRM traces. When requested, GRM collects trace data from all machines where the application is running and transfers it to the machine where the trace is visualized by PROVE. GRM uses the Mercury Grid monitoring system to deliver trace data to the host undertaking the visualization process. PROVE visualizes trace information online during the execution of the Grid applications. Figure 5.18 shows the architecture of GRM, which is based on Mercury.

- A LM is running on each host where application processes are executed. It is responsible for handling trace events from processes on the same host. It creates a shared memory buffer where processes place event records directly. Thus even if the process terminates abnormally, all trace events are available for the user

5.5 OTHER MONITORING SYSTEMS

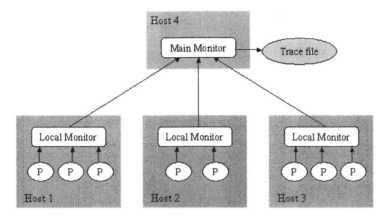

P: Process

Figure 5.18 The architecture of GRM

up to the point of failure. In statistics collection mode, the shared memory buffer is used to store the counters and LM is responsible for generating the final statistics data in an appropriate form.

- The MM is coordinating the work of the Local Monitors. It collects trace data from them when the user asks or a trace buffer on a local host becomes full. The trace is written into a text file in Tape/PVM format, which is a record-based format for trace events in ASCII representation.

5.5.4 Nagios

Nagios [28] is a system and network monitoring application operating through external "plugins" which publish status information to Nagios. Some of the Nagios features include:

- Monitoring of network services (SMTP, POP3, HTTP, NNTP and ICMP).
- Monitoring of host resources (processor load, disk usage, etc.).
- Contact notifications when service or host problems occur and get resolved (via email, pager or a user-defined method).
- Optional Web interface for viewing current network status, notification, problem history and log files.

5.5.5 NetLogger

Networked Application Logger (NetLogger) [48] is a set of tools for monitoring the behaviour of all the elements of the application-to-application communication path, applications, operating systems, hosts and networks. It includes tools for generating time-stamped event logs to provide detailed end-to-end application- and system-level monitoring; and tools for visualizing the log data and real-time state of the distributed system. System monitoring is based on standard UNIX and networking tools, which are first instrumented.

NetLogger consists of four components:

- An API and library of functions to simplify the generation of application-level event logs;
- A set of tools for collecting and sorting log files;
- A set of host and network monitoring tools;
- A tool for visualization and analysis of the log files.

In order to instrument an application to produce event logs, the application developer inserts calls to the NetLogger API at all the critical points in the code, then links the application with the NetLogger library. Figure 5.19 shows the architecture of NetLogger.

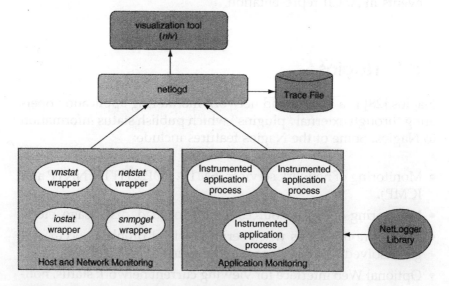

Figure 5.19 The architecture of NetLogger

5.5 OTHER MONITORING SYSTEMS

Components in NetLogger are described below. NetLogger uses IETF ULM (Universal Logger Message) format for logging and exchanging messages.

To instrument an application to produce event logs, the application developer inserts NetLogger calls at critical points in the code, and then links the application with the NetLogger library. This API is currently available in several languages, including Java, C, C++, Python and Perl.

NetLogger facilitates the collection of event logs from distributed applications by providing automatic logging to a single host and port. A server daemon, called *netlogd*, receives the log entries and writes them into a file on the local disk. Thus, applications can transparently log events in real time to a single destination over the wide-area network.

NetLogger includes wrappers for several standard UNIX system and network monitoring tools. These wrappers take the output from the tool and generate NetLogger-formatted monitoring events. Current wrappers include `vmstat` (CPU and memory monitoring), `netstat` (network interface monitoring), `iostat` (disk monitoring) and `snmpget` (remote access to a variety of host and network monitoring information).

The NetLogger visualization tool is used for interactive graphical representation of system-level and application-level events. The visualization tool can display several types of events at once. It is user configurable and can, for example, play, pause, rewind, provide slow motion and zoom. It can run post-mortem or in real time.

5.5.6 SCALEA-G

SCALEA-G [90] is a monitoring and performance analysis system for the Grid. SCALEA-G is based on GMA and is implemented as a set of OGSA [19]-based services that conduct online monitoring and performance analysis of a variety of computational and network resources, as well as applications. Source code and dynamic instrumentation are exploited to perform profiling and tracing of applications. Figure 5.20 shows the architecture of SCALEA-G.

SCALEA-G consists of:

- A Directory Service is used for publishing and searching information about producers and consumers as well as information about the types and characteristics of that data they produce.

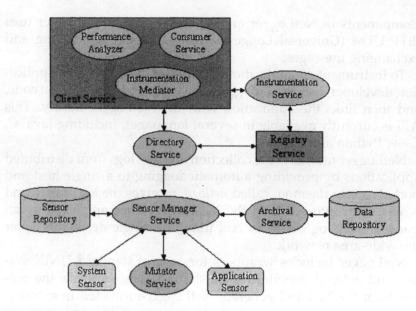

Figure 5.20 The architecture of SCALEA-G

- An Archival Service is a data repository which is used to store monitored data and performance results collected and analysed by other components.
- Sensor Manager Service whose role is to control and manage activities of sensors in the sensor repository, to register information about sensors that send data to it with a directory service, to receive and buffer data sensors produce, to support data subscription and query and to forward instrumentation requests to the Instrumentation Service.
- An Instrumentation Service is used to instrument applications, either at the source code level by using the Source Code Instrumentation Service or dynamically at run time through the Mutator Service.
- The Client Service provides interfaces for administrating other SCALEA-G services and accessing data in these services. In addition, it provides facilities for analysing performance data.
- System Sensors monitor computational services and network services such as network links, hard disks, memory usage and CPU availability.
- Application Sensors are used to measure execution behaviour of code regions and to monitor events in applications. Sensor

instances deliver data they collect to Sensor Manager Services. All sensors are associated with some common properties such as sensor identifier, data schema and parameters.

- The Consumer Service controls the activities of sensor manager services, such as when sensors subscribe, unsubscribe or query the directory service.
- The Instrumentation Mediator acts as an intermediary agent in communicating between users/tools with the Source Code Instrumentation Service (based on SCALEA Instrumentation Service)
- The Performance Analyzer is used to analyse collected data provided by the Consumer Service and provide the result to the user.
- The Registry Service is used to register information about service instances.

5.6 SUMMARY

Many systems are available in the field, each with their own particular focus. This review concentrated on a number of representative systems for resource and service monitoring that had software available at the time of writing, and met the high-level criteria presented in (Section 5.3 Review Criteria). This section summarizes our findings and highlights the trends that have emerged.

5.6.1 Resource categories

The majority of the projects reviewed (CODE, GPIR, GridRM, JAMM, MDS, Mercury, R-GMA and visPerf) provide access to host, service and network resource information. Of the remaining projects, GridICE and MapCenter provide computer and service monitoring, Hawkeye and NWS provide computer and network monitoring, while Autopilot is focused on computer host monitoring.

5.6.2 Native agents

With the exception of GridRM, MapCenter, MDS3 and R-GMA, all systems require their own agents to be installed, before monitoring can take place. GridRM and MapCenter interact with resources using a range of native protocols. For example, GridRM drivers are provided to interact with SNMP, Ganglia and NWS agents. MapCenter uses sensors installed on its server to probe remote MDS2, Tomcat and ICMP host availability. Generally MDS3 interacts with Grid services installed on monitored resources; however, the MDS3 Index Service contains mechanisms that allow locally executing code to insert GLUE-formatted information which could be used to interact with Ganglia, for example. R-GMA provides a general-purpose transport mechanism for publishing, locating and consuming monitoring information; developers are free to implement producers and therefore can potentially connect to native agents on monitored resources to retrieve information. An example might be the Canonical Producer retrieving Ganglia data from a remote resource.

5.6.3 Architecture

The projects reviewed can be categorized by their architecture into GMA-like and non-GMA-like systems. Nine of the projects (Autopilot, CODE, GridRM, JAMM, MDS3, Mercury, NWS, R-GMA and visPerf) provide a GMA-like architecture and therefore should be scalable, fault tolerant and ideally, interoperable. Even though this is the case, the overall implementation of these systems may actually prevent them meeting these goals. The non-GMA-like systems include GPIR, GridICE, Hawkeye and MapCenter.

5.6.4 Interoperability

Although the majority of systems have GMA-like architectures, interoperability is unlikely without considerable effort. GMA does not provide standard wire protocols or sufficient details of the registry APIs. Therefore while projects can follow the GMA in a coarse-grained manner, they define their own particular protocol formats and registry functionality.

5.6.5 **Homogeneous data presentation**

Systems may directly provide support to transform raw resource data into a standard format, or they may simply require that all data to be used in the system must first be converted explicitly by the resource producer. The reviewed projects are categorized into systems that directly provide normalization to achieve standard information, those that require the input of standard information, and those that operate with raw data, regardless of the heterogeneity of resources.

CODE, GridRM, Mercury and R-GMA provide mechanisms for normalization of heterogeneous data. The MapCenter server provides partial transformation via plugins. GPIR, GridICE and MDS3 require user-defined code or sensors to gather data and format it according to a predefined schema before submitting resource information into the system. GridICE and MDS3 use GLUE. While GLUE is the MDS3 default, custom providers are free to use other naming schemas. GPIR uses a restricted set of XML schemas to determine the format and layout of information. Hawkeye, JAMM, NWS and visPerf do not provide any mechanisms for normalizing data from the heterogeneous resources they monitor.

5.6.6 Intrusiveness of monitoring

Systems are categorized by their ability to potentially reduce monitoring intrusiveness. This is determined by the manner in which resource information is polled, either on demand, or periodically according to predefined intervals.

5.6.6.1 Periodic monitoring

- Autopilot sensors periodically gather information and cache it locally regardless of client interest. Client requests are fulfilled from the sensor's cache.
- CODE sensor managers can be set to perform periodic polling of resources in order to fulfil client subscription requests. The sensor manager could potentially use caching to service multiple client requests.
- GPIR Information Providers are executed remotely, and publish their information into the GPIR database. Clients then query the database for updated information. It is likely that a remote scheduler, for example `cron`, is used to periodically gather resource information from a range of monitored hosts.

- GridICE performs periodic polling of resources and publishes latest readings into the MDS2. Despite automated periodic polling, monitoring intrusiveness can be reduced based on the rate at which sensor information is expected to change, for example dynamic information can be queried more frequently than static.
- GridRM gateways are designed to provide the ability of periodically polling resources, based on client subscription.
- Hawkeye sensors are executed periodically on monitored hosts by external scheduling mechanisms. For example, the condor_startd daemon can be used on Condor hosts and cron on UNIX hosts.
- JAMM sensors can be configured to execute continuously or in response to specified conditions, for example when network activity is detected on a given port. Event Gateways cache data received from sensors and use the cache to respond to client requests.
- MapCenter services are probed according to predefined frequencies for specific service instances. The results of probes are cached in the gateway and used to respond to client queries.
- MDS3 resources are queried periodically by Service Data Managers; monitoring is independent of a client's interest. Polling frequency is defined separately for each Service Data Element. Administrators have the option to modify the rate a resource is polled in order to reflect the frequency with which underlying information is expected to change, or to meet client demand for up-to-date information.
- NWS sensors periodically transmit time-stamped local resource information to memory components. Clients retrieve cached information from memory components, which operate a circular buffer and only hold most-recent information.
- R-GMA user codes are free to execute periodically or in response to a local system event. Latest information produced by the user code is cached in the producer agent for retrieval by clients.
- visPerf sensors periodically gather local host performance information which is cached within the sensor. Sensors operate either in slave mode, where preconfigured monitoring duties are performed, or in a proxy mode, where the sensor aggregates information from slave sensors and makes this available to clients.

The use of caching at sensors can reduce overheads from client requests, as can the aggregation of information using proxy sensors.
- visPerf monitoring of Grid middleware is achieved through the parsing of application log files, or directly from the application if it provides an API for performance observations at run time. A direct connection between the sensor and its clients is used to return information.

5.6.6.2 Request-based monitoring

- CODE sensors gather resource information only in response to a direct request from a Sensor Manager. Sensor Managers can request a sensor to gather information in response to a direct client request.
- GridRM drivers query resources directly in response to a client request for specific subsets of information. Caching is provided at gateways and clients can control the cache policy on a per-request basis specifying the maximum age of information they are willing to tolerate. In addition, driver developers have the choice of implementing their own caching policies within drivers, as appropriate for a particular native agent.
- MapCenter probes service status in direct response to a particular client request.
- Mercury queries resources in response to a client request; the Main Monitor instructs appropriate Local Monitors to take measurements and pass back their results. As well as providing information from Local Monitors, the Local Monitoring Architecture can potentially aggregate information from multiple resources.
- The R-GMA CanonicalProducer executes user code to retrieve information in direct response to a client query.

5.6.6.3 Event-based monitoring

- GridRM is designed to respond to events issued by native agents, by passing event information to subscribed clients.
- Hawkeye Managers are capable of executing tasks in response to periodic updates received from agents, for example to notify a user by email if a host is running low on disk space.

- JAMM Process Sensors are used to generate events that describe a process' life cycle. For example, when the process starts, when it terminates, if termination was normal or abnormal. Events are transmitted to and cached by Event Gateways, which respond to client requests.
- MDS3 supports a subscription/notification protocol, for notifying clients of changes in Service Data Elements. Clients register their interest in services, which in return send notification messages to the client, when specified information is updated.
- NWS can provide information in response to certain events occurring on the monitored host. For example, the `nws_insert` command can be used in conjunction with locally executing processes to transmit an event in response to a specific local condition.
- R-GMA user codes are free to execute periodically or in response to a local system event. Clients cache any information produced by the user code in the producer's agent for later retrieval. R-GMA also provides event notification by utilizing "continuous" producers to stream information as it is generated, to subscribed consumers.
- visPerf sensors can be configured to stream real-time information to a predefined client.

5.6.6.4 Remote monitoring control

- Autopilot clients can interact with actuators to start, stop or modify the polling frequency of associated sensors.
- GridRM uses SQL to interact with native agents executing on resources. In addition to querying for resource values, SQL can be used to transmit data to drivers. In response, suitably equipped drivers can potentially control the behaviour of the resources they are monitoring; for example, to remotely kill a process or to restart a daemon.
- GPIR Information Providers (GIP) gather information only when remotely executed. This implies that the frequency with which GIPs publish information can be modified to meet expected client requirements for up-to-date information.
- JAMM sensors can be stopped and started remotely, therefore it may be possible to tailor monitoring overhead to client request

patterns; for example, by stopping a continuously executing sensor when interested clients are not present.
- NWS sensors can start and stop monitoring activities in response to client commands to the sensor control API.
- The visPerf Main component is used for browsing and controlling the status of remote sensors. This component periodically transmits synchronization messages to registered sensors in order to determine if any are unreachable or experiencing operational problems. Sensors can be started, stopped and their status viewed. Monitoring intrusiveness can potentially be controlled, in a coarse-grained fashion, by remotely instructing sensors to halt their monitoring activities when it is clear that clients are not interested in certain information.

5.6.7 Information searching and retrieval

- In Autopilot clients locate sensors based on a register's keyword and knowledge of the information associated. Given a match, clients connect to the sensors and retrieve the available information.
- In CODE clients locate Observers using the registry and then connect directly to a suitable one. To ascertain the sensors supported by an Observer, the registry can either be searched, or a given Observer can be queried directly. If a consumer executes a subscription query to an Observer, then it is possible for the Sensor Manager to return only those results that match a consumer-specified criteria, e.g. CPU load greater than 50%.
- GPIR can be queried by resource or by virtual organisation name. Query requests are formulated using the name of one of the GPIR XML naming schemas, for example "load" represents machine load and "services" represents the availability of services executing on a system. GPIR performs filtering and fusing of data by providing summary information for a VO and statistics for a range of resources.
- GridICE publishes information into the Globus MDS2. Clients query the MDS2 directly or use the GridICE Web portal. Clients searching the MDS2 can locate information based on name, category and value comparisons; for example, to return all Linux hosts with load greater than a specified threshold.

- GridRM gateways are registered in the jGMA directory and publish monitored resource and driver naming schemas; clients can locate resources by GridRM gateway, by name or by category. Clients submit SQL-like queries to appropriate gateways. For example, a gateway can be queried for all GLUE hosts with memory greater than a specified value. Information is filtered and fused based on the client's SQL query.
- In Hawkeye, clients query Hawkeye Managers for information based on ClassAd attribute names. For example, clients are able to return a list of all hosts that have a load average greater than some threshold. The Hawkeye Manager provides filtering and fusing operations, which allow clients to select a subset of information from selected hosts.
- On start-up in JAMM, sensors register with a directory service. Sensors gather monitoring data and submit it to an Event Gateway (EG) as the data is generated at the monitored resource. Clients can request to receive all data cached by an EG, or subscribe to receive data as the EG receives it from sensors. The EG performs filtering functions whereby clients can select particular types of data. For example, by taking 1-, 10- and 15-minute CPU load averages.
- MapCenter does not natively support information searching by resource category, functionality, capability or load. Event subscription and notification are also not supported. Probed service status is fused and filtered to provide a number of map views for the user to navigate. MapCenter could potentially populate service data into an MDS2 and then clients can use the LDAP query interface for advanced or custom queries.
- MDS3 clients can use the subscription mechanism in order to be notified when specified information is updated. MDS3 supports both simple and complex client resource queries. Query-by-name is performed by specifying a service and one or more Service Data Element names. Optionally clients may also provide an XPath expression to refine the information returned from a query. The approaches used allow information to be returned by resource type and by the comparison of resource values, for example CPULoad greater than a specified value. Filtering and fusing of data is provided by aggregation and indexing operations performed by the MDS3 Index Service.

- Mercury clients can request real-time information by request–response or subscription mechanisms. Templates are used to define the information produced by sensors and to assign a unique name to individual items of information. Clients may be required to provide query parameters, for example a resource network address and perhaps a specific processor or network interface to be queried.
- The NWS Name Server (NS) is used to locate categories of resource information; searches can be by hostname, sensor, monitoring activity or clique. The NS does not contain details of performance information so, for example, searching for a host by CPU load is not possible. Client event notification is not supported; clients have to initiate requests to retrieve data from NWS memory and forecaster components.
- R-GMA provides SQL queries for locating resource information. Queries can be refined by naming schema and by predicate, for example the name of the resource's controlling site. Data from separate producers can be filtered and fused using producer-consumer components, such as an Archiver. Customized user code is needed to provide the filtering and fusing functionality for a given set of producers.
- visPerf sensors can be located using the directory service and queried directly to provide a list of the monitoring services they provide. Sensors support simple request/response (client pull) mechanisms to respond on demand to client requests.

5.7 CHAPTER SUMMARY

Monitoring is critical for providing a robust, high-performance Grid environment. Without monitoring mechanisms it will be impossible to determine the status or health of the environment, and thus difficult to use it efficiently and effectively. Monitoring data can be used for performance analysis, performance diagnosis, performance tuning, fault detection and scheduling. A basic monitoring system has the following components:

- Producers (sensors) that generate monitoring data (called events);
- Consumers that sink events;
- One or more directory services for registration and discovery of sensors/events/consumers.

We have reviewed a set of representative monitoring systems. While they differ in their implementations, all of them can provide basic monitoring functionalities. Due to the complexity and dynamics of the Grid, a monitoring system should have the following features:

- GMA compliance,
- Scalable,
- Resources monitored include network resources, host resources and jobs,
- Resource performance forecasting,
- Resource performance analysis,
- Various presentation views for resource monitoring,
- Directory service for events subscription and notification.

Since OGSA is the *de facto* standard for developing service-oriented Grid systems, therefore, a compliant monitoring system should potentially expose monitoring information via as Service Data Elements (SDEs). A monitoring tool can then be implemented as a Grid monitoring service that provides interfaces to interact with a variety of existing monitoring systems as shown in Figure 5.21. Monitoring data can be queried from the service's SDEs for resource performance forecasting, analysis and presentation purposes. An overview of Grid monitoring systems features is shown in Table 5.3.

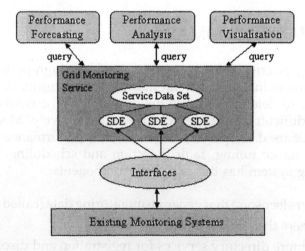

Figure 5.21 An OGSA-compliant Grid monitoring system

5.7 CHAPTER SUMMARY

Table 5.3 Overview of Grid monitoring systems features

Category	Autopilot	CODE	GPIR	GridICE	GridRM	Hawkeye	JAMM	MapCenter	MDS3	Mercury	NWS	R-GMA	visPerf
Scalable + wide area	1	1	1	1	Y	N, 1	Y	Unclear	Y	Y	Y	Y	Y
Monitor (C, S, N)	C	All	All	C, S	All	C, N	All	S	All	All	C, N	All	All
Cross-API monitoring	ζ	ζ	?	ζ	Y	ζ	ζ	Y	ζ	ζ	ζ	Y	ζ
Homogeneous data	N	Y	Θ	Θ	Y	N	N	Partly	Θ	Y	N	Y	N
Info search by resource class	Y	Y	Y	Y	Y	Y	Y	N	Y	Y	Y	Y	N
Run-time extensibility	N	Y	Y	Y	Y	Y	Y	Y	Y	Y	N	Y	N
Filtering/fusing of data	Y	Filter	Y	N	Y	Y	Y	N	Y	Y	N	Y	Y
Open standards	N	Y	Y	N	Y	N	N	N	Y	N	N	N	N
Security	N	Y	N	N	Y	N	N	N	Y	Y	N	Y	N
Software available	Y	Y	Y	Y	Y	Y	Y	Y	Y	Y	Y	Y	Y
Software dependency	Y	Y	Y	Y	Y	N	Y	Y	Y	N	N	Y	Y
Active and supported	Y	Y	Y	Y	Y	Y	N	Y	Y	Y	Y	Y	Y
Open-source license	Y	Y	Y	Y	Y	Y	Y	Y	Y	Y	Y	Y	Y

Notes: C, S, N – Computer, Service, Network; ζ – Custom agent or sensor must be installed on monitored resources; Θ – Custom sensors at monitored resource are required to format data into GLUE before submitting to the monitoring system.

5.8 FURTHER READING AND TESTING

A wide variety of monitoring systems have been reviewed in the chapter. Many can be freely downloaded from their Web sites cited in the text. Readers can download one to experience how a monitoring system works. Based on that, you can start with implementing an OGSA compliant Grid monitoring system as proposed in Figure 5.21.

5.9 KEY POINTS

- Monitoring is critical for a robust, effective and efficient Grid environment.
- Monitoring data can be used to tune an environment's performance, detect faults and schedule jobs.
- The three fundamental components in a monitoring system are sensors (producers), clients (consumers) and directory services.
- A good monitoring system should potentially have a scalable GMA-like architecture and support a variety of sensors, and for analysis to provide a number of different views of the monitored resources.
- With the introduction of OGSA, a monitoring system can be implemented as a Grid Service in which monitoring data can be exposed as SDEs that can be queried.

5.10 REFERENCES

[1] Global Grid Forum (GGF), Grid Monitoring Architecture (GMA) Working Group, http://www-didc.lbl.gov/GGF-PERF/GMA-WG/.
[2] GMA White Paper, http://www-didc.lbl.gov/GGF-PERF/GMA-WG/.
[3] Grid Security Infrastructure (GSI) documentation, May 2004, http://www-unix.globus.org/toolkit/docs/3.2/gsi/.
[4] Generic Security Services Application Programming Interface, June 2004, http://www.faqs.org/rfcs/rfc1508.html.
[5] Autopilot, June 2003, http://vibes.cs.uiuc.edu/Software/Autopilot/autopilot.htm.
[6] Vetter, J.S. and Reed, D.A., *Real-time Performance Monitoring, Adaptive Control, and Interactive Steering of Computational Grids*, 14(4): 357–366, 2000.

5.10 REFERENCES

[7] Grid Application Development Software (GrADS), August 2003, http://nhse2.cs.rice.edu/grads/.

[8] Vraalsen, F., Aydt, R.A., Mendes, C.L. and Reed, D.A., *Performance Contracts: Predicting and Monitoring Grid Application Behavior*, GRID 2001 Proceedings of the Second International Workshop on Grid Computing, Denver, CO. Lecture Notes in Computer Science, Springer-Verlag, November 2001, pp. 154–165.

[9] SDDF, June 2004, http://www-pablo.cs.uiuc.edu/Project/SDDF/SDDFOverview.htm.

[10] Shaffer, E., Reed, D.A., Whitmore, S. and Schaeffer, B., *Virtue: Performance Visualization of Parallel and Distributed Applications*. Los Alamitos, CA, USA, IEEE Computer Society Press, December 1999, ISSN 0018-9162, 32(12), pp. 44–51.

[11] The Globus Alliance, March 2004, http://www.globus.org/.

[12] Pablo Project Software License Agreement, August 2002, http://www.pablo.cs.uiuc.edu/Software/license.htm.

[13] Smith, W., *A System for Monitoring and Management of Computational Grids*, International Conference on Parallel Processing (ICPP '02), 18–21 August 2002, Vancouver, BC, Canada.

[14] Smith, W.W., Code: A Framework for Control and Observation in Distributed Environments, March 2004, http://www.nas.nasa.gov/Research/Software/Open-Source/CODE/.

[15] NASA Ames Research Center, April 2004, http://www.arc.nasa.gov/.

[16] NASA Information Power Grid, April 2004, http://www.nas.nasa.gov/About/IPG/ipg.html.

[17] NASA Open Source Agreement Version 1.0. May 2004, http://opensource.arc.nasa.gov/agreement.jsp?id=9.

[18] Andreozzi, S., De Bortoli, N., Fantinel, S., Ghiselli, A., Tortone, G. and Vistoli, C., *GridICE: A Monitoring Service for the Grid*, Proceedings of the 3rd Cracow Grid Workshop, October 2003.

[19] Andreozzi, S., GridICE – The Eyes of the Grid. Technical Report, Instituto Nazionale di Fisica Nucleare, 2004.

[20] GridICE: The Eyes of the Grid, February 2004, http://server11.infn.it/gridice/.

[21] Czajkowski, K., Fitzgerald, S., Foster, I. and Kesselman, C., *Grid Information Services for Distributed Resource Sharing*. 10th IEEE International Symposium on High Performance Distributed Computing (HPDC-10 '01), 7–9 August 2001, San Francisco, California.

[22] MDS2.4 Features in the Globus Toolkit v2.4. August 2003, http://www.globus.org/mds/mds2/.

[23] Lemon: Fabric Monitoring Toolkit, February 2004, http://lemon.web.cern.ch/lemon/.

[24] INFN Grid, February 2004, http://server11.infn.it/grid/.

[25] DataTAG, February 2004, http://datatag.web.cern.ch/datatag/.

[26] LHC Computing Grid (LCG) project, February 2004, http://lcg.web.cern.ch/LCG/.

[27] INFN Production Grid, February 2004, http://grid-it.cnaf.infn.it/.

[28] Nagios, June 2004, http://www.nagios.org.

[29] Ganglia distributed monitoring and execution system, March 2004, http://ganglia.sourceforge.net/.
[30] INFN Software License, July 2002, http://www.cnaf.infn.it/license.html.
[31] Grid Portals Information Repository (GPIR), February 2004, http://www.tacc.utexas.edu/projects/gpir/.
[32] Gridport, February 2004, http://gridport.net/.
[33] NPACI Hotpage Grid Computing Portal, September 2003, http://hotpage.npaci.edu/.
[34] JBoss, February 2004, http://www.jboss.org.
[35] PostgresSQL, February 2004, http://www.postgresql.org/.
[36] Baker, M.A. and Smith, G., *GridRM: An Extensible Resource Management System*. Proceedings of IEEE International Conference on Cluster Computing (Cluster 2003) Hong Kong, IEEE Computer Society Press, 2003, ISBN 0-7695-2066-9, pp. 207, 215.
[37] GridRM, February 2004, http://gridrm.org.
[38] Wolski, R., Spring, N. and Hayes, J., The Network Weather Service: A Distributed Resource Performance Forecasting Service for Metacomputing. *Journal of Future Generation Computing Systems*, 15(5–6): 757–768, October 1999, http://nws.cs.ucsb.edu/publications.html.
[39] DSG, June 2004, http://dsg.port.ac.uk.
[40] Baker, M.A. and Grove, M., *jGMA: A Lightweight Implementation of the Grid Monitoring Architecture.* UKUUG LISA/Winter Conference, February 2004.
[41] *jGMA: A Lightweight Implementation of the Grid Monitoring Architecture.* May 2004, http://dsg.port.ac.uk/projects/jGMA/.
[42] The GNU General Public License (GPL), version 2, June 1991. February 2004, http://www.opensource.org/licenses/gpl-license.php.
[43] Hawkeye: A Monitoring and Management Tool for Distributed Systems, April 2004, http://www.cs.wisc.edu/condor/hawkeye/.
[44] Condor, April 2004, http://www.cs.wisc.edu/condor/.
[45] Classified Advertisements, November 2003, http://www.cs.wisc.edu/condor/classad/.
[46] Tierney, B., Crowley, B., Gunter, D., Holding, M., Lee, J. and Thompson, M. *A Monitoring Sensor Management System for Grid Environments*, High Performance Distributed Computing (HPDC-9), Pittsburgh, Pennsylvania, August 2000, pp. 97–104.
[47] Java Agents for Monitoring and Management (JAMM), July 2000, http://www-didc.lbl.gov/JAMM.
[48] The NetLogger Toolkit, April 2004, http://www-didc.lbl.gov/NetLogger/.
[49] The MATISSE Project, April 2000, http://www.cnri.net/matisse/.
[50] The Data Intensive Distributed Computing Research Group (DIDC), February 2004, http://www-didc.lbl.gov/.
[51] Gunter, D., Tierney, B.L., Tull, C.E. and Virmani, V., *On Demand Grid Application Tuning and Debugging with the NetLogger Activation Service*, 4th International Workshop on Grid Computing, 17 November 2003, Phoenix, Arizona.
[52] Universal Format for Logger Messages, June 2004, http://www.hsc.fr/gulp/draft-abela-ulm-05.txt.
[53] Bonnassieux, F., Harakaly, R. and Primet, P., Automatic Services Discovery, Monitoring and Visualization of Grid Environments: The Mapcenter

5.10 REFERENCES

Approach. Across Grid Conference, 13–14 February 2003, Santiago de Compostela, Spain.

[54] Mapcenter, February 2004, http://mapcenter.in2p3.fr/.

[55] Bonnassieux, F., Harakaly, R. and Primet, P., *Mapcenter: An Open Grid Status Visualization Tool*. Proceedings of ISCA 15th International Conference on Parallel and Distributed Computing Systems, Louisville, Kentucky, USA, 19–21 September 2002.

[56] Bonnassieux, F., MapCenter v2.3.0 Administration Guide: Installation and Configuration Instructions, Technical Report, European Data Grid, February 2004.

[57] The datagrid project. February 2004, http://eu-datagrid.web.cern.ch/eudatagrid/.

[58] DataTAG, June 2004, http://datatag.web.cern.ch/datatag/.

[59] The CrossGrid project, February 2004, http://www.eu-crossgrid.org/.

[60] PlanetLab, February 2004, http://www.planet-lab.org/.

[61] L-Bone, February 2004, http://loci.cs.utk.edu/lbone/.

[62] Atlas, February 2004, http://atlas.web.cern.ch/Atlas/GROUPS/SOFTWARE/OO/grid/.

[63] Information services in the Globus Toolkit 3.0 release, August 2003, http://www.globus.org/mds/.

[64] Foster, I., Kesselman, C., Nick, J. and Tuecke, S., The Physiology of the Grid: An Open Grid Services Architecture for Distributed Systems Integration, 2002.

[65] Tuecke, S., Czajkowski, K., Foster, I., Frey, J., Graham, S., Kesselman, C., Maquire, T., Sandholm, T., Snelling, D. and Vanderbilt, P., Open Grid Services Infrastructure (OGSI) version 1.0. Technical Report, Global Grid Forum, June 2003.

[66] MDS2.4 features in the Globus Toolkit release 2.4. August 2003, http://www.globus.org/mds/mds2/.

[67] Gridlab work package 11: Mercury, February 2004, http://www.gridlab.org/WorkPackages/wp-11/.

[68] Balaton, Z. and Gombas, G., Detailed Architecture Specification. Technical Report, GridLab Project, Information Society Technologies, September 2002.

[69] Balaton, Z. and Gombas, G., Extended Architecture Specification. Technical Report D11.4, GridLab WP11, Information Society Technologies, 2004.

[70] Gridlab: A Grid Application Toolkit and Testbed, February 2004, http://www.gridlab.org/.

[71] Ludwig, T., On-line Monitoring Interface Specification 2.0. August 1998, http://wwwbode.cs.tum.edu/omis/.

[72] RFC 1014 – XDR: External Data Representation Standard, June 1987, http://www.faqs.org/rfcs/rfc1014.html.

[73] The Network Weather Service, July 2003, http://nws.cs.ucsb.edu/.

[74] NSF Middleware Initiative, February 2004, http://www.nsfmiddleware.org.

[75] NPACI Grid, August 2003, http://npacigrid.npaci.edu/.

[76] Cooke, A., Gray, A.J.G., Ma, L., Nutt, W., Magowan, J., Oevers, M., Taylor, P., Byrom, R., Field, L., Hicks, S., Leake, J., Soni, M., Wilson, A., Cordenonsi, R., Cornwall, L., Djaoui, A., Fisher, S., Podhorszki, N., Coghlan, B.A., Kenny, S.

and O'Callaghan, D., *R-GMA: An Information Integration System for Grid Monitoring*. Robert Meersman, Zahir Tari and Douglas C. Schmidt (eds), COOPIS 2003, Lecture Notes in Computer Science, Springer, 2003, 2888, pp. 462–481.
[77] R-GMA: Relational Grid Monitoring Architecture, December 2003, http://rgma.org.
[78] GLUE, June 2004, http://www.cnaf.infn.it/~sergio/datatag/glue/.
[79] DataGrid Information and Monitoring Services Architecture: Design, Requirements, and Evaluation Criteria. Technical Report DataGrid-03-NOT-???-x-y, Data Grid, January 2004, http://hepunx.rl.ac.uk/edg/wp3/documentation/doc/arch.pdf.
[80] EGEE, June 2004, http://public.eu-egee.org/.
[81] visPerf monitor, February 2004, http://icl.cs.utk.edu/netsolvedev/monitor/.
[82] NetSolve, June 2004, http://icl.cs.utk.edu/netsolve/.
[83] Netsolve/gridsolve-2.0, March 2004, http://icl.cs.utk.edu/netsolve/.
[84] Godard, S., Sysstat: System performance tools for Linux OS, February 2004, http://perso.wanadoo.fr/sebastien.godard/.
[85] RRDTool, June 2004, http://www.caida.org/tools/utilities/rrdtool/.
[86] GridMon, June 2004, http://gridmon.dl.ac.uk/.
[87] Iperf, June 2004, http://dast.nlanr.net/Projects/Iperf/.
[88] Balaton, Z., Kacsuk, P., Podhorszki, N. and Vajda, F., *From Cluster Monitoring to Grid Monitoring based on GRM*, Proceedings of EuroPar 2001, Manchester, UK.
[89] Kacsuk, P., de Kergommeaux, J.C., Maillet, É. and Vincent, J.M., The Tape/PVM Monitor and the PROVE Visualisation Tool, In: Parallel Program Development for Cluster Computing, Methodology, Tools and Integrated Environments (eds: Cunha, C., Kacsuk, P. and Winter, S.C.), Nova Science Publishers, Inc., pp. 291–303, 2001.
[90] Truong, H.-L. and Fahringer, T., SCALEA-G: A Unified Monitoring and Performance Analysis System for the Grid, 2nd European Across Grids Conference, Nicosia, Cyprus, January 2004.

Part Three

Job Management and User Interaction

6

Grid Scheduling and Resource Management

LEARNING OBJECTIVES

In this chapter, we will study Grid scheduling and resource management, which play a critical role in building an effective and efficient Grid environment. From this chapter, you will learn:

- What a scheduling system is about and how it works.
- Scheduling paradigms.
- Condor, SGE, PBS and LSF.
- Grid scheduling with quality-of-services (QoS) support, e.g. AppLeS, Nimrod/G, Grid rescheduling.
- Grid scheduling optimization with heuristics.

CHAPTER OUTLINE

6.1 Introduction
6.2 Scheduling Paradigms
6.3 How Scheduling Works
6.4 A Review of Condor, SGE, PBS and LSF

The Grid: Core Technologies Maozhen Li and Mark Baker
© 2005 John Wiley & Sons, Ltd

6.5 Grid Scheduling with QoS
6.6 Chapter Summary
6.7 Further Reading and Testing

6.1 INTRODUCTION

The Grid is emerging as a new paradigm for solving problems in science, engineering, industry and commerce. Increasing numbers of applications are utilizing the Grid infrastructure to meet their computational, storage and other needs. A single site can simply no longer meet all the resource needs of today's demanding applications, and using distributed resources can bring many benefits to application users. The deployment of Grid systems involves the efficient management of heterogeneous, geographically distributed and dynamically available resources. However, the effectiveness of a Grid environment is largely dependent on the effectiveness and efficiency of its schedulers, which act as localized resource brokers. Figure 6.1 shows that user tasks, for example, can be submitted via Globus to a range of resource management and job scheduling systems, such as Condor [1], the Sun Grid Engine (SGE) [2], the Portable Batch System (PBS) [3] and the Load Sharing Facility (LSF) [4].

Grid scheduling is defined as the process of mapping Grid jobs to resources over multiple administrative domains. A Grid job can

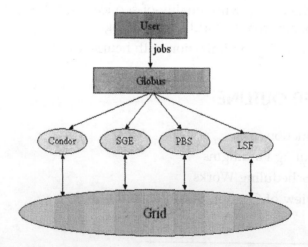

Figure 6.1 Jobs, via Globus, can be submitted to systems managed by Condor, SGE, PBS and LSF

be split into many small tasks. The scheduler has the responsibility of selecting resources and scheduling jobs in such a way that the user and application requirements are met, in terms of overall execution time (throughput) and cost of the resources utilized.

This chapter is organized as follows. In Section 6.2, we present three scheduling paradigms – centralized, hierarchical and decentralized. In Section 6.3, we describe the steps involved in the scheduling process. In Section 6.4, we give a review of the current widely used resource management and job scheduling such as Condor and SGE. In Section 6.5, we discuss some issues related to scheduling with QoS. In Section 6.6, we conclude the chapter and in Section 6.7, provide references for further reading and testing.

6.2 SCHEDULING PARADIGMS

Hamscher *et al.* [5] present three scheduling paradigms – centralized, hierarchical and distributed. In this section, we give a brief review of the scheduling paradigms. A performance evaluation of the three scheduling paradigms can also be found in Hamscher *et al.* [5].

6.2.1 Centralized scheduling

In a centralized scheduling environment, a central machine (node) acts as a resource manager to schedule jobs to all the surrounding nodes that are part of the environment. This scheduling paradigm is often used in situations like a computing centre where resources have similar characteristics and usage policies. Figure 6.2 shows the architecture of centralized scheduling.

In this scenario, jobs are first submitted to the central scheduler, which then dispatches the jobs to the appropriate nodes. Those jobs that cannot be started on a node are normally stored in a central job queue for a later start.

One advantage of a centralized scheduling system is that the scheduler may produce better scheduling decisions because it has all necessary, and up-to-date, information about the available resources. However, centralized scheduling obviously does not scale well with the increasing size of the environment that it manages. The scheduler itself may well become a bottleneck, and if

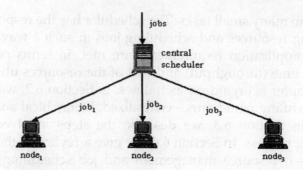

Figure 6.2 Centralized scheduling

there is a problem with the hardware or software of the scheduler's server, i.e. a failure, it presents a single point of failure in the environment.

6.2.2 Distributed scheduling

In this paradigm, there is no central scheduler responsible for managing all the jobs. Instead, distributed scheduling involves multiple localized schedulers, which interact with each other in order to dispatch jobs to the participating nodes. There are two mechanisms for a scheduler to communicate with other schedulers – direct or indirect communication.

Distributed scheduling overcomes scalability problems, which are incurred in the centralized paradigm; in addition it can offer better fault tolerance and reliability. However, the lack of a global scheduler, which has all the necessary information on available resource, usually leads to sub-optimal scheduling decisions.

6.2.2.1 Direct communication

In this scenario, each local scheduler can directly communicate with other schedulers for job dispatching. Each scheduler has a list of remote schedulers that they can interact with, or there may exist a central directory that maintains all the information related to each scheduler. Figure 6.3 shows the architecture of direct communication in the distributed scheduling paradigm.

If a job cannot be dispatched to its local resources, its scheduler will communicate with other remote schedulers to find resources

6.2 SCHEDULING PARADIGMS

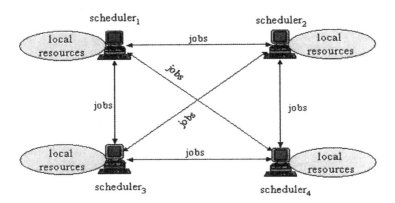

Figure 6.3 Direct communications in distributed scheduling

appropriate and available for executing its job. Each scheduler may maintain a local job queue(s) for job management.

6.2.2.2 Communication via a central job pool

In this scenario, jobs that cannot be executed immediately are sent to a central job pool. Compared with direct communication, the local schedulers can potentially choose suitable jobs to schedule on their resources. Policies are required so that all the jobs in the pool are executed at some time. Figure 6.4 shows the architecture of using a job pool for distributed scheduling.

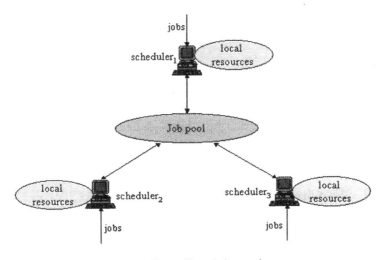

Figure 6.4 Distributed scheduling with a job pool

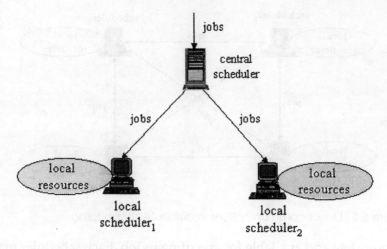

Figure 6.5 Hierarchical scheduling

6.2.3 Hierarchical scheduling

In hierarchical scheduling, a centralized scheduler interacts with local schedulers for job submission. The centralized scheduler is a kind of a meta-scheduler that dispatches submitted jobs to local schedulers. Figure 6.5 shows the architecture of this paradigm.

Similar to the centralized scheduling paradigm, hierarchical scheduling can have scalability and communication bottlenecks. However, compared with centralized scheduling, one advantage of hierarchical scheduling is that the global scheduler and local scheduler can have different policies in scheduling jobs.

6.3 HOW SCHEDULING WORKS

Grid scheduling involves four main stages: resource discovery, resource selection, schedule generation and job execution.

6.3.1 Resource discovery

The goal of resource discovery is to identify a list of authenticated resources that are available for job submission. In order to cope with the dynamic nature of the Grid, a scheduler needs to have

6.3 HOW SCHEDULING WORKS

some way of incorporating dynamic state information about the available resources into its decision-making process.

This decision-making process is somewhat analogous to an ordinary compiler for a single processor machine. The compiler needs to know how many registers and functional units exist and whether or not they are available or "busy". It should also be aware of how much memory it has to work with, what kind of cache configuration has been implemented and the various communication latencies involved in accessing these resources. It is through this information that a compiler can effectively schedule instructions to minimize resource idle time. Similarly, a scheduler should always know what resources it can access, how busy they are, how long it takes to communicate with them and how long it takes for them to communicate with each other. With this information, the scheduler optimizes the scheduling of jobs to make more efficient and effective use of the available resources.

A Grid environment typically uses a pull model, a push model or a push–pull model for resource discovery. The outcome of the resource discovery process is the identity of resources available ($R_{available}$) in a Grid environment for job submission and execution.

6.3.1.1 The pull model

In this model, a single daemon associated with the scheduler can query Grid resources and collect state information such as CPU loads or the available memory. The pull model for gathering resource information incurs relatively small communication overhead, but unless it requests resource information frequently, it tends to provide fairly stale information which is likely to be constantly out-of-date, and potentially misleading. In centralized scheduling, the resource discovery/query process could be rather intrusive and begin to take significant amounts of time as the environment being monitored gets larger and larger. Figure 6.6 shows the architecture of the model.

6.3.1.2 The push model

In this model, each resource in the environment has a daemon for gathering local state information, which will be sent to a centralized scheduler that maintains a database to record each resource's

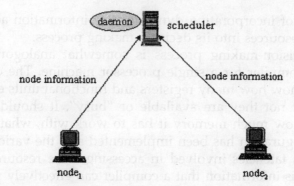

Figure 6.6 The pull model for resource discovery

activity. If the updates are frequent, an accurate view of the system state can be maintained over time; obviously, frequent updates to the database are intrusive and consume network bandwidth. Figure 6.7 shows the architecture of the push model.

6.3.1.3 The push–pull model

The push–pull model lies somewhere between the pull model and the push model. Each resource in the environment runs a daemon that collects state information. Instead of directly sending this information to a central scheduler, there exist some intermediate nodes running daemons that aggregate state information from different sub-resources that respond to queries from the scheduler.

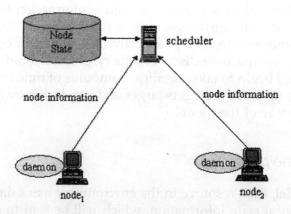

Figure 6.7 The push model for resource discovery

6.3 HOW SCHEDULING WORKS

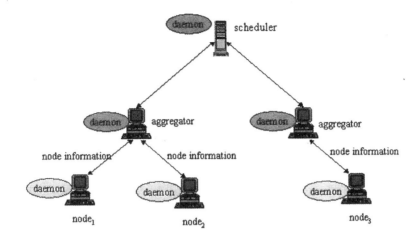

Figure 6.8 The push–pull model for resource discovery

A challenge of this model is to find out what information is most useful, how often it should be collected and how long this information should be kept around. Figure 6.8 shows the architecture of the push–pull model.

6.3.2 Resource selection

Once the list of possible target resources is known, the second phase of the scheduling process is to select those resources that best suit the constraints and conditions imposed by the user, such as CPU usage, RAM available or disk storage. The result of resource selection is to identify a resource list ($R_{selected}$) in which all resources can meet the minimum requirements for a submitted job or a job list. The relationship between resources available ($R_{available}$) and resources selected ($R_{selected}$) is:

$$R_{selected} \subseteq R_{available}$$

6.3.3 Schedule generation

The generation of schedules involves two steps, selecting jobs and producing resource selection strategies.

6.3.3.1 Job selection

The resource selection process is used to choose resource(s) from the resource list ($R_{selected}$) for a given job. Since all resources in the list $R_{selected}$ could meet the minimum requirements imposed by the job, an algorithm is needed to choose the best resource(s) to execute the job. Although random selection is a choice, it is not an ideal resource selection policy. The resource selection algorithm should take into account the current state of resources and choose the best one based on a quantitative evaluation. A resource selection algorithm that only takes CPU and RAM into account could be designed as follows:

$$\text{Evaluation}_{resource} = \frac{\text{Evaluation}_{CPU} + \text{Evaluation}_{RAM}}{W_{CPU} + W_{RAM}} \quad (6.1)$$

$$\text{Evaluation}_{CPU} = W^*_{CPU}(1 - \text{CPU}_{load}) * \frac{\text{CPU}_{speed}}{\text{CPU}_{min}} \quad (6.2)$$

$$\text{Evaluation}_{RAM} = W^*_{RAM}(1 - \text{RAM}_{usage}) * \frac{\text{RAM}_{size}}{\text{RAM}_{min}} \quad (6.3)$$

where W_{CPU} – the weight allocated to CPU speed; CPU_{load} – the current CPU load; CPU_{speed} – real CPU speed; CPU_{min} – minimum CPU speed; W_{RAM} – the weight allocated to RAM; RAM_{usage} – the current RAM usage; RAM_{size} – original RAM size; and RAM_{min} – minimum RAM size.

Now we give an example to explain the algorithm used to choose one resource from three possible candidates. The assumed parameters associated with each resource are given in Table 6.1.

Let us suppose that the total weighting used in the algorithm is 10, where the CPU weight is 6 and the RAM weight is 4. The minimum CPU speed is 1 GHz and minimum RAM size is 256 MB.

Table 6.1 The resource information matrix

	CPU speed (GHz)	CPU load (%)	RAM size (MB)	RAM usage (%)
Resource$_1$	1.8	50	256	50
Resource$_2$	2.6	70	512	60
Resource$_3$	1.2	40	512	30

6.3 HOW SCHEDULING WORKS

Then, evaluation values for resources can be calculated using the three formulas:

$$\text{Evaluation}_{resource_1} = \frac{5.4+2}{10} = 0.74$$

$$\text{Evaluation}_{resource_2} = \frac{4.68+3.2}{10} = 0.788$$

$$\text{Evaluation}_{resource_3} = \frac{4.32+5.6}{10} = 0.992$$

From the results we know Resource$_3$ is the best choice for the submitted job.

6.3.3.2 Resource selection

The goal of job selection is to select a job from a job queue for execution. Four strategies that can be used to select a job are given below.

- *First come first serve*: The scheduler selects jobs for execution in the order of their submissions. If there is no resource available for the selected job, the scheduler will wait until the job can be started. The other jobs in the job queue have to wait. There are two main drawbacks with this type of job selection. It may waste resources when, for example, the job selected needs more resources to be available before it can start, which results in a long waiting time. And jobs with high priorities cannot get dispatched immediately if a job with a low priority needs more time to complete.
- *Random selection*: The next job to be scheduled is randomly selected from the job queue. Apart from the two drawbacks with the first-come-first-serve strategy, jobs selection is not fair and job submitted earlier may not be scheduled until much later.
- *Priority-based selection*: Jobs submitted to the scheduler have different priorities. The next job to be scheduled is the job with the highest priority in the job queue. A job priority can be set when the job is submitted. One drawback of this strategy is that it is hard to set an optimal criterion for a job priority. A job with the highest priority may need more resources than available and may also result in a long waiting time and inability to make good use of the available resources.

- *Backfilling selection [6]*: The backfilling strategy requires knowledge of the expected execution time of a job to be scheduled. If the next job in the job queue cannot be started due to a lack of available resources, backfilling tries to find another job in the queue that can use the idle resources.

6.3.4 Job execution

Once a job and a resource are selected, the next step is to submit the job to the resource for execution. Job execution may be as easy as running a single command or as complicated as running a series of scripts that may, or may not, include set up or staging.

6.4 A REVIEW OF CONDOR, SGE, PBS AND LSF

In this section, we give a review on Condor/Condor-G, the SGE, PBS and LSF. The four systems have been widely used for Grid-based resource management and job scheduling.

6.4.1 Condor

Condor is a resource management and job scheduling system, a research project from University of Wisconsin–Madison. In this section we study Condor based on its latest version, Condor 6.6.3.

6.4.1.1 Condor platforms

Condor 6.6.3 supports a variety of systems as follows:

- HP systems running HPUX10.20
- Sun SPARC systems running Solaris 2.6/2.7/8/9
- SGI systems running IRIX 6.5 (not fully supported)
- Intel x86 systems running Redhat Linux 7.1/7.2/7.3/8.0/9.0, Windows NT4.0, XP and 2003 Server (the Windows systems are not fully supported)
- ALPHA systems running Digital UNIX 4.0, Redhat Linux 7.1/7.2/7.3 and Tru64 5.1 (not fully supported)

6.4 A REVIEW OF CONDOR, SGE, PBS AND LSF

- PowerPC systems running Macintosh OS X and AIX 5.2L (not fully supported)
- Itanium systems running Redhat 7.1/7.2/7.3 (not fully supported)
- Windows systems (not fully supported).

UNIX machines and Windows machines running Condor can co-exist in the same Condor pool without any problems, e.g. a job submitted from a Windows machine can run on a Windows machine or a UNIX machine, a job submitted from a UNIX machine can run on a UNIX or a Windows machine. There is absolutely no need to run more than one Condor central manager, even if you have both UNIX and Windows machines. The Condor central manager itself can run on either UNIX or Windows machines.

6.4.1.2 The architecture of a Condor pool

Resources in Condor are normally organized in the form of Condor pools. A pool is an administrated domain of hosts, not specifically dedicated to a Condor environment. A Condor system can have multiple pools of which each follows a flat machine organization.

As shown in Figure 6.9, a Condor pool normally has one Central Manager (master host) and an arbitrary number of Execution (worker) hosts. A Condor Execution host can be configured as a job Execution host or a job Submission host or both. The Central Manager host is used to manage resources and jobs in a Condor

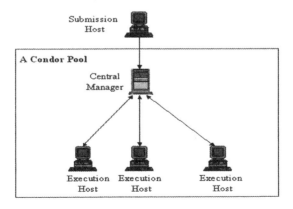

Figure 6.9 The architecture of a Condor pool

pool. Host machines in a Condor pool may not be dedicated to Condor.

If the Central Manager host in a Condor pool crashes, jobs that are already running will continue to run unaffected. Queued jobs will remain in the queue unharmed, but they cannot begin running until the Central Manager host is restarted.

6.4.1.3 Daemons in a Condor pool

A daemon is a program that runs in the background once started. To configure a Condor pool, the following Condor daemons need to be started. Figure 6.10 shows the interactions between Condor daemons.

condor_master
The *condor_master* daemon runs on each host in a Condor pool to keep all the other daemons running in the pool. It spawns daemons such as *condor_startd* and *condor_schedd*, and periodically checks if there are new binaries installed for any of these daemons. If so, the *condor_master* will restart the affected daemons. In addition, if any daemon crashes, the master will send an email to the administrator

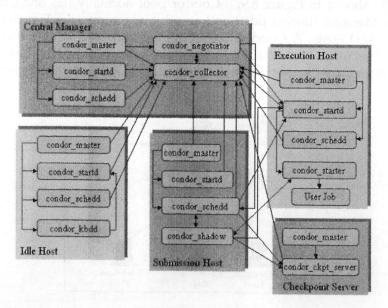

Figure 6.10 Daemons in a Condor pool

of the Condor pool and restart the daemon. The *condor_master* also supports various administrative commands, such as starting, stopping or reconfiguring daemons remotely.

condor_startd
The *condor_startd* daemon runs on each host in a Condor pool. It advertises information related to the node resources to the *condor_collector* daemons running on the Master host for matching pending resource requests. This daemon is also responsible for enforcing the policies that resource owners require, which determine under what conditions remote jobs will be started, suspended, resumed, vacated or killed. When the *condor_startd* is ready to execute a Condor job on an Execution host, it spawns the *condor_starter*.

condor_starter
The *condor_starter* daemon only runs on Execution hosts. It is the *condor_starter* that actually spawns a remote Condor job on a given host in a Condor pool. The *condor_starter* daemon sets up the execution environment and monitors the job once it is running. When a job completes, the *condor_starter* sends back job status information to the job Submission node and exits.

condor_schedd
The *condor_schedd* daemon running on each host in a Condor pool deals with resource requests. User jobs submitted to a node are stored in a local job queue managed by the *condor_schedd* daemon. Condor command-line tools such as *condor_submit*, *condor_q* or *condor_rm* interact with the *condor_schedd* daemon to allow users to submit a job into a job queue, and to view and manipulate the job queue. If the *condor_schedd* is down on a given machine, none of these commands will work.

The *condor_schedd* advertises the job requests with resource requirements in its local job queue to the *condor_collector* daemon running on the Master hosts. Once a job request from a *condor_schedd* on a Submission host has been matched with a given resource on an Execution host, the *condor_schedd* on the Submission host will spawn a *condor_shadow* daemon to serve that particular job request.

condor_shadow
The *condor_shadow* daemon only runs on Submission hosts in a Condor pool and acts as the resource manager for user job

submission requests. The *condor_shadow* daemon performs remote system calls allowing jobs submitted to Condor to be checkpointed. Any system call performed on a remote Execution host is sent over the network, back to the *condor_shadow* daemon on the Submission host, and the results are also sent back to the Submission host. In addition, the *condor_shadow* daemon is responsible for making decisions about a user job submission request, such as where checkpoint files should be stored or how certain files should be accessed.

condor_collector

The *condor_collector* daemon only runs on the Central Manager host. This daemon interacts with *condor_startd* and *condor_schedd* daemons running on other hosts to collect all the information about the status of a Condor pool such as job requests and resources available. The *condor_status* command can be used to query the *condor_collector* daemon for specific status information about a Condor pool.

condor_negotiator

The *condor_negotiator* daemon only runs on the Central Manager host and is responsible for matching a resource with a specific job request within a Condor pool. Periodically, the *condor_negotiator* daemon starts a negotiation cycle, where it queries the *condor_collector* daemon for the current state of all the resources available in the pool. It interacts with each *condor_schedd* daemon running on a Submission host that has resource requests in a priority order, and tries to match available resources with those requests. If a user with a higher priority has jobs that are waiting to run, and another user claims resources with a lower priority, the *condor_negotiator* daemon can preempt a resource and match it with the user job request with a higher priority.

condor_kbdd

The *condor_kbdd* daemon only runs on an Execution host installing Digital Unix or IRIX. On these platforms, the *condor_startd* daemon cannot determine console (keyboard or mouse) activity directly from the operating system. The *condor_kbdd* daemon connects to an X Server and periodically checks if there is any user activity. If so, the *condor_kbdd* daemon sends a command to the *condor_startd* daemon running on the same host. In this way, the *condor_startd* daemon knows the machine owner is using the machine again and it can perform whatever actions are necessary, given the policy it has been configured to enforce. Therefore, Condor can

6.4 A REVIEW OF CONDOR, SGE, PBS AND LSF

be used in a non-dedicated computing environment to scavenge idle computing resources.

condor_ckpt_server
The *condor_ckpt_server* daemon runs on a checkpoint server, which is an Execution host, to store and retrieve checkpointed files. If a checkpoint server in a Condor pool is down, Condor will revert to sending the checkpointed files for a given job back to the job Submission host.

6.4.1.4 Job life cycle in Condor

A job submitted to a Condor pool will go through the following steps as shown in Figure 6.11.

1. *Job submission*: A job is submitted by a Submission host with *condor_submit* command (Step 1).
2. *Job request advertising*: Once it receives a job request, the *condor_schedd* daemon on the Submission host advertises the request to the *condor_collector* daemon running on the Central Manager host (Step 2).
3. *Resource advertising*: Each *condor_startd* daemon running on an Execution host advertises resources available on the host to the

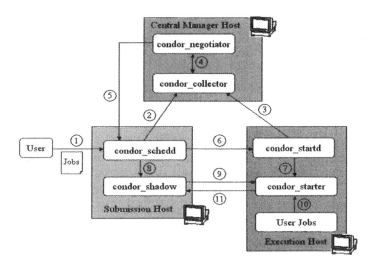

Figure 6.11 Job life cycle in Condor

condor_collector daemon running on the Central Manager host (Step 3).

4. *Resource matching*: The *condor_negotiator* daemon running on the Central Manager host periodically queries the *condor_collector* daemon (Step 4) to match a resource for a user job request. It then informs the *condor_schedd* daemon running on the Submission host of the matched Execution host (Step 5).

5. *Job execution*: The *condor_schedd* daemon running on the job Submission host interacts with the *condor_startd* daemon running on the matched Execution host (Step 6), which will spawn a *condor_starter* daemon (Step 7). The *condor_schedd* daemon on the Submission host spawns a *condor_shadow* daemon (Step 8) to interact with the *condor_starter* daemon for job execution (Step 9). The *condor_starter* daemon running on the matched Execution host receives a user job to execute (Step 10).

6. *Return output*: When a job is completed, the results will be sent back to the Submission host by the interaction between the *condor_shadow* daemon running on the Submission host and the *condor_starter* daemon running on the matched Execution host (Step 11).

6.4.1.5 Security management in Condor

Condor provides strong support for authentication, encryption, integrity assurance, as well as authorization. A Condor system administrator using configuration macros enables most of these security features.

When Condor is installed, there is no authentication, encryption, integrity checks or authorization checks in the default configuration settings. This allows newer versions of Condor with security features to work or interact with previous versions without security support. An administrator must modify the configuration settings to enable the security features.

Authorization
Authorization protects resource usage by granting or denying access requests made to the resources. It defines who is allowed to do what. Authorization is granted based on specified access levels, e.g. if you want to view the status of a Condor pool, you need

READ permission; if you want to submit a job, you need WRITE permission.

Authentication

Authentication provides an assurance of an identity. Through configuration macros, both a client and a daemon can specify whether authentication is required. For example, if the macro defined in the configuration file for a daemon is

$$SEC\_WRITE\_AUTHENTICATION = REQUIRED$$

then the daemon must authenticate the client for any communication that requires the WRITE access level. If the daemon's configuration contains

$$SEC\_DEFAULT\_AUTHENTICATION = REQUIRED$$

and does not contain any other security configuration for AUTHENTICATION, then this default configuration defines the daemon's needs for authentication over all access levels.

If no authentication methods are specified in the configuration, Condor uses a default authentication such as Globus GSI authentication with x.509 certificates, Kerberos authentication or file system authentication as we have discussed in Chapter 4.

Encryption

Encryption provides privacy support between two communicating parties. Through configuration macros, both a client and a daemon can specify whether encryption is required for further communication.

Integrity checks

An integrity check assures that the messages between communicating parties have not been tampered with. Any change, such as addition, modification or deletion, can be detected. Through configuration macros, both a client and a daemon can specify whether an integrity check is required of further communication.

6.4.1.6 Job management in Condor

Condor manages jobs in the following aspects.

Job

A Condor job is a work unit submitted to a Condor pool for execution.

Job types
Jobs that can be managed by Condor are executable sequential or parallel codes, using, for example, PVM or MPI. A job submission may involve a job that runs over a long period, a job that needs to run many times or a job that needs many machines to run in parallel.

Queue
Each Submission host has a job queue maintained by the *condor_schedd* daemon running on the host. A job in a queue can be removed and placed on hold.

Job status
A job can have one of the following status:

- *Idle*: There is no job activity.
- *Busy*: A job is busy running.
- *Suspended*: A job is currently suspended.
- *Vacating*: A job is currently checkpointing.
- *Killing*: A job is currently being killed.
- *Benchmarking*: The *condor_startd* is running benchmarks.

Job run-time environments
The Condor universe specifies a Condor execution environment. There are seven universes in Condor 6.6.3 as described below.

- The default universe is the *Standard Universe* (except where the configuration variable DEFAULT_UNIVERSE defines it otherwise), and tells Condor that this job has been re-linked via *condor_compile* with Condor libraries and therefore supports checkpointing and remote system calls.
- The *Vanilla Universe* is an execution environment for jobs which have not been linked with Condor libraries; and it is used to submit shell scripts to Condor.
- The *PVM Universe* is used for a parallel job written with PVM 3.4.
- The *Globus Universe* is intended to provide the standard Condor interface to users who wish to start Globus jobs from Condor. Each job queued in the job submission file is translated into the Globus Resource Specification Language (RSL) and subsequently submitted to Globus via the Globus GRAM protocol.

- The *MPI Universe* is used for MPI jobs written with the MPICH package.
- The *Java Universe* is used for programs written in Java.
- The Scheduler Universe allows a Condor job to be executed on the host where the job is submitted. The job does not need matchmaking for a host and it will never be preempted.

Job submission with a shared file system
If Vanilla, Java or MPI jobs are submitted without using the file transfer mechanism, Condor must use a shared file system to access input and output files. In this case, the job must be able to access the data files from any machine on which it could potentially run.

Job submission without a shared file system
Condor also works well without a shared file system. A user can use the file transfer mechanism in Condor when submitting jobs. Condor will transfer any files needed by a job from the host machine where the job is submitted into a temporary working directory on the machine where the job is to be executed. Condor executes the job and transfers output back to the Submission machine.

The user specifies which files to transfer, and at what point the output files should be copied back to the Submission host. This specification is done within the job's submission description file. The default behavior of the file transfer mechanism varies across the different Condor universes, which have been discussed above and it differs between UNIX and Windows systems.

Job priority
Job priorities allow the assignment of a priority level to each submitted Condor job in order to control the order of execution. The priority of a Condor job can be changed.

Chirp I/O
The Chirp I/O facility in Condor provides a sophisticated I/O functionality. It has two advantages over simple whole-file transfers.

- First, the use of input files is done at run time rather than submission time.
- Second, a part of a file can be transferred instead of transferring the whole file.

Job flow management

A Condor job can have many tasks of which each task is an executable code. Condor uses a Directed Acyclic Graph (DAG) to represent a set of tasks in a job submission, where the input/output, or execution of one or more tasks is dependent on one or more other tasks. The tasks are nodes (vertices) in the graph, and the edges (arcs) identify the dependencies of the tasks. Condor finds the Execution hosts for the execution of the tasks involved, but it does not schedule the tasks in terms of dependencies.

The Directed Acyclic Graph Manager (DAGMan) [7] is a meta-scheduler for Condor jobs. DAGMan submits jobs to Condor in an order represented by a DAG and processes the results. An input file is used to describe the dependencies of the tasks involved in the DAG, and each task in the DAG also has its own description file.

Job monitoring

Once submitted, the status of a Condor job can be monitored using *condor_q* command. In the case of DAG, the progress of the DAG can also be monitored by looking at the log file(s), or by using *condor_q–dag*.

Job recovery: The rescue DAG

DAGMan can help with the resubmission of uncompleted portions of a DAG when one or more nodes fail. If any node in the DAG fails, the remainder of the DAG is continued until no more forward progress can be made based on the DAG's dependencies. When a node in the DAG fails, DAGMan automatically produces a file called a Rescue DAG, which is a DAG input file whose functionality is the same as the original DAG file. The Rescue DAG file additionally contains indication of successfully completed nodes using the *DONE* option. If the DAG is re-submitted using this Rescue DAG input file, the nodes marked as completed will not be re-executed.

Job checkpointing mechanism

Checkpointing is normally used in a Condor job that needs a long time to complete. It takes a snapshot of the current state of a job in such a way that the job can be restarted from that checkpointed state at a later time.

Checkpointing gives the Condor scheduler the freedom to reconsider scheduling decisions through preemptive-resume scheduling. If the scheduler decides to no longer allocate a host to a job, e.g. when the owner of that host starts using the host, it can checkpoint the job and preempt it without losing the work the job has already accomplished. The job can be resumed later when the scheduler allocates it a new host. Additionally, periodic checkpointing provides fault tolerance in Condor.

Computing On Demand
Computing On Demand (COD) extends Condor's high throughput computing abilities to include a method for running short-term jobs on available resources immediately.

COD extends Condor's job management to include interactive, computation-intensive jobs, giving these jobs immediate access to the computing power they need over a relatively short period of time. COD provides computing power *on demand*, switching predefined resources from working on Condor jobs to working on the COD jobs. These COD jobs cannot use the batch scheduling functionality of Condor since the COD jobs require interactive response time.

Flocking
Flocking means that a Condor job submitted in a Condor pool can be executed in another Condor pool. Via configuration, the *condor_schedd* daemon running on Submission hosts can implement job flocking.

6.4.1.7 Resource management in Condor

Condor manages resources in a Condor pool in the following aspects.

Tracking resource usage
The *condor_startd* daemon on each host reports to the *condor_collector* daemon on the Central Manager host about the resources available on that host.

User priority
Condor hosts are allocated to users based upon a user's priority. A lower numerical value for user priority means higher priority, so a user with priority 5 will get more resources than a user with priority 50.

6.4.1.8 Job scheduling policies in Condor

Job scheduling in a Condor pool is not strictly based on a first-come-first-server selection policy. Rather, to keep large jobs from draining the pool of resources, Condor uses a unique up-down algorithm [8] that prioritizes jobs inversely to the number of cycles required to run the job. Condor supports the following policies in scheduling jobs.

- *First come first serve*: This is the default scheduling policy.
- *Preemptive scheduling*: Preemptive policy lets a pending high-priority job take resources away from a running job of lower priority.
- *Dedicated scheduling*: Dedicated scheduling means that jobs scheduled to dedicated resources cannot be preempted.

6.4.1.9 Resource matching in Condor

Resource matching [9] is used to match an Execution host to run a selected job or jobs. The *condor_collector* daemon running on the Central Manager host receives job request advertisements from the *condor_schedd* daemon running on a Submission host and resource availability advertisements from the *condor_startd* daemon running on an Execution host. A resource match is performed by the *condor_negotiator* daemon on the Central Manager host by selecting a resource based on job requirements. Both job request advertisements and resource advertisements are described in Condor Classified Advertisement (ClassAd) language, a mechanism for representing the characteristics and constraints of hosts and jobs in the Condor system.

A ClassAd is a set of uniquely named expressions. Each named expression is called an attribute. ClassAds use a semi-structured data model for resource descriptions. Thus, no specific schema is required by the matchmaker, allowing it to work naturally in a heterogeneous environment.

The ClassAd language includes a query language, allowing advertising agents such as the *condor_startd* and *condor_schedd* daemons to specify the constraints in matching resource offers and user job requests. Figure 6.12 shows an example of a ClassAd job request advertisement and a ClassAd resource advertisement.

6.4 A REVIEW OF CONDOR, SGE, PBS AND LSF

Job ClassAd	Host ClassAd
[MyType="job" TargetType="Machine" Requirements= ((other.Arch == "INTEL"&& other OpSys == "LINUX")&& Other.Disk>my.DiskUsage) Rank=(Memory*10000)+Kflops CMD="/home/eestmml/bin/test-exe Department="ECE" Owner="eestmml" DiskUsage=8000]	[MyType = "Machine" TargetType = "Job" Machine = "s140n209.brunel.ac.uk" Arch = "INTEL" OpSys = "LINUX" Disk = 35882 KeyboardIdle = 173 LoadAvg = 0.1000 Rank=other.Department == self.Department Requirements = TARGET.Owner == "eestmml" || LoadAvg<= 0.3 & & KeyboardIdle> 15*60]

Figure 6.12 Two ClassAd samples

These two ClassAds will be used by the *condor_negotiator* daemon running on the Central Manager host to check whether the host can be matched with the job requirements.

6.4.1.10 Condor support in Globus

Jobs can be submitted directly to a Condor pool from a Condor host, or via Globus (GT2 or earlier versions of Globus), as shown in Figure 6.13. The Globus host is configured with *Condor jobmanager* provided by Globus. When using a *Condor jobmanager*, jobs are submitted to the Globus resource, e.g. using *globus_job_run*. However, instead of forking the jobs on the local machine, jobs are re-submitted by Globus to Condor using the *condor_submit* tool.

6.4.1.11 Condor-G

Condor-G is a version of Condor that has the ability to maintain contact with a Globus gatekeeper, submitting and monitoring jobs to Globus (GT2 or earlier versions of Globus). Condor-G allows users to write familiar Condor job-submission scripts with a few changes and run them on Grid resources managed by Globus, as shown in Figure 6.14.

To use Condor-G, we do not need to install a Condor pool. Condor-G is only the job management part of Condor. Condor-G can be installed on just one machine within an organization and

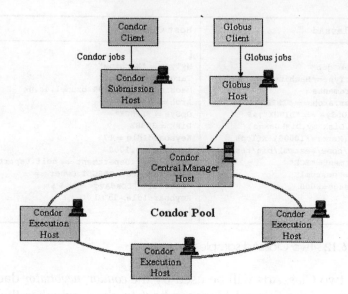

Figure 6.13 Submitting jobs to a Condor pool via Condor or Globus

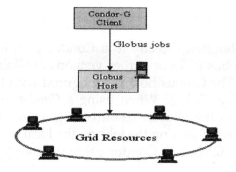

Figure 6.14 Submitting jobs to Globus via Condor-G

the access to remote Grid resources using a Globus interface can be done through it.

Submitting Globus jobs using Condor-G provides a much higher level of service than simply using *globus_job_run* command provided by Globus.

- First, jobs submitted to Globus with Condor-G enter a local Condor queue that can be effectively managed by Condor.

6.4 A REVIEW OF CONDOR, SGE, PBS AND LSF

- Secondly, jobs remain in the Condor queue until they are completed. Therefore, should the job crash while running remotely, Condor-G can re-submit it again without user intervention.

In a word, Condor-G provides a level of service guarantee that is not available with *globus_job_run* and other Globus commands.

Note: Condor-G does not have a GUI (the "G" is for Grid). However, the following graphic tools can be used with both Condor and Condor-G:

- *CondorView*: Shows a graphical history of the resources in a pool.
- *Condor UserLogViewer*: Shows a graphical history of a large set of jobs submitted to Condor or Condor-G.

6.4.2 Sun Grid Engine

The SGE is a distributed resource management and scheduling system from Sun Microsystems that can be used to optimize the utilization of software and hardware resources in a UNIX-based computing environment. The SGE can be used to find a pool of idle resources and harnesses these resources; also it can be used for normal activities, such as managing and scheduling jobs onto the available resources. The latest version of SGE is Sun N1 Grid Engine (N1GE) version 6 (see Table 6.2). In this section, we focus on SGE 5.3 Standard Edition because it is freely downloadable.

6.4.2.1 The SGE architecture

Hosts (machines or nodes) in SGE are classified into four categories – master, submission, execution, administration and shadow. Figure 6.15 shows the SGE architecture.

- *Master host*: A single host is selected to be the SGE master host. This host handles all requests from users, makes job-scheduling decisions and dispatches jobs to execution hosts.
- *Submit host*: Submit hosts are machines configured to submit, monitor and administer jobs, and to manage the entire cluster.
- *Execution host*: Execution hosts have the permission to run SGE jobs.

Table 6.2 A note of the differences between N1 Grid Engine and Sun Grid Engine

N1GE 6 differs from the version 6 of Sun Grid Engine Open Source builds in the following aspects:

- Sun support only available for N1GE (for allmajor UNIX platforms and Windows soon).
- Accounting and Reporting Console (database for storing of accounting, metering and statistics data, plus a Web UI forqueries and generating reports).
- Binaries for MS Windows execution/submit functionality (to be delivered in the second half of 2005).
- Grid Engine Management Model for 1-click deployment of execution hosts on an arbitrary number of hosts (to be delivered in the second quarter of 2005).

The basic software components underneath N1GE and SGE are identical. In fact, the open-source project is the development platform for those components. Proprietary Sun code only exists for the differentiators listed above (where applicable). Note that some of those differentiators use other Sun products or technologies, which are not open source themselves.

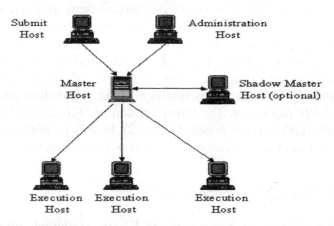

Figure 6.15 The architecture of the SGE

- *Administration host*: SGE administrators use administration hosts to make changes to the cluster's configuration, such as changing distributed resource management parameters, configuring new nodes or adding or changing users.
- *Shadow master host*: While there is only one master host, other machines in the cluster can be designated as shadow master hosts to provide greater availability. A shadow master host continually monitors the master host, and automatically and

6.4 A REVIEW OF CONDOR, SGE, PBS AND LSF

transparently assumes control in the event that the master host fails. Jobs already in the cluster are not affected by a master host failure.

6.4.2.2 Daemons in an SGE cluster

As shown in Figure 6.16, to configure an SGE cluster, the following daemons need to be started.

sge_qmaster – The Master daemon
The *sge_qmaster* daemon is the centre of the cluster's management and scheduling activities; it maintains tables about hosts, queues, jobs, system load and user permissions. It receives scheduling decisions from *sge_schedd* daemon and requests actions from *sge_execd* daemon on the appropriate execution host(s). The *sge_qmaster* daemon runs on the Master host.

sge_schedd – The Scheduler daemon
The *sge_schedd* is a scheduling daemon that maintains an up-to-date view of the cluster's status with the help of *sge_qmaster* daemon. It makes the scheduling decision about which job(s) are dispatched to which queue(s). It then forwards these decisions to

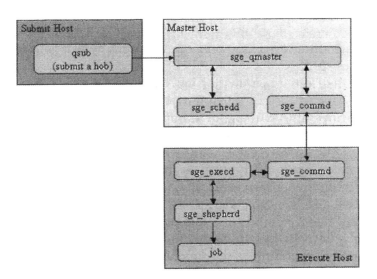

Figure 6.16 Daemons in SGE

the *sge_qmaster* daemon, which initiates the requisite actions. The *sge_schedd* daemon also runs on the Master host.

sge_execd – The Execution daemon
The *sge_execd* daemon is responsible for the queue(s) on its host and for the execution of jobs in these queues by starting *sge_shepherd* daemons. Periodically, it forwards information such as job status or load on its host, to the *sge_qmaster* daemon. The *sge_execd* daemon runs on an Execute host.

sge_commd – The Communication daemon
The *sge_commd* daemon communicates over a well-known TCP port and is used for all communication among SGE components. The *sge_commd* daemon runs on each Execute host and the Master host in an SGE cluster.

sge_shepherd – The Job Control daemon
Started by the *sge_execd* daemon, the *sge_shepherd* daemon runs for each job being actually executed on a host. The *sge_shepherd* daemon controls the job's process hierarchy and collects accounting data after the job has completed.

6.4.2.3 Job management in SGE

SGE supports four job types – batch, interactive, parallel and array. The first three have obvious meanings, the fourth type – array job – is where a single job can be replicated a specified number of times, each differing only by its input data set, which is useful for parameter studies.

Submitted jobs are put into job queues. An SGE queue is a container for a class of jobs allowed to execute on a particular host concurrently. A queue determines certain job attributes; for example, whether it may be migrated. Throughout their lifetimes, running jobs are associated with their queues. Association with a queue affects some of the actions that can happen to a job. For example, if a queue is suspended, all the jobs associated with that queue will also be suspended.

In SGE, there is no need to submit jobs directly to a queue. A user only needs to specify the requirement profile of the job (such as memory, operating system and available software) and SGE will dispatch the job to a suitable queue on a lightly loaded host automatically. If a job is submitted to a particular queue, the job

6.4 A REVIEW OF CONDOR, SGE, PBS AND LSF

will be bound to this queue and to its host, and thus SGE daemons will be unable to select a lightly loaded or better-suited resource.

6.4.2.4 Job run-time environments in SGE

SGE supports three execution modes – batch, interactive and parallel. Batch mode is used to run straightforward sequential programs. In interactive mode, users are given shell access (command line) to some suitable host via, for example, X-windows. In a parallel mode, parallel programs using the likes of MPI and PVM are supported.

6.4.2.5 Job selection and resource matching in SGE

Jobs submitted to the Master host in an SGE cluster are held in a spooling area until the scheduler determines that the job is ready to run. SGE matches the available resources to a job's requirements; for example, matching the available memory, CPU speed and available software licences, which are periodically collected by Execution hosts. The requirements of the jobs may be very different and only certain hosts may be able to provide the corresponding services. Once a resource becomes available for execution of a new job, SGE dispatches the job with the highest priority and matching requirements.

Fundamentally, SGE uses two sets of criteria to schedule jobs – job priorities and equal share.

Job priorities
This criterion concerns the order of the scheduling of different jobs, a *first-in-first-out* (FIFO) rule is applied by default. All *pending* (not yet scheduled) jobs are inserted in a list, with the first submitted job being at the head of the list, followed by the second submitted job, and so on. SGE will attempt to schedule the FIFO queue of jobs. If at least one suitable queue is available, the job will be scheduled. SGE will try to schedule the second job afterwards no matter whether the first has been dispatched or not.

The cluster administrator via a priority value being assigned to a job may overrule this order of precedence among the pending jobs. The actual priority value can be displayed by using the *qstat* command (the priority value is contained in the last column of the pending jobs display titled "P"). The default priority value that is assigned to a job at submission time is 0. The priority values are positive and negative integers and the pending job list is sorted correspondingly in the order of descending priority values. By assigning a relatively high-priority value to a job, it is moved to the top of the pending list. A job will be given a negative priority value after the job is just submitted. If there are several jobs with the same priority value, the FIFO rule is applied to these jobs.

Equal-share scheduling
The FIFO rule sometimes leads to problems, especially when users tend to submit a series of jobs at almost the same time (e.g. via a shell script issuing a series of job submissions). All the jobs that are submitted in this case will be designated to the same group of queues and will have to potentially wait a very long time before executing. *equal-share scheduling* avoids this problem by sorting the jobs of a user already owning an executing job to the end of the precedence list. The sorting is performed only among jobs within the same priority value category. Equal-share scheduling is activated if the SGE scheduler configuration entry *user_sort* switch is set to TRUE.

Jobs can be directly submitted to an SGE cluster on a Submit host or via Globus, as shown in Figure 6.17. It should be noted that the newer version of N1GE 6 has more sophisticated scheduling criteria than those mentioned above. There is "old" Enterprise Edition policy system [10], and in N1GE 6, there is an urgency control scheme combined with resource reservation [11]. Also, the Equal-share scheduling mentioned above has been replaced in N1GE 6 by a combination of other more advanced scheduling facilities.

6.4.3 The Portable Batch System (PBS)

The PBS is a resource management and scheduling system. It accepts batch jobs (shell scripts with control attributes), preserves and protects the job until it runs; it executes the job, and delivers the output back to the submitter. A batch job is a program that executes on the backend computing environment without further user interaction.

6.4 A REVIEW OF CONDOR, SGE, PBS AND LSF

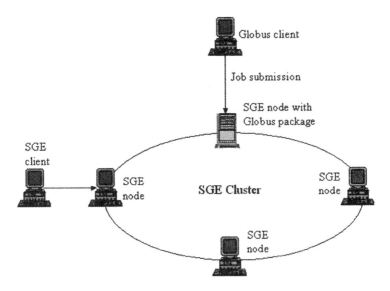

Figure 6.17 Submitting jobs to an N1GE cluster via N1GE or Globus

PBS may be installed and configured to support jobs executing on a single computer or on a cluster-based computing environment. PBS is capable of allowing its nodes to be grouped into many configurations. OpenPBS is available for free source download; however, there is no support for OpenPBS, nor has it been improved in a number of years. PBS Pro [12] is fully supported and is undergoing continuous development and improvement.

6.4.3.1 The PBS architecture

As Figure 6.18 shows, PBS uses a Master host and an arbitrary number of Execution and job Submission hosts. The Master host is the central manager of a PBS cluster; a host can be configured as a Master host and an Execution host.

6.4.3.2 Daemons in a PBS cluster

As shown in Figure 6.19, to configure a PBS cluster, the following daemons need to be started.

- *pbs_server*: The *pbs_server* daemon only runs on the PBS Master host (server). Its main function is to provide the basic batch

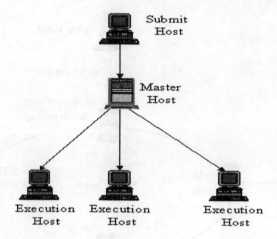

Figure 6.18 The architecture of a PBS cluster

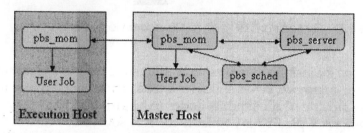

Figure 6.19 The daemons in a PBS cluster

services, such as receiving/creating a batch job, modifying a job, protecting the job against system crashes and executing the job.
- *pbs_mom*: The *pbs_mom* daemon runs on each host and is used to start, monitor and terminate jobs, under instruction from the *pbs_server* daemon.
- *pbs_sched*: The *pbs_sched* daemon runs on the Master host and determines when and where to run jobs. It requests job state information from *pbs_server* daemon and resource state information from *pbs_mom* daemon, and then makes decisions for scheduling jobs.

6.4.3.3 Job selection in PBS

Jobs submitted to PBS are put in job queues. Jobs can be sequential or parallel codes using MPI. A server can manage one or more

6.4 A REVIEW OF CONDOR, SGE, PBS AND LSF

queues; a batch queue consists of a collection of zero or more jobs and a set of queue attributes. Jobs reside in the queue or are members of the queue. In spite of the name, jobs residing in a queue need not be ordered with FIFO. Access to a queue is limited to the server that owns the queue. All clients gain information about a queue or jobs within a queue through batch requests to the server. Two main queue types are defined: routing and execution queues. When a job resides in an execution queue, it is a candidate for execution. A job being executed is still a member of the execution queue from which it is selected. When a job resides in a routing queue, it is a candidate for routing to a new destination. Each routing queue has a list of destinations to which jobs may be routed. The new destination may be a different queue within the same server or a queue under a different server. A job submitted to PBS can

- be batch or interactive
- define a list of required resources such as CPU, RAM, hostname, number of nodes or any of site-defined resources
- define a priority
- define the time of execution
- send a mail to a user when execution starts, ends or aborts
- define dependencies (such as *after, afterOk afterNotOk or before*)
- be checkpointed (if the host OS has provision for it)
- be suspended and later resumed.

Jobs can be directly submitted to a PBS cluster or via Globus, as shown in Figure 6.20.

6.4.3.4 Resource matching in PBS

In PBS, resources can be identified either explicitly through a job control language, or implicitly by submitting the job to a particular queue that is associated with a set of resources. Once a suitable resource is identified, a job can be dispatched for execution. PBS clients have to identify a specific queue to submit to in advance, which then fixes the set of resources that may be used; this hinders further dynamic and qualitative resource discovery. Furthermore,

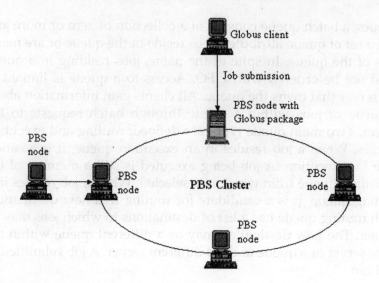

Figure 6.20 Jobs can be submitted to or from a PBS cluster to Globus

system administrators have to anticipate the services that will be requested by clients and set up queues to provide these services. Additional PBS Pro services include:

- *Cycle harvesting*: PBS Pro can run jobs on idle workstations and suspend or re-queue the jobs when the workstation becomes used, based on either load average or keyboard/mouse input.
- *Site-defined resources*: A site can define one or more resources which can be requested by jobs. If the resource is "consumable", it can be tracked at the server, queue and/or node level.
- *"Peer to Peer" scheduling*: A site can have multiple PBS Pro clusters (each cluster has its server, scheduler and one or more execution systems). A scheduler in any given cluster can be configured to move jobs from other clusters to its cluster when the resources required by the job are available locally.
- *Advance reservations*: Resources, such as nodes or CPUs, can be reserved in advance with a specified start and end time/date. Jobs can be submitted against the reservation and run in the time period specified. This ensures the required computational resources are available when time-critical work must be performed.

6.4.4 LSF

The LSF is a resource management and workload scheduling system from Platform Computing Corporation. LSF utilizes computing resources that include desktops, servers and mainframes to ensure policy-driven, prioritized service levels for access to resources. In this section we discuss LSF version 6 (v6).

6.4.4.1 LSF platforms

LSF v6 supports a variety of computer architectures and operating systems, including HP, IBM, Intel, SGI, Sun and NEC.

6.4.4.2 The architecture of an LSF cluster

An LSF cluster has one Master host and an arbitrary number of Execution hosts as shown in Figure 6.21. The Master host acts as the overall coordinator of the cluster. It is responsible for job scheduling and dispatching. If the Master host fails, another LSF server in the cluster becomes the Master host. An Execution host is used to run jobs. A Submission host is responsible for submitting jobs to an LSF cluster. In addition, a Submission host can also be an Execution host. A Server host is capable of job submission and execution.

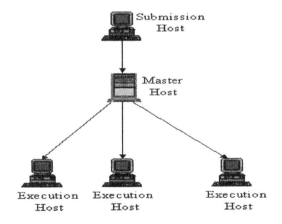

Figure 6.21 The architecture of an LSF cluster

6.4.4.3 LSF daemons

Table 6.3 shows the daemons used in LSF.

6.4.4.4 Job life cycle in LSF

Figure 6.22 shows their interaction within an LSF cluster. As shown in Figure 6.22, a job submitted to an LSF cluster will go through the following steps.

Table 6.3 LSF daemons

Daemon	Function
lim	The load information manager (*lim*) daemon runs on the Master host and each Execution host. It collects host state information and forwards it to the master *lim* daemon on the Master host.
res	The resource execution server (*res*) daemon runs on the Master host and each Execution host to provide transparent and secure remote execution of jobs.
mbatchd	The master batch daemon runs on the Master host. It is responsible for the overall state of jobs in an LSF cluster. Started by *sbatchd*, the *mbatchd* daemon receives job submissions and information query requests, and manages jobs held in queues. It also dispatches jobs to hosts as determined by *mbschd*.
mbschd	The master batch scheduler daemon runs on the Master host. It starts and works with *mbatchd*. This daemon makes scheduling decisions based on job requirements and policies.
sbatchd	The slave batch daemon runs on the Master host and each Execution host. It receives the request to run the job from *mbatchd* daemon and manages the local execution of the job. It is responsible for enforcing local policies and maintaining the state of jobs on the host. The *sbatchd* daemon forks a child *sbatchd* for every job. The child *sbatchd* runs an instance of *res* to create the execution environment in which the job runs. The child *sbatchd* exits when the job is complete.
pim	The process information manager (*pim*) daemon, started by *lim*, running on the Master host and on each Execution host, is used to collect statistics about job processes executing on a host. The statistics may be, for example, how much CPU time or the amount of memory being used; this is reported back to the *sbatchd* daemon.

6.4 A REVIEW OF CONDOR, SGE, PBS AND LSF

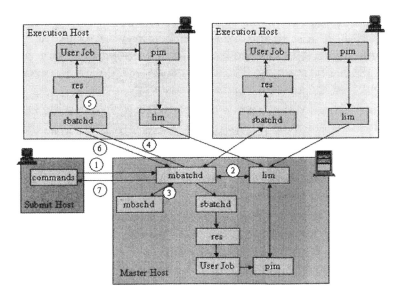

Figure 6.22 The interaction between LSF daemons

1. *Job submission*: A job is submitted from a Submission host with *bsub* command to the Master host in an LSF cluster (Step 1).
2. *Resource collection*: The *lim* daemon on the Master host collects resource information from other *lim* daemons at present time intervals (Step 2). Then the master *lim* daemon communicates with the *mbatchd* daemon, which in turn communicates with the *mbschd* daemon to support scheduling decisions.
3. *Job scheduling*: The *mbatchd* daemon on the Master host looks at jobs in the queue and sends the jobs to the *mbschd* daemon on the Master host for scheduling at a set time interval (Step 3). The *mbschd* daemon evaluates jobs and makes scheduling decisions based on job priority, scheduling policies and available resources. The *mbschd* daemon selects the most appropriate host to execute a job and sends its decisions back to the *mbatchd* daemon.
4. *Job dispatching*: Upon receiving scheduling decisions, the *mbatchd* daemon immediately dispatches the jobs to hosts.
5. *Job execution*: The *sbatchd* daemon on an Execution host handles the job execution. It receives the request from the *mbatchd* daemon (Step 4), creates a child *sbatchd* process for the job, creates the execution environment and starts the job using the *res*

daemon (Step 5). The execution environment is copied from the Submission host to the Execution host and includes the environment variables needed by the job's working directory where the job begins running. The job runs under the user account that submitted the job and has the status RUN.

6. *Return output*: When a job is completed, it is assigned the DONE status if it has been completed without any problems. The job is assigned the EXIT status if errors had prevented it from finishing. The *sbatchd* daemon on the Execution host sends job information including errors and output to the *mbatchd* daemon on the Master host (Step 6).

7. *Send email to client*: The *mbatchd* daemon on the Master host returns the job output, job error and job information to the Submission host through email (Step 7). A job report is sent by email to the LSF client and includes job information such as CPU use, memory use, name of the account that submitted the job, job output and errors.

6.4.4.5 Security management in LSF

LSF provides secure access to local and remote data files. It uses standard 56-bit (and optional 128-bit) encryption for authentication. LSF provides three levels of authentications – user level, host level and daemon level.

- *User level authentication*: LSF recognizes UNIX and Windows authentication environments, including different Windows domains and individual workgroup hosts.
- *Host authentication*: Upon receiving a job request, LSF first determines the user's identity. Once the user is identified, LSF decides whether it can trust the host from which the request is submitted. Users must have valid accounts on all hosts. This allows any user to run a job with their own permissions on any host in the cluster. Remote execution requests and batch job submissions are rejected if they come from a host not in the LSF cluster. A site can configure an external executable to perform additional user or host authorization.
- *Daemon authentication*: LSF can use the *eauth* program to authenticate the communications between daemons, e.g. *mbatchd* requests to *sbatchd*, *sbatchd* updates to *mbatchd*.

6.4.4.6 Job management in LSF

LSF provides the following support for job management.

- *Job and job slot*: An LSF job is a unit of work submitted for execution. A job slot is a sink into which a single unit of work is assigned. Hosts are configured to have a number of job slots available and queues dispatch jobs to fill job slots.
- *Queue*: A queue is a cluster-level container for Execution hosts. Queues do not correspond to individual hosts; each queue can use all nodes in the cluster or a configured subset of the nodes. Queues implement different job scheduling and control policies. LSF can have multiple queues. LSF can automatically choose a suitable queue from a list of candidate default queues for a job submitted without specifying a queue name.
- *Preemptive and preemptable queues*: Jobs in a preemptive queue can preempt jobs in any queue of lower priority, even if the low-priority queues are not specified as preemptable. Jobs in a preemptable queue can be preempted by jobs from any queue of a higher priority, even if the high-priority queues are not specified as preemptive.
- *Job types*: LSF supports batch and interactive jobs, which can be sequential or parallel, using MPI or PVM.
- *Job deadline constraint*: Deadline constraints will suspend or terminate running jobs at a certain time. A deadline constraint can be specified at the queue level to suspend a running job, or can be specified at the job level to terminate a running job.
- *Job re-queue*: LSF provides a way to automatically recover from temporary errors in executing jobs. Users can configure certain exit values, and in that case the job will be automatically re-queued. It is also possible for users to configure the queue such that a re-queued job will not be scheduled to hosts on which the job had previously failed to run.
- *Job checkpointing and restarting*: Checkpointing a job involves capturing the state of an executing job, and the data necessary to restart the job. In LSF, checkpointing can be configured at kernel level, user level and application level.
  - *Kernel-level checkpointing* is provided by the operating system and can be applied to arbitrary jobs running on the system. This approach is transparent to the application; there are no

source code changes and no need to re-link the application with checkpoint libraries.
- *User-level checkpointing* requires that the application object file (.o) be re-linked with a set of LSF libraries.
- *Application-level checkpointing* requires that checkpoint and restart routines are embedded in the source code.

LSF can restart a checkpointed job on a host other than the original execution host using the information saved in the checkpoint file to recreate the execution environment.

- *Job grouping and chunking*: A collection of jobs can be organized into a job group, which is a container for jobs, e.g. a payroll application may have one group of jobs that calculates weekly payments, another job group for calculating monthly salaries, and a third job group that handles the salaries of part-time or contract employees. Users can submit, view and control jobs according to their groups rather than looking at individual jobs. LSF supports *job chunking*, where jobs with similar resource requirements submitted by the same user are grouped together for dispatch.
- *Job submission with no shared file system*: When shared file space is not available, the *bsub -f* command copies the needed files to the Execution host before running the job, and copies result files back to the Submission host after the job completes.
- *LSF MultiCluster*: LSF MultiCluster extends an organization's reach to share virtualized resources beyond a single LSF cluster to span geographical locations. With LSF MultiCluster, local ownership and control is maintained, ensuring priority access to any local cluster while providing global access across an enterprise Grid. Organizations using LSF MultiCluster complete workload processing faster with increased computing power, enhancing productivity and speed.

6.4.4.7 Resource management in LSF

LSF provides the following support for resource management.

- *Tracking resource usage*: LSF tracks resource availability and usage; Jobs are scheduled according to the available resources.

Resources executing jobs are monitored by LSF; information, such as total CPU time consumed by all processes in the job, total resident memory usage in KB of all currently running processes in a job and currently active processes in a job, is collected.

- *Load indices and threshold*: Load indices are used to measure the availability of dynamic, non-shared resources on hosts in an LSF cluster. LSF also supports load threshold. Two load thresholds can be configured; each specifies a load index value as follows:
  - *loadSched* determines the load condition for dispatching pending jobs. If a host's load is beyond any defined *loadSched* value, a job will not be started on the host. This threshold is also used as the condition for resuming suspended jobs.
  - *loadStop* determines when the running jobs should be suspended.
- *Resource reservation*: When submitting a job, resource reservation requirements can be specified as part of the resource requirements, or they can be configured into the queue level resource requirements.
- *Memory reservation for pending jobs*: In LSF, resources are not reserved for pending jobs, therefore, some memory-intensive jobs could be pending indefinitely because smaller jobs take the resources immediately before the larger jobs can start running. Memory reservation solves this problem by reserving memory as it becomes available, until the total required memory specified is accumulated and the job can start.
- *Advanced resource reservation*: Advance reservations ensure access to specific hosts during specified times. Each reservation consists of the number of processors to reserve, a list of hosts for the reservation, a start time, an end time and an owner.

6.4.4.8 Job scheduling policies in LSF

Job scheduling is not strictly based on a first-come-first-serve selection policy. Rather, each queue has a priority number; LSF tries to start jobs in the highest priority queue first. The LSF administrator sets the queue priority when the queue is configured. LSF has the job schedule policies shown in Table 6.4.

Table 6.4 LSF scheduling polices

Policy	Features
First come first serve	The default scheduling policy.
Fair share scheduling	The fair share policy divides the processing resources of a cluster among users and groups to provide fair access to resources. A fair share policy can be configured at either the queue level or the host level.
Preemptive scheduling	A preemptive policy lets a pending high-priority job take resources away from a executing job of lower priority.
Deadline constraint scheduling	Deadline constraints will suspend or terminate running jobs at a certain time. There are two kinds of deadline constraints.
Exclusive scheduling	Exclusive policy gives a job exclusive use of the host that it executes on. LSF dispatches the job to a host that has no other jobs executing, and does not place any more jobs on the host until the exclusive job is finished.
Job dependency scheduling	Specifies that the running of a job is dependent on the completion of another job, which can be specified in the job submission.
Goal-oriented SLA scheduling	It helps users configure workload so that user jobs are completed on time and reduce the risk of missed deadlines. A Service-Level Agreement (SLA) defines how a service is delivered and the parameters for the delivery of a service.

6.4.4.9 Resource matching in LSF

Resource matching is used to match a host and its resources to execute a selected job or jobs. Each job has its resource requirements. Each time LSF attempts to dispatch a job, it checks to see which hosts are eligible to run the job. Hosts that match the resource requirements are the candidate hosts. When LSF dispatches a job, it uses the load index values of all the candidate hosts. The load values of each host are compared with the scheduling conditions. Jobs are only dispatched to a host if all the load values of the host are within the scheduling thresholds. If a job is queued and there is an eligible host for that job, the job is placed on that host. If more

6.4 A REVIEW OF CONDOR, SGE, PBS AND LSF

than one host is eligible, the job is started on the best host based on both the job and the queue resource requirements.

6.4.4.10 LSF support in Globus

Jobs can be submitted directly to a cluster from an LSF host, or via Globus (GT2 or earlier versions of Globus), as shown in Figure 6.23. The Globus host is configured with LSF *jobmanager* provided by Globus. When using an *LSF jobmanager*, jobs are submitted to the Globus resource, e.g. using *globus_job_run*. However, instead of forking the jobs on the local machine, jobs are re-submitted by Globus to LSF using the *bsub* tool.

6.4.4.11 Platform Globus Toolkit 3.0

Platform Globus Toolkit 3.0 (PGT3), as shown in Figure 6.24, is a commercial distribution of the Globus Toolkit 3.0 (GT3). In addition to the core services available in GT3, PGT3 includes extensions based upon the Community Scheduler Framework (CSF), an open-source implementation of a number of Grid services built on the GT3. CSF provides components for implementing meta-schedulers.

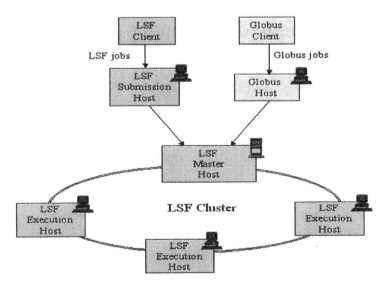

Figure 6.23 Submitting jobs to an LSF cluster via LSF or Globus

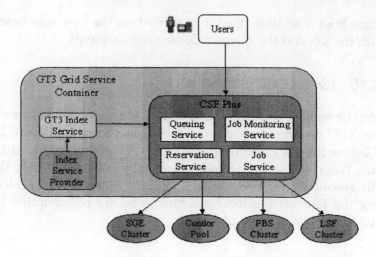

Figure 6.24 The PGT3 architecture

PGT3 provides transparent interactivity and interoperability among various resource management systems such as SGE, Condor, LSF and PBS. It allows an LSF client to submit jobs to an SGE-managed cluster. The core component of PGT3 is CSF Plus, which provides a meta-scheduler infrastructure that allows end users to interact with underlying resource management systems.

6.4.5 A comparison of Condor, SGE, PBS and LSF

Condor, SGE, PBS and LSF are single administrative domain resource management and scheduling systems. These systems can be interconnected across administrative boundaries, using the Globus toolkit, specifically GT2. While they differ in their implementation and functionality, they all share the following features.

- They are master–worker-based scheduling systems;
- There is one master host (central manager) per system;
- A system can have an arbitrary number of worker machines that can be used for job submission, job execution or both;
- Scheduling is centralized;
- Priority-based job scheduling;

6.4 A REVIEW OF CONDOR, SGE, PBS AND LSF

- Support batch jobs;
- Support a variety of platforms;
- Authentication and authorization support.

However, the four systems differ in many ways. Table 6.5 gives a comparison of the four systems in terms of availability, usability, jobs supported, work-load management, fault tolerance and accounting.

- *Availability*: Condor and SGE 5.3 can be freely downloaded. However, PBS Pro and LSF are commercial products.
- *Windows platform support*: PBS Pro and LSF fully support Windows platform. Condor partially supports Windows platform. SGE currently does not support Windows.
- *GUI support*: The use of Condor is based on command-line interfaces. However, the following graphic tools can be used with Condor:
  - *Condor view*: Shows a graphical history of the resources in a pool.

Table 6.5 A comparison of Condor, SGE, PBS and LSF

	Condor 6.6.3	SGE 5.3	PBS Pro 5.4	LSF 6.0
Availability	Freely downloadable	Freely downloadable	Commercial	Commercial
Windows platform support	Partial	No	Yes	Yes
GUI support	No	Yes	Yes	Yes
Jobs supported				
Batch jobs	Yes	Yes	Yes	Yes
Interactive jobs	Yes	Yes	Yes	Yes
Parallel jobs	Yes	Yes	Yes	Yes
Resource reservation	No	Yes	Yes	Yes
Job checkpointing	Yes	Yes	Yes	Yes
Job flocking	Yes	No	Yes	Yes
Job scheduling				
Preemptive scheduling	Yes	Yes	Yes	Yes
Deadline scheduling	No	Yes	No	Yes
Resource matching	Yes	Yes	Yes	Yes
Job flow management	Yes	Yes	Yes	Yes

– *Condor UserLogViewer*: Shows a graphical history of a large set of jobs submitted to Condor or Condor-G.

All the other systems have their own GUIs.

- *Jobs supported*: The four systems all support batch and parallel jobs using MPI and PVM. Only Condor does not support interactive jobs, the other three systems support interactive jobs.
- *Resource reservation*: Apart from Condor, all the other three systems support resource reservations.
- *Job checkpointing*: All the four systems support job checkpointing and fault recovery.
- *Job flocking*: Condor, PBS and LSF currently support job flocking, by which a job submitted to one cluster can be executed in another cluster.
- *Job scheduling*: All the four systems support preemptive scheduling. SGE and LSF support deadline constraint scheduling, but Condor and PBS do not support this feature.
- *Resource matching*: Each of the four systems has its own mechanism to match resources based on the resource requirements of a job submission and resources available.
- *Job flow management*: The four systems support inter-job dependency descriptions for complex applications.

6.5 GRID SCHEDULING WITH QOS

In Section 6.4, we have reviewed Condor, SGE, PBS and LSF. One major problem with the four systems is their lack of QoS support in scheduling jobs; such a system should take the following issues into account when scheduling jobs:

- Job characteristics
- Market-based scheduling model
- Planning in scheduling
- Rescheduling
- Scheduling optimization
- Performance prediction.

6.5.1 AppLeS

AppLeS [13] is an adaptive application-level scheduling system that can be applied to the Grid. Each application submitted to the Grid can have its own AppLeS. The design philosophy of AppLeS is that all aspects of system performance and utilization are experienced from the perspective of an application using the system. To achieve application performance, AppLeS measures the performance of the application on a specific site resource and utilizes this information to make resource selection and scheduling decisions. Figure 6.25 shows the architecture of AppLeS.

The AppLeS components are:

- *Network Weather Service* (NWS) [14]: Dynamic gathering of information of system state and forecasting of resource loads.
- *User specifications*: This is the information about user criterion for aspects such as performance, execution constraints and specific request for implementation.
- *Model*: This is a repository of default models, populated by similar classes of applications and specific applications that can be used for performance estimation, planning and resource selection.

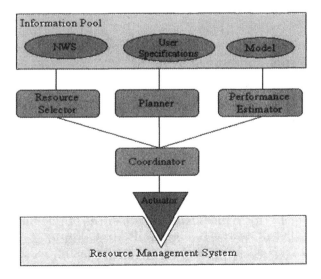

Figure 6.25 The architecture of AppLeS

- *Resource selector*: Choose and filter different resource combinations.
- *Planner*: This component is used to generate a description of a resource-dependent schedule from a given resource combination.
- *Performance estimator*: This component is used to generate an estimate for candidate schedules according to the user's performance metric.
- *Coordinator*: This component chooses the "best" schedule.
- *Actuator*: This component implements the "best" schedule on the target resource management system.

When AppLeS is used, the following steps are performed.

1. The user provides information to AppLeS via a Heterogeneous Application Template (HAT) and user specifications. The HAT provides information for the structure, characteristics and implementation of an application and its tasks.
2. The coordinator uses this information to filter out infeasible/possibly bad schedules.
3. The resource selector identifies promising sets of resources, and prioritizes them based on the logical "distance" between resources.
4. The planner computes a potential schedule for each viable resource configuration.
5. The performance estimator evaluates each schedule in terms of the user's performance objective.
6. The coordinator chooses the best schedule and then implements it with the actuator.

AppLeS differs from other scheduling systems in that the resource selection and scheduling decisions are based on the specific needs and exhibited performance characteristics of an application. AppLeS targets parallel master–slave applications. Condor, SGE, PBS and LSF do not take the application-level attributes into account when scheduling a job. Note that AppLeS is not a resource management system, it is a Grid application-level scheduling system.

6.5.2 Scheduling in GrADS

AppLeS focuses on per-job scheduling. Each application has its own AppLeS. When scheduling a job, AppLeS assumes that there is only one job to use the resources. One problem with AppLeS is that it does not have resource managers that can negotiate with applications to balance their interests. The absence of these negotiating mechanisms in the Grid can lead to variety of problems in which focus will be on the improvement of the performance of individual AppLeS. However, there will be many AppLeS agents in a system simultaneously, each working on behalf of its own application. A worst-case scenario is that all of the AppLeS agents may identify the same resources as "best" for their applications and seek to use them simultaneously. Recognizing that the targeted resources are no longer optimal or available, they all might seek to reschedule their applications on another resource. In this way, multiple unconstrained AppLeS might exhibit "thrashing" behaviour and achieve good performance neither for their own applications nor from the system's perspective. This is called the Bushel of AppLeS Problem.

The Grid Application Development Software (GrADS) project [15] seeks to provide a comprehensive programming environment that explicitly incorporates application characteristics and requirements in application design decisions. The goal of this project is to provide an integrated Grid application development solution that incorporates activities such as compilation, scheduling, staging of binaries and data, application launching, and monitoring during execution. The meta-scheduler in GrADS receives candidate schedules of different application-level schedulers and implements scheduling policies for balancing the interests of different applications as shown in Figure 6.26.

6.5.3 Nimrod/G

Nimrod/G [16] is a Grid-enabled resource management and scheduling system that supports deadline and economy-based computations for parameter sweep applications. It supports a simple *declarative* parametric modelling language for expressing parametric experiments. The domain experts (application experts or users) can create a *plan* for parameter studies and use the Nimrod/G broker to handle all the issues related to management issues

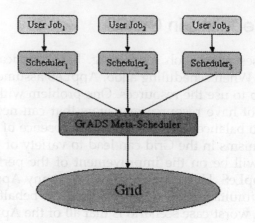

Figure 6.26 Job scheduling in GrADS

including resource discovery, mapping of jobs to the appropriate resources, data and code staging and gathering results from multiple nodes. Figure 6.27 shows the main components in Nimrod/G.

- *Nimrod/G client*: This component acts as a user-interface for controlling and supervising the experiment under consideration. The user can vary parameters related to time and cost that influence the decisions the scheduler takes while selecting resources. It also serves as a monitoring console and will list the status of all jobs, which a user can view and control.

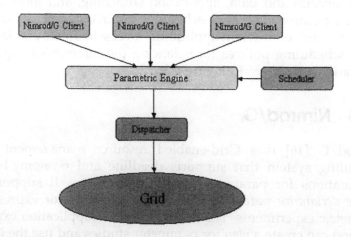

Figure 6.27 The architecture of Nimrod/G

- *Parametric engine*: The parametric engine acts as a persistent job control agent and is the central component from which the whole experiment is managed and maintained. It is responsible for the parameterization of an experiment and the actual creation of jobs, maintenance of job status, and interaction with clients, the schedule advisor, and dispatcher. The parametric engine takes the experiment's plan as input described by using a *declarative* parametric modelling language and manages the experiment under the direction of the schedule advisor. It then informs the dispatcher how to map an application's tasks to the selected resource.
- *Scheduler*: The scheduler is responsible for resource discovery, resource selection and job assignment. The resource discovery system interacts with Grid information services, such as the MDS in Globus, and it identifies the list of authorized machines, and keeps track of resource status information. The resource selection algorithm is used to select those resources that meet some criterion, such as a deadline or minimize the cost of a computation.
- *Dispatcher*: The dispatcher primarily initiates the execution of a task on the selected resource as instructed by the scheduler. It periodically updates the status of the task's execution to the parametric-engine.

6.5.4 Rescheduling

On the Grid, long-running jobs must be able to adapt to nodes leaving the Grid due to failure, scheduled downtime, or nodes becoming overloaded. A Grid scheduler should have the ability to reschedule a task [17], whether this means acknowledging failure and re-sending lost work to a live node, or stopping and moving a job away from a heavily loaded node.

Condor's Master–Worker (MW) library [18] supports limited rescheduling. The AppLeS Parameter Sweep Template (APST) is capable of limited rescheduling in that it can adjust future schedules, but cannot move current jobs. APST is able to assume the existence of many independent tasks in each application because it is designed for parameter sweep applications. It computes a schedule for all its tasks and may adapt the schedule at preconfigured and tunable intervals to compensate for resource variations.

A promising project that incorporates rescheduling is the Cactus [19], which dynamically checks the performance of an application to determine if it has violated a contract, and moves the application to a different set of processors to improve performance. The contract is defined as not degrading a particular metric, such as iterations per second, and detects aspects like a new load on a processor.

Rescheduling is an important issue for Grid schedulers, and no such system currently implements a complete set of rescheduling features. A mature scheduler with rescheduling should be able to adjust current schedules, move jobs from poorly performing nodes and recover from failures.

6.5.5 Scheduling with heuristics

Scheduling optimization is the process to find an optimal or near-optimal schedule in terms of a set of tasks and resources. The mapping of tasks to resources is known as an NP-hard problem. Heuristics such as genetic algorithms and simulated annealing can be used for Grid scheduling optimization.

Genetic algorithms (GAs)
GAs are adaptive methods that can be used to solve optimization problems. Spooner *et al.* [20] apply a GA to explore the solution space to find optimal schedules. Their performance evaluation has shown that GA-based schedules are optimal compared with other scheduling strategies, such as first come first serve, deadline sort and deadline sort with node limitations. The approach used in this work first generates a set of initial schedules, then it evaluates the schedules to obtain a measure of fitness, and finally it selects the most appropriate and combines them together using operators (crossover and mutation) to formulate a new set of solutions. This is repeated using the current schedule as a basis, rather than restarting the process, allowing the system to capitalize on the organization that has occurred previously.

Simulated annealing (SA)
SA is based on the analogy to the physical process of cooling and re-crystallization of metals. It is a generalization of the Monte Carlo method used for optimization of multi-variate problems. Possible solutions are generated randomly and then accepted or discarded

based on the differences in their benefits in comparison to a currently selected solution, consisting of either a maximum number of new solutions being accepted, or a maximum number of solutions being considered. Work on applying SA to Grid scheduling can be found in Young et al. [21] and YarKhan and Dongarra [22].

6.6 CHAPTER SUMMARY

In this chapter, we have studied the core techniques involved in scheduling, e.g. paradigms, the steps involved in implementing a scheduling system, scheduling with QoS and scheduling optimization with heuristics. We have reviewed four resource management and job scheduling systems – Condor, SGE, PBS and LSF. The reason we choose to review these four systems is because they have been widely used on the Grid. The four systems usually work locally within a single administrative domain, they manage a range of computer-based systems and they generally operate in a type of master–worker paradigm in which there is only one master host and arbitrary number of execution hosts. Clusters based on the four systems can be connected by the Globus Toolkit, which thus provides a multi-domain Grid-enabled system. The process of scheduling a job on the Grid involves resource selection, resource filtering, schedule generation, optimal schedule selection, and finally job dispatching. Scheduling systems can work at the resource level, like Condor, SGE, PBS and LSF, or at the application level, like AppLeS. A job-scheduling algorithm can be implemented by using very simple strategies, such as first come first serve or complex heuristics using GAs and SA to achieve an optimal schedule. Apart from information related to jobs and system resources, a Grid scheduler, such as Nimrod/G, also takes the cost of using a resource into account when scheduling jobs.

Scheduling has been playing a crucial role in the construction of effective Grid environments. Grid scheduling will continue to be a research focus as the Grid evolves. A good scheduling system on the Grid should have the following features:

- Efficiency in generating schedules;
- Adaptability, where resources may join or leave dynamically;
- Scalability in managing resources and jobs;

- Ability to predict and estimate performance;
- Ability to coordinate the competition and collaboration of different cluster-level schedulers;
- Ability to reserve resources for scheduling;
- Ability to take the cost of resources into account when scheduling;
- Ability to take user preferences and site policies into account.

6.7 FURTHER READING AND TESTING

To help understand this chapter more completely, you can start with the installation of a resource management system such as Condor, SGE, PBS or LSF. Condor, SGE and PBS can be freely downloaded from their Web sites and detailed information can be found for installation and execution. However, LSF needs to be purchased. To test a meta-scheduler, a Grid environment can be instantiated by first installing Globus, Condor and SGE, then connecting a Condor pool and an SGE cluster with Globus. Jobs submitted to a meta-scheduler can be dispatched in the Condor pool or the SGE cluster or both.

6.8 KEY POINTS

- Grid scheduling involves resource selection, resource filtering, schedule generation, optimal schedule selection, job dispatching and execution.
- A scheduler is a software component in a Grid system.
- Condor, SGE, PBS and LSF are cluster-level resource management and job scheduling systems that can utilize the Grid. They normally work within one administrative domain.
- AppLeS is an application-level scheduling system.
- Nimrod/G is a scheduling system that can take the cost of resources into account.
- Grid scheduling can be optimized with more complex heuristics such as generic algorithms and simulated annealing.

6.9 REFERENCES

[1] Condor, http://www.cs.wisc.edu/condor/.
[2] Sun Grid Engine, http://wwws.sun.com/software/gridware/.
[3] PBS, http://www.openpbs.org/.
[4] LSF, http://www.platform.com/products/LSF/.
[5] Hamscher, V., Schwiegelshohn, U., Streit, A. and Yahyapour, R. *Evaluation of Job-Scheduling Strategies for Grid Computing*. GRID 2000, 191–202, 17–20 December 2000, Bangalore, India. Lecture Notes in Computer Science, Springer-Verlag.
[6] Srinivasan, S., Kettimuthu, R., Subramani, V. and Sadayappan, P. *Characterization of Backfilling Strategies for Parallel Job Scheduling*. ICPP Workshops 2002, 514–522, August 2002, Vancouver, BC, Canada. CS Press.
[7] DAGManager, http://www.cs.wisc.edu/condor/dagman/.
[8] Ghare, G. and Leutenegger, S. *Improving Small Job Response Time for Opportunistic Scheduling*. Proceedings of 8th International Workshop on Modeling, Analysis, and Simulation of Computer and Telecommunication Systems (MASCOTS 2000), San Francisco, CA, USA. CS Press.
[9] Raman, R., Livny, M. and Solomon, M. *Matchmaking: Distributed Resource Management for High Throughput Computing*. Proceedings of the 7th IEEE International Symposium on High Performance Distributed Computing, July 1998, Chicago, IL, USA. CS Press.
[10] Enterprise Edition policy, http://www.sun.com/blueprints/0703/817-3179.pdf.
[11] N1GE 6 Scheduling, http://docs.sun.com/app/docs/doc/817-5678/6ml4alis7?a=view.
[12] PBS Pro, http://www.pbspro.com/.
[13] Figueira, M., Hayes, J., Obertelli, G., Schopf, J., Shao, G., Smallen, S., Spring, N., Su, A. and Zagorodnov, D. Adaptive Computing on the Grid Using AppLeS. *IEEE Transactions on Parallel and Distributed Systems*, 14(4): 369–382 (2003).
[14] NWS, http://nws.cs.ucsb.edu/.
[15] Dail, H., Berman, F. and Casanova, H. A Decoupled Scheduling Approach for Grid Application Development Environments. *Journal of Parallel Distributed Computing*, 63(5): 505–524 (2003).
[16] Abramson, D., Giddy, J. and Kotler, L. *High Performance Parametric Modeling with Nimrod/G: Killer Application for the Global Grid?* Proceedings of the International Parallel and Distributed Processing (IPDPS 2000), May 2000, Cancun, Mexico. CS Press.
[17] Gerasoulis, A. and Jiao, J. *Rescheduling Support for Mapping Dynamic Scientific Computation onto Distributed Memory Multiprocessors*. Proceedings of the Euro-Pa '97, August 1997, Passau, Germany. Lecture Notes in Computer Science, Springer-Verlag.
[18] Goux, Jean-Pierre, Kulkarni, Sanjeev, Yoder, Michael and Linderoth, Jeff. Master-Worker: An Enabling Framework for Applications on the Computational Grid. *Cluster Computing*, 4(1): 63–70 (2001).
[19] Cactus, http://www.cactuscode.org/.

[20] Spooner, D., Jarvis, S., Cao, J., Saini, S. and Nudd, G. Local Grid Scheduling Techniques using Performance Prediction, *IEE Proc. – Comp. Digit. Tech.*, 150(2): 87–96 (2003).

[21] Young, L., McGough, S., Newhouse, S. and Darlington, J. *Scheduling Architecture and Algorithms within the ICENI Grid Middleware*. Proceedings of the UK e-Science All Hands Meeting, September 2003, Nottingham, UK.

[22] YarKhan, A. and Dongarra, J. *Experiments with Scheduling Using Simulated Annealing in a Grid Environment*. Proceedings of the 3rd International Workshop on Grid Computing (GRID 2002), November 2002, Baltimore, MD, USA. CS Press.

7

Workflow Management for the Grid

LEARNING OUTCOMES

In this chapter, we will study Grid workflow management. From this chapter, you will learn:

- What a workflow management system is and the roles it will play in the Grid.
- The techniques involved in building workflow systems.
- The state-of-the-art development of workflow systems for the Grid.

CHAPTER OUTLINE

7.1 Introduction
7.2 The Workflow Management Coalition
7.3 Web Services-Oriented Flow Languages
7.4 Grid Services-Oriented Flow Languages
7.5 Workflow Management for the Grid

The Grid: Core Technologies   Maozhen Li and Mark Baker
© 2005 John Wiley & Sons, Ltd

7.6 Chapter summary

7.7 Further reading and testing

7.1 INTRODUCTION

As we have discussed in Chapter 2, OGSA is becoming the *de facto* standard for building service-oriented Grid systems. OGSA defines Grid services as Web services with additional features and attributes. A Web service itself is a software component with a specific WSDL interface that completely describes the service and how to interact with it. Information about a particular Web service can be published in a registry, such as UDDI. A client interacts with the registry to search and discover the services available. SOAP is a protocol for message exchanging between a client and a service. Apart from that, an important feature of Web services is service composition in which a compound service can be composed from other services.

The main goal of OGSA is to make compliant Grid services interoperable. Grid services can be used in the following two ways: independent pre-OGSA Grid services and interdependent OGSA compliant Grid services.

Independent pre-OGSA Grid services
As shown in Figure 7.1, a user makes use of independent pre-OGSA Grid services to access the Grid. These services normally interact with a pre-OGSA Grid middleware toolkit such as the GT2 to access Grid resources.

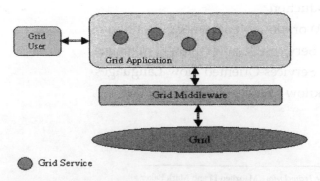

Figure 7.1 Accessing the Grid via independent Grid services

7.2 THE WORKFLOW MANAGEMENT COALITION

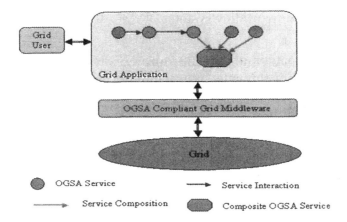

Figure 7.2 Accessing the Grid via interdependent OGSA services

Interdependent OGSA compliant Grid services
OGSA compliant Grid services are interoperable and can be composed in a Grid application. The execution of a Grid application may involve the running of a number of interdependent Grid services. These services then interact with an OGSA compliant Grid middleware toolkit such as the GT3 to access Grid resources. As shown in Figure 7.2, interdependent OGSA compliant Grid services are the one where the output of one service can be an input of another service. Services can also be composed into an amalgamated service accessed directly by users. The interactions and executions of services are managed by a workflow management system, specifically a workflow engine, which will be described in this chapter.

This chapter is organized as follows. In Section 7.2, we introduce the Workflow Management Coalition (WfMC) [1], a workflow standard body to promote the interoperability of heterogeneous workflow systems. In Section 7.3, we describe workflow management in the context of Web services. In Section 7.4, we review the state-of-the-art of workflow development for the Grid. In Section 7.5, we conclude the chapter and provide further readings in Section 7.6.

7.2 THE WORKFLOW MANAGEMENT COALITION

Founded in August 1993, now with more than 300 members from both industry and academia, WfMC aims to identify the common workflow management functional areas and develop appropriate specifications for workflow systems. WfMC defines a workflow as follows:

> The automation of a business process, in whole or part, during which documents, information or tasks are passed from one participant to another for action, according to a set of procedural rules [2].

Figure 7.3 shows the mapping from a business process in the real world to a workflow process in the world of computer systems. A workflow process is a coordinated (parallel and/or sequential) set of process activities that are connected in order to achieve a common business goal. A process activity is defined as a logical step or description of a piece of work that contributes towards the accomplishment of a process. A process activity may be a manual process activity and/or an automated process activity. A workflow process is first specified using a process definition language and then executed by a Workflow Management System (WFMS), which is defined by WfMC as follows:

> A system that defines, creates and manages the execution of workflows through the use of software, running on one or more workflow engines, which is able to interpret the process definition, interact with workflow participants and, where required, invoke the use of information technology tools and applications [2].

WfMC defines a reference model, as shown in Figure 7.4, to identify the interfaces within a generic WFMS. The reference model specifies a framework for workflow systems, identifying their

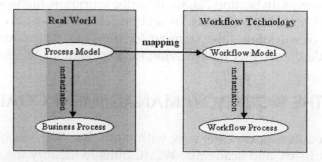

Figure 7.3 Mapping a business process to a workflow process

7.2 THE WORKFLOW MANAGEMENT COALITION

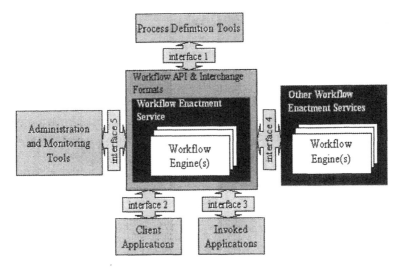

Figure 7.4 The WfMC reference model

characteristics, functions and interfaces. A major focus of WfMC has been on specifying the five interfaces that surround the workflow engine. These interfaces provide a standard means of communication between workflow engines and clients, including other workflow components such as process definition and monitoring tools.

7.2.1 The workflow enactment service

A workflow enactment service provides the run-time environment in which one or more workflow processes can be executed; which may involve more than one actual workflow engine. A workflow enactment service can be a homogeneous or a heterogeneous service. A homogeneous service consists of one or more compatible workflow engines which provide the run-time execution environment for workflow processes with a defined set of process definition attributes. On the other hand, a heterogeneous service consists of two or more heterogeneous services which follow common standards for interoperability at a defined conformance level. When heterogeneous services are involved, a standardized interchange format is necessary between workflow engines. Using interface 4 (which will be described later in this section), the enactment service may transfer activities or sub-processes to other enactment services for execution.

7.2.2 The workflow engine

A workflow engine provides the run-time environment for activating, managing and executing workflow processes. The WfMC focuses on a paradigm in which the workflow engine instantiates a workflow specification defined by a flow language, decomposes it into smaller activities and then allocates activities to processing entities for execution. This approach distinguishes between the process definition, which describes the processes to be executed, and the process instantiation, which is the actual enactment (execution) of the process. This paradigm is referred to as the scheduler-based paradigm [3].

7.2.2.1 A scheduler-based paradigm

The implementation and deployment of the scheduler-based approach to a workflow engine can be described in terms of a state transition machine. Individual process or activity instances change state in response to workflow engine decisions or external events, such as the completion of an activity. A process instance may be *initiated* once selected for enactment; it is *active* after at least one of its activities has been started; *suspended*, when perhaps waiting for some events or *completed*. Similarly, an activity may be *inactive*, *active*, *suspended* or *completed*. It is the role of the workflow engine to manage this state transition, selecting processes to be instantiated, initiating activities by scheduling them to processing components, and controlling and monitoring the resulting state transitions. The workflow engine must also implement the rules that govern the transitions between tasks, updating the processes as tasks complete or fail, and taking appropriate actions in response.

The scheduler-based paradigm has been widely used. However, there are two alternative paradigms, namely *data-flow* and *information pull*:

- The *data-flow* paradigm views the workflow as a repository of data that is passed between processing activities according to sets of rules, the current state and history information related to the workflow.
- The *information pull* paradigm originated with the network and information management fields, where the requirement for information drives the creation and enactment of workflow processes.

7.2.2.2 Workflow engine tasks

A workflow engine normally performs the following tasks.

Process selection
One key responsibility of the workflow engine is to manage the selection and instantiation of process templates. The engine will respond to some stimulus (i.e. a *triggering event*) by selecting a suitable process from the library of templates. Examples of possible triggering events include the arrival of a new user request, the generation of a product by an already active process or even the passage of time. The workflow engine manages the instantiation of the relevant process. There may be alternative and applicable processes that must be compared with the triggering conditions and selected as appropriate. In many existing WFMSs this task is trivial, as there is none or little choice among processes, given the predefined stimulus for enactment. But there are domains where there may be many, or even no, directly applicable and valid processes for a given stimulus, thus requiring process selection, adaptation or even dynamic process creation.

Task allocation
Once a process is selected and instantiated, the workflow engine forwards activities to an *activity list manager* to allocate the activities to processing entities. An activity is assigned to a processing entity according to its capability, availability and the temporal and sequencing constraints of the activity. This allocation of tasks can be treated as a scheduling problem. Thus, the workflow engine takes a centralized role in coordinating the operation of processing entities.

Scheduling techniques within workflow management systems have employed straightforward enumerative or heuristic-based algorithms to date. As the complexity of WFMS domains increases, more sophisticated approaches that provide robust reactive scheduling will be critical to accommodate processing entities.

Enactment control, execution monitoring and failure recovery
The workflow engine must maintain all the knowledge and internal control data to identify the state of each of the individually instantiated activities, transition conditions, connections among processes (e.g. parent/child relationships) and performance

metrics. The WfMC defines two types of data relevant to the control and monitoring of workflow processes:

- *Workflow control data* encompass state information about processes, activities, and possibly performance criteria. It is internal information managed directly by a workflow engine.
- *Workflow relevant data* is used by the WFMS to determine when to enact new processes and when the transition among states within enacted processes should be performed.

7.2.3 WfMC interfaces

The WfMC has identified five functional interfaces (Figure 7.4) that are described below.

Interface 1
This interface defines a common meta-model for describing workflow process definitions, a textual grammar in Workflow Process Definition Language (WPDL) for the interchange of process definitions and a set of APIs for the manipulation of process definition data. The WPDL has been replaced by XML Process Definition Language (XPDL) [4] which allows the definition of processes in a standardized format via XML.

XPDL is conceived as a graph-structured language with additional concepts to handle blocks of workflow processes. In XPDL, process definitions cannot be nested and routing is handled by the specification of transitions between activities. The activities in a process can be thought of as the nodes of a directed graph, with the transitions being the edges. Conditions associated with the transitions determine at execution time which activity or activities should be executed next.

Interface 2
Interface 2 defines how client applications interact with different workflow systems. It was specified as a series of Workflow APIs to allow the control of process, activity and worklist handling functions. These APIs were originally defined in "C" and subsequently re-expressed in CORBA IDL and Microsoft's Object Linking and Embedding (OLE).

Interface 3
Interface 3 defines a set of APIs for invoking third-party applications.

7.2 THE WORKFLOW MANAGEMENT COALITION

Interface 4

Interface 4 defines the interoperability of workflow engines. It comprises an interchange protocol covering five basic operations, specified in abstract terms and with separate concrete bindings. The initial version was defined as a MIME body part for use with email; subsequent versions have been specified in XML (Wf-XML) [5], which is an interoperability specification defined by WfMC. It combines the elementary concept of Simple Workflow Access Protocol (SWAP) [6] with the abstract commands defined by the WfMC Interface 4. Wf-XML defines a set of request/response messages that are exchanged between an observer, which may or may not be a WFMS, and a WFMS that controls the execution of a remote workflow instance. Figure 7.5 shows the interaction between two workflow engines (A and B) via Wf-XML. Ongoing work has lead to version 2 of Wf-XML, layered over SOAP and Asynchronous Service Access Protocol (ASAP) [7].

Interface 5

Interface 5 allows several workflow services to share a range of common management and monitoring functions. The proposed interface provides a complete view of the status of a workflow in an organization.

7.2.4 Other components in the WfMC reference model

- *Process definition tools* provide users with the ability to analyse and model actual business processes and generate corresponding

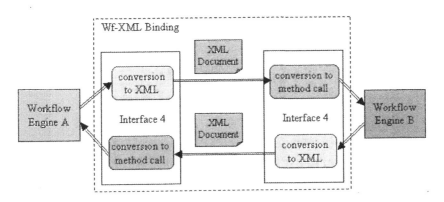

Figure 7.5 The interoperation of workflow engines via Wf-XML

representations. The design of a process definition can be separated from the run time of the process, which makes it possible for a process definition to be executed by an arbitrary workflow system implementing this interface at run time.
- *Client applications* interact with a workflow engine, requesting facilities and services from the engine. Client applications may perform some common functions such as work list handling, process instance initiation and process state control functions.
- *Invoked applications* are applications that are invoked by a WFMS to fully or partly perform an activity, or to support a workflow participant in processing a work-item. Usually these invoked applications are server based and do not have any user interfaces. The Interface 3 defines the semantics and syntax of the APIs for standardized invocation, which includes session establishment, activity management and data handling functions.
- *Administration and monitoring tools* are used to manage and monitor workflows. A management and monitoring tool may exist as an independent application interacting with different workflow engines. In addition, it may be implemented as an integral part of a workflow enactment service with the additional functionality to manage other workflow engines.

7.2.5 A summary of WfMC reference model

The WfMC reference model is a general model that provides guidelines for developing interoperable WFMSs. However, at present, most of the workflow management systems in the marketplace do not implement all the interfaces defined by the reference model. Usually, they implement a subset of interfaces and functionality that is defined in the model.

7.3 WEB SERVICES-ORIENTED FLOW LANGUAGES

Web services aim to exploit XML technology and the HTTP protocol by integrating applications that can be published, located and invoked over the Web. To integrate processes across multiple business enterprises, traditional interaction using standard messages and protocols is insufficient. Business interactions require long-running exchanges that are driven by an explicit process model. This raises the need for composition languages, which for Web services are flow languages that are the means to manage the orchestration of Web services, the instantiation and execution of workflows. In this section, we give a brief overview of representative Web services flow languages that build on WSDL. These languages are either block structured, graph based or both. Whereas a block-structured workflow language specifies a predefined order in executing services, a graph-based workflow language uses graphs to specify the data and control flows between services.

7.3.1 XLANG

XLANG [8], initially developed by Microsoft, is used to describe how a process works as part of a business flow. It is a block-structured language with basic control flow structures: <sequence> and <switch> for conditional routing; <while> for looping; <all> for parallel routing; and <pick> for race conditions based on timing or external triggers. XLANG focuses on the creation of business processes and the interactions between Web service providers. It also includes a robust exception handling facility, with support for long-running transactions through compensation.

An XLANG service is a WSDL service with a behaviour. Instances of XLANG services are started either implicitly by specially marked operations or explicitly by some background functionality. As shown in Figure 7.6, the XLANG sample specifies the execution sequence of the two services: ServiceA and ServiceB. The two services use WSDL to describe their interfaces.

7.3.2 Web services flow language

Web Services Flow Language (WSFL) [9], initially developed by IBM, is a graph-based language that defines a specific order of activities and data exchanges for a particular process. It defines both the execution sequence and the mapping of each step in the flow to specific operations, referred to as flow models and global models.

```
<definition>

ServiceA WSDL description
ServiceB WSDL description

<xlang:behavior>
  <xlang:body>
    <xlang:sequence>
      <xlang:action operation="OpA" port="ServiceA"activation="true"/>
      <xlang:action operation="OpB" port="ServiceB"/>
    </xlang:sequence>
  </xlang:body>
</xlang:behavior>

</definition>
```

Figure 7.6 An XLANG sample

Flow model
The flow model in WSFL specifies the execution sequence of the composed Web services and defines the flow of control and data exchange between Web services involved. Figure 7.7 shows a flow model sample in WSFL to define how the two service providers can collaborate. *controlLink* and *dataLink* are used to separate data from control in service interactions.

```
<flowModel name="myWorkflow" serviceProvierType="">
  <serviceProvider name="Provider A" type="">
    <locator type="static" service="Provider A.com"/>
  </serviceProvider>
  <serviceProvider name="Provider B" type="">
    <locator type="static" service="Provider B.com"/>
  </serviceProvider>
  <activity name="Activity A">
    <performedBy serviceProvider="Provider A"/>
    <implement><export><target portType="" operation=" OpA"/>
    </export></implement>
  </activity>
  <activity name=ActivityB>
    <performedBy serviceProvider="Provider A"/>
    ...
  </activity>
  <controlLink source="Activity A" target="ActivityB">
  <dataLink source="Activity A "target="Activity B"/>
    <map sourceMessage="" targetMessage=""/>
  </dataLink>
</flowModel>
```

Figure 7.7 A flow model sample in WSFL

Global model

The global model in WSFL describes how the composed Web services interact with each other. The interactions are modelled as links between endpoints of the Web services' interfaces in terms of WSDL, with each link corresponding to the interaction of one Web service with another's interface.

A WSFL definition can also be exposed with a WSDL interface, allowing for recursive decomposition. WSFL supports the handling of exceptions but has no direct support for transactions. In contrast to XLANG, WSFL is not limited to block structures and allows for directed graphs. The graphs in WSFL can be nested but need to be acyclic. Iteration in WSFL is only supported through exit conditions, i.e. an activity or a sub-process is iterated until its exit condition is met.

7.3.3 WSCI

Web Services Choreography Interface (WSCI) [10], initially developed by Sun, SAP, BEA and Intalio, is a block-structured language that describes the messages exchanged between Web services participating in a collaborative exchange. WSCI was recently published as a W3C note. As shown in Figure 7.8, a WSCI choreography would include a set of WSCI interfaces associated with Web services, one for each partner involved in the collaboration. In WSCI, there is no single controlling process managing the interaction between collaborative parties.

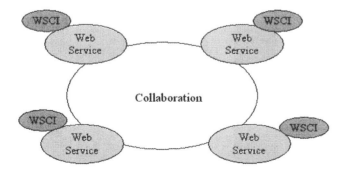

Figure 7.8 A view of WSCI

Each action in WSCI represents a unit of work, which typically would map to a specific WSDL operation. While WSDL describes the entry points for each Web service, WSCI describes the interactions among these WSDL operations. WSCI supports both basic and structured activities. For example, <action> is used for defining a basic request or response message; <call> for invoking external services; <all> for indicating that the specific actions have to be performed, but not in any particular order.

Each activity specifies the WSDL operation involved and the role being played by this participant. WSCI supports the definition of the following types of choreographies:

- *Sequential execution*: The activities must be executed in a sequential order.
- *Parallel execution*: All activities must be executed, but they may be executed in any order.
- *Looping*: The activities are repeatedly executed based on the evaluation of a condition or an expression. WSCI supports *for-each*, *while* and *repeat-until* style loops.
- *Conditional execution*: One out of several sets of activities is executed based on the evaluation of conditions (<switch>) or based on the occurrence of an event (<choice>).

Figure 7.9 shows a WSCI example. An ordering process is created containing two sequential activities, "Receive Order" and "Confirm Order". Each activity maps to a WSDL portType, and a correlation is established between the two steps.

A key aspect of WSCI is that it only describes the observable or visible behaviour between Web services. WSCI does not address

```
<process name="Order" instantiation="message">
  <sequence>
   <action name="ReceiveOrder" role=" Agent"operation="tns:Order" />
   <action name="ConfirmOrder" role=" Agent" operation="tns:Confirm"/>
    <correlate correlation="tns:ordered"/>
    <call process="tns:Order"/>
   </action>
  </sequence>
</process>
```

Figure 7.9 A WSCI example

the definition of executable business processes as defined by (BPEL4WS) which will be described below. Furthermore, a single WSCI definition can only describe one partner's participation in a message exchange. For example, the WSCI definition as shown in Figure 7.9 is the WSCI document from the perspective of the Agent. The buyer and the supplier involved in the process also have their own WSCI definitions.

7.3.4 BPEL4WS

The Business Process Execution Language for Web Services (BPEL4WS) [11], proposed by IBM, Microsoft and BEA, builds on WSFL and XLANG and combines the features of the block-structured XLANG and the graph-based WSFL. BPEL4WS is replacing XLANG and WSFL. Unlike a traditional programming language implementation of a Web service, each operation of each WSDL portType in the service does not map to a separate piece of logic in BPEL4WS. Instead, the set of WSDL portTypes of the Web service is implemented by one single BPEL4WS process. BPEL4WS is intended to support the modelling of two types of processes: executable and abstract processes. An abstract process specifies the message exchange behaviour between different parties without revealing the internal behaviour for anyone of them. An executable process specifies the execution order between a number of activities constituting the process, the partners involved in the process, the messages exchanged between these partners and the fault and exception handling specifying the behaviour in cases of errors and exceptions. Figure 7.10 shows the components in BPEL4WS.

The BPEL4WS itself is like a flow chart in which each step involved is called an activity. An activity is either primitive or structured. There are a collection of primitive activities: *<invoke>* for invoking an operation on a Web service; *<receive>* for waiting for a message from an external source; *<reply>* for generating the response of an input/output operation; *<wait>* for waiting for some time; *<assign>* for copying data from one place to another; *<throw>* for indicating exceptions in the execution; *<terminate>* for terminating the entire service instance; and *<empty>* for doing nothing. Structured activities prescribe the order in which a collection of activities take place: *<sequence>* for defining an execution order; *<switch>* for conditional routing; *<while>* for looping; *<pick>*

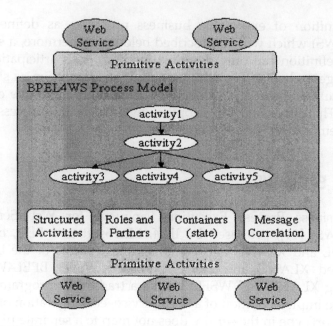

Figure 7.10 The architecture of BPEL4WS

for race conditions based on timing or external triggers; <*flow*> for parallel routing; and <*scope*> for grouping activities to be treated by the same fault-handler.

While standard Web services are stateless, workflows in BPEL4WS are stateful with persistent containers. A BPEL4WS container is a typed data structure which stores messages associated with a workflow instance. A partner could be any Web service that a process invokes or any Web service that invokes the process. Each partner is mapped to a specific role that it fills within the business process. A specific partner might play one role in one business process but a completely different role in another process. Message correlation is used to link messages and specific workflow instances. A general structure of BPEL4WS is shown in Figure 7.11.

Within the BPEL4WS model, data is accessed and manipulated using XML standards. Transformations within <*assign*> activities are expressed with XSLT [12] and XPath [13]. The use of XML as the data format and XML Schema [14] as the associated type system follows from the use of these standards in the WSDL specification.

While BPEL4WS supports the notion of "abstract processes", most of its focus is aimed at BPEL4WS executable processes.

7.3 WEB SERVICES-ORIENTED FLOW LANGUAGES

```
<process ...>

    <partners> ... </partners>
        <!--Web services that the process interacts with -->
    <containers> ... </containers>
        <!-- Stateful data used by the process -->
    <correlationSets> ... </correlationSets>
        <!-- Used to support asynchronous interactions -->
    <faultHandlers> ... </faultHandlers>
        <!--Alternate execution path to deal with faulty conditions -->
    <compensationHandlers> ... </compensationHandlers>
        <!--Code to execute when "undoing" an action -->
    (activities)*
        <!-- What the process actually does -->

</process>
```

Figure 7.11 The BPEL4WS structure

BPEL4WS describes an executable process from the perspective of one of the partners. WSCI takes more of a collaborative and choreographed approach, requiring each participant in the message exchange to define a WSCI interface.

7.3.5 BPML

Business Process Modelling Language (BPML) [15] is a meta-language for modelling business processes. The specification was developed by Business Process Management Initiative (BPMI.org), an independent organization chartered by Intalio, Sterling Commerce, Sun, CSC and others. BPML defines basic activities for sending, receiving and invoking services available, along with structured activities that handle conditional choices, sequential and parallel activities, joins and looping. BPML also supports the scheduling of tasks at specific times.

BPML is conceived as a block-structured flow language. Recursive block structure plays a significant role in scoping issues that are relevant for declarations, definitions and process execution. Flow control (routing) is handled entirely by the block-structure concepts, e.g. execute all the activities in the block sequentially.

BPML provides transactional support and exception handling mechanisms. Both short and long running transactions are supported, with compensation techniques used for more complex transactions. BPML uses a scoping technique similar to BPEL4WS to manage the compensation rules. It also provides the ability to nest processes and transactions, a feature that BPEL4WS currently does not provide. BPML and BPEL4WS both have capabilities to define a business process. WSCI is now considered a part of BPML, with WSCI defining the interactions between the services and BPML defining the business processes behind each service [16].

7.3.6 A summary of Web services flow languages

XLANG, WSFL, BPEL4WS, WSCI and BPML are flow languages for Web services orchestration in terms of WSDL. References [17, 18] compare their features and functionality in more depth. Commercial products have been available to support these flow languages. For example, Microsoft BizTalk Server [19] supports XLANG; Collaxa Orchestration Server [20] support BPEL4WS; IBM BPWS4J [21] supports BPEL4WS.

7.4 GRID SERVICES-ORIENTED FLOW LANGUAGES

Built on Web services, OGSA promotes service-oriented Grid systems. We define a workflow in a service-oriented system as follows:

> The automation of service composition, in whole or part, during which documents, information or tasks are passed from one service to another for action, according to a set of procedural rules.

While there are a number of flow languages available for Web services composition, they cannot be directly used for the composition of Grid services due to the complexity and dynamic nature of the Grid. In the following sections, we review some representative flow languages for the composition of Grid services.

7.4.1 GSFL

Built on WSFL, Grid Services Flow Language (GSFL) [22] is a flow language for the composition of OGSA compliant Grid services.

7.4 GRID SERVICES-ORIENTED FLOW LANGUAGES

Figure 7.12 shows a general GSFL definition. The features of GSFL includes:

Service provider (<serviceProvider>)
All the services that are part of the workflow have to be specified in the <serviceProvider> list which includes a *name* to uniquely identify a service provider and a *type* to specify the type of a service to be provided. Service providers can be located using the <locator> element. Services can be located statically, via a URL, which would point to an already running service. Factories in GSFL can also be used to start services.

Activity model (<activityModel>)
For each of the services provided by a service provider, their operations have to be defined in the <activityModel> list. It contains a list of activities, each of which has a *name* to identify the service and a *source* which is a reference to an operation in a Web service defined by an <endPointType> element. The <endPointType> element contains the *names* of the operation: *portType*, *portName* and *serviceName* for a particular operation.

Composition model (<compositionModel>)
The <compositionModel> describes how a composite service can be composed from different Grid services. It describes the control and data flow between various operations of the services, and also the direct communication between them in a peer-to-peer fashion. The <compositionModel> consists of an export model (<exportModel>) and a notification model (<notificationModel>).

- The <exportModel> contains the list of activities that have to be exported as operations of the workflow process. Any client can invoke these operations in the workflow instance using standard mechanisms. Since the workflow instance can also be viewed as a standard Grid service, it can be used recursively as part of another workflow process. For each activity exported, the control and data flow are described by the <controlModel> and the <dataModel> respectively in the <exportModel>.

- The <notificationModel> is used for the communication between a service source and sink, which helps peer services directly transmit large amounts of data without going through the workflow engine, as done by normal Web services flow languages such as WSFL.

```xml
<definitions ...>
  <!--List of Service Providers -->
  <serviceProvider name="" type=""> <locator type="" handle=""/> </serviceProvider>*
  <!-- List of activities -->
  <activityModel>
    <activity name=""> <source serviceName="" portType="" portName="" operation=""/> </activity>*
  </activityModel>
  <!-- The Composition Model -->
  <compositionModel>
    <exportModel>
      <exportedActivity>
        <exportedActivityInfo name="" portType=""/>
        <controlModel controlIn=""> <controlLink label="" source="" target=""/> </controlModel>
        <dataModel dataInfo="" dataOutFrom=""> <dataLink label="" source="" sink=""/> </dataModel>
      </exportedActivity>*
    </exportModel>
    <notificationModel> <notificationLink label="" source="" sink="" topic=""/> </notificationModel>
  </compositionModel>
  <!-- Lifecycle for the Services -->
  <lifecycleModel>
    <serviceLifecycleModel> <precedenceLink label="">
      <parent> <element serviceName=""/> <element serviceName=""/> </parent>
      <child> <element serviceName=""/> </child>
      </precedenceLink> </serviceLifecycleModel>
  </lifecycleModel>
</definitions>
```

Figure 7.12 A general structure of GSFL

Lifecycle model (<lifeModel>)
The <lifecycleModel> addresses the order in the services and the activities to be executed. The <serviceLifecycleModel> contains a list of precedence links describing the order in which the services are meant to execute. Hence, all services need not be instantiated at start-up, but only once the preceding services have stopped executing.

7.4.2 SWFL

Built on WSFL, Service Workflow Language (SWFL) [23] is a flow language for the composition of Grid services. It extends WSFL in three ways:

- SWFL improves the representation of conditional and loop control constructs. Currently WSFL can handle *if-then-else*, *switch* and *do-while* constructs and permits only one service within each conditional clause or loop body. SWFL also handles *while* and *for* loops, as well as permitting sequences of services within conditional clauses and loop bodies.
- SWFL permits more general data mappings than WSFL. For example, SWFL can describe data mappings for arrays and compound objects, as well as permitting an activity to receive input data from more than one preceding activity.
- SWFL provides an *assignment* activity which allows the assignment of one variable to another.

7.4.3 GWEL

Grid Workflow Execution Language (GWEL) [24] is a flow language based on BPEL4WS. Similar to the BPEL4WS's process model, GWEL's process model represents a peer-to-peer interaction between services described in WSDL. Both the Grid workflow and its partners are modelled as Grid services with WSDL interfaces.

A GWEL workflow definition can be seen as a template for creating Grid service instances, performing a set of operations on the instances and finally destroying them. The creation of instances is always implicit which means that an occurrence of an activity can result in the creation of an instance. The elements of GWEL include <factoryLinks> for specifying services, <dataLinks> for specifying data sources or storage locations, variables for mapping WSDL <messageTypes>, <faultHandlers> for dealing with faults, <lifecycle> for lifetime management of instances and <onNotification>, <destroyInstanc>, <onAlarm>, <controlflow> for controlling flow data.

7.4.4 GALE

The Grid Access Language for High-Performance Computing (HPC) Environments (GALE) [25] is a flow language that defines a HPC workflow vocabulary to specify complex task sequences that are Grid site independent. The features of GALE are discussed below.

Resource Query (<ResourceQuery>)
<ResourceQuery> is used to query the availability of Grid resources. All Grid resources advertised in a Grid Information Service (GIS) can be queried.

Computation (<Computation>)
A <Computation> is usually a computationally intensive activity (like a physics simulation). A computation has any number of environment settings, any number of application arguments and any number of computation attributes, which specify things like working directory or the name of a submission queue. An attribute named *input* is used to specify the target properties of a <ResourceQuery> activity. The workflow engine uses this *input* attribute to resolve information like executable paths, host contacts, default environments and any other information that is required by the computation. Finally, the *id* attribute is the unique identifier of the Grid activity within the GALE interface document.

Data transfer (<DataTransfer>)
The data transfer directive in a workflow script instructs the workflow engine to execute the appropriate data transfer program. The minimal arguments required are source and destination. The data transfer activity recognizes Grid output variables set by resolving query activities.

New features are being added to GALE, these include, e.g. <DataTransfer> for indicating high-bandwidth tools and interfaces to mass storage systems like High-Performance Storage System (HPSS).

7.4.5 A summary of Grid services flow languages

GSFL, SWFL, GWEL and GALE are XML-based languages for the composition of Grid services. While GALE is designed specifically for HPC systems, GSFL, SWFL and GWEL are based on Web services composition languages, e.g. GSFL and SWFL are based on WSFL, and GWEL is based on BPEL4WS. Since GSFL, SWFL and GWEL are Web services based flow languages, they can be leveraged for the composition of OGSA compliant Grid services.

While WSFL provides a flexible and effective basis for representing a Grid application, BPEL4WS is replacing WSFL. A standard Grid services composition language is needed for the Grid, which may be defined as a subset of a Web services composition language such as BPEL4WS.

7.5 WORKFLOW MANAGEMENT FOR THE GRID

In this section, we review some current activities being carried out in the area of workflow management for the Grid.

7.5.1 Grid workflow management projects

7.5.1.1 BioPipe

BioPipe [26] is a cluster-level workflow framework that seeks to address some of the complexity involved in carrying out large-scale bioinformatics analyses. The main idea behind BioPipe is to allow users to integrate data from disparate sources into a common analysis framework. BioPipe is implemented as a collection of Perl modules for constructing workflows from BioPerl applications. BioPipe uses XML to define the pipeline including a workflow definition, inputs/outputs. It provides a GUI for the construction of workflows. With BioPipe, processes in a workflow can be scheduled in a cluster environment using PBS or LSF.

7.5.1.2 myGrid

myGrid [27] is a project targeted at developing open source high-level service-oriented middleware to support data intensive *in silico* experiments in biology on a Grid [28]. The Taverna project [29], a component in the myGrid project, provides a workflow workbench to visually build, edit and browse workflows. The workbench includes easy import of external Web services and workflow definitions and can submit workflows directly to the workflow enactor for execution. The enactor coordinates execution of the parallel and sequential activities in the workflow and supports data iteration and nested workflows. The enactor can invoke arbitrary Web services as well as more specific bioinformatic services such as Talisman [30] (an open source rapid application development platform with a particular focus on the bioinformatics domain) and SoapLab [31]. In Taverna, workflows are represented in the Scufl language [32].

7.5.1.3 IT Innovation Workflow Enactment Engine

The IT Innovation Workflow Enactment Engine [33] is one component within the myGrid project. It is a WSFL-based workflow orchestration tool for Web services. It can search a standard UDDI registry, given preferences from the workflow author, to obtain actual service instances to invoke. It provides a GUI for the construction of workflows.

7.5.1.4 Triana

Triana [34] is an open-source problem-solving environment written in Java, which allows the user to compose workflows from existing codes in a variety of problem domains. Triana uses XML and contains a workflow engine for coordinating and invoking a set of Grid services. Triana provides a GUI for the construction of workflows. BPEL4WS readers and writers have been integrated with Triana to handle BPEL4WS graphs.

7.5.1.5 JISGA

Jini-based Service-oriented Grid Architecture (JISGA) [35] is a Jini-based service-oriented workflow management system for the Grid. JISGA provides the following support for workflow management:

- It defines the SWFL as described in Section 7.4.2 for Grid services composition.
- It provides the Visual Service Composition Environment (VSCE) which is a graphical user interface that aids services composition. A composite service is described in SWFL.
- It includes a Workflow Engine that provides a run-time environment for executing composite services described in SWFL. The SWFL2Java tool is used to automatically generate Java codes from a SWFL definition.

7.5.1.6 ICENI

Imperial College e-Science Networked Infrastructure (ICENI) [36] is middleware that provides a dynamic service management framework to aid resource administrators, application developers and end users to manage and use Grid environments. ICENI provides a GUI-based workflow tool for service composition; that supports spatial and temporal composition. In spatial composition, all the components that make up an application are represented simultaneously, with information representing how they relate and interact with each other. There is no ordering between the components. In temporal composition, all components are ordered with respect to their temporal dependence. Concurrency, where it exists, is explicit.

Each component has attached workflow information, which consists of a graph in which the directed arcs represent temporal dependence, e.g. a node's behaviour occurs after those which have an arc directed to it. Each node represents some behaviour, and the behaviour happens in an ordered fashion, beginning with the *Start* nodes, and finishing with the *Stop* nodes.

Once the components and links of an application are determined using spatial composition, a textual "execution plan" will be generated and defined in an XML-based language derived from YAWL (Yet Another Workflow Language) [37]. The ICENI workflow system supports conditionals, loops and parallel execution of workflow activities.

7.5.1.7 BioOpera

BioOpera [38] provides a programming environment for multiple step computations over heterogeneous applications, including program invocations, the management of control and data flow. It also provides a run-time environment for processes navigation, distribution, load balancing and failure recovery. It provides monitoring tools at process and task levels. BioOpera provides a GUI tool for the construction of workflows from Web services and Grid components. BioOpera represents a workflow "process template" as a Directed Acyclic Graph (DAG), and translates this set of activities into an executable Condor-G script.

7.5.1.8 GridFlow

GridFlow [39] is a Grid workflow management system based on ARMS [40] (an agent-based resource management system), Titan [41] (a resource scheduling system) and PACE [42] (a performance prediction tool). Workflow management in GridFlow is performed at multiple layers:

- *Tasks*: Tasks are the smallest elements in a workflow. In general, workflow tasks are MPI and PVM jobs running on multiple processors, data transfers to visualization servers or archiving of large data sets to mass storage. Task scheduling is implemented using Titan, which focuses on the sub-workflow and workflow levels of management and scheduling.
- *Sub-workflow*: A sub-workflow is a flow of closely related tasks that is to be executed in a predefined sequence on resources in a local cluster environment.
- *Workflow*: A Grid application can be represented as a flow of several different activities, with each activity represented by a sub-workflow. These activities are loosely coupled and may require multi-sited Grid resources.

The user portal in GridFlow provides a GUI to facilitate the composition of workflow elements and the access to additional services. Grid workflows are described in XML, which will be parsed by a global workflow management system in GridFlow for simulation, execution and monitoring.

7.5.1.9 CCA

The Common Component Architecture (CCA) [43] project defines a common architecture for building large-scale scientific applications from well-tested software components that run on both parallel and distributed systems. XCAT [44] is a Web services-based CCA implementation. XCAT allows components to be connected to each other dynamically using ports, making it possible to build complex applications from simple components described in XML.

7.5.1.10 Geodise

The Geodise project [45] aims to assist engineers in the design process by making available a suite of design search and optimization tools, Computational Fluid Dynamics (CFD) analysis packages, resources, including computer, databases and knowledge management technologies. The workflow construction environment (WCE) in Geodise provides a GUI for workflow construction and validation, execution and visualization [46]. Process activities in a Geodise workflow are Matlab scripts whose interactions are defined in XML.

7.5.1.11 GridAnt

GridAnt [47] provides supports for mapping complex client-side workflows, but also works as a simplistic client to test the functionality of different Grid services. GridAnt helps Grid applications make a smooth transition from GT2 to GT3. GridAnt essentially consists of four components: a workflow engine, a run-time environment, a workflow vocabulary and a workflow monitor.

- *Workflow engine*: The workflow engine, which is based on Apache Ant [48], is responsible for directing the flow of control and data through multiple GridAnt activities. It is the central controller that handles task dependencies, failure recovery, performance analysis and process synchronization.

- *Run-time environment*: Apache Ant lacks the functionality to support workflow compositions. In order to overcome the deficiencies of Ant in the context of an advanced workflow system, the GridAnt architecture introduces a run-time environment that offers a globally accessible whiteboard-style communication model. The run-time environment is capable of hosting arbitrary data structures that can be read and written by individual GridAnt tasks. Additionally, the run-time environment supports important constructs such as constants, arithmetic expressions, global variables, array references and literals.
- *Workflow vocabulary*: The workflow vocabulary specifies a set of predefined activities or tasks upon which complex workflows can be developed.
- *Workflow monitor*: A workflow in GridAnt is specified in XML, which can be edited with any text editor or loaded in a visualization tool for monitoring the execution of elements in the workflow.

7.5.1.12 Symphony

Symphony [49] is a Java-based composition and manipulation framework for the Grid. The framework has two principal elements: a composition and control environment in which a meta-program is constructed and a backend execution environment in which the described computation is performed. Symphony provides a GUI in which components are instantiated and customized to describe specific programs and files. These individual components can then be connected by links indicating data-flow relationships. Individual components and complete meta-programs can be saved, restored for later use and shared with other users. During execution, the components initiate and monitor the operations performed in the backend execution environment. The operations are performed in terms of the defined data-flow relationships. The backend execution environment uses Globus or proprietary mechanisms to access remote resources.

7.5.1.13 Discovery Net

Discovery Net [50] provides knowledge discovery services that allow users to conduct and manage complex data analysis and knowledge discovery activities on the Grid. The workflow system in the Discovery Net is based on data-flow dependencies. Each element (node) of the workflow can describe some constraints it has in terms of execution, e.g. the execution must occur on a particular resource, the execution is done through a script or the execution can be done on any resource available at run time. Data analysis tasks in a workflow are defined with Discovery Process Markup Language (DPML). In addition, Discovery Net workflows can coordinate the execution of OGSA services through a Grid service interface. Any of these workflows can then be published as a new Grid service for programmatic access from other systems.

7.5.1.14 P-GRADE

P-GRADE [51] is graphical programming environment that provides an integrated workflow support to enable the construction, execution and monitoring of complex applications on Grid environments built with Condor-G/Globus or pure Condor. An application workflow describes both the control flow and the data flow of the jobs involved, which can be sequential, MPI, PVM or GRAPNEL jobs generated by the P-GRADE programming environment. The execution of a workflow in P-GRADE is managed by Condor DAGMan.

7.5.2 A summary of Grid workflow management

Table 7.1 summarizes the Grid workflow management efforts as described in Section 7.5.1 from the aspects of the support of GUI, areas focused, languages for workflow definition, computing environments. Apart from the efforts mentioned above, the Pegasus [52] workflow manager in the GriPhyN [53] project can map and execute complex workflows on the Grid. It introduces planning for mapping abstract workflows to Grid resources. The notion of an abstract workflow is to separate abstract entities such as the actors, data and operations from their concrete instances.

Table 7.1 A comparison of Grid workflow management projects

Grid workflow projects	GUI support	Workflow definition language	Areas focused	Computing environments
BioPipe	Yes	XML	Bioinformatics	PBS, LSF
myGrid	Yes	XML/WSFL/Sculf	Bioinformatics	Web services
IT Innovation	Yes	XML/WSFL	General	Web services
Triana	Yes	XML/Part BPEL4WS	General	Web services/JXTA
JISGA	Yes	XML/SWFL	General	Jini
ICENI	Yes	A subset of YAWL	General	Jini/OGSA
BioOpera	Yes	No	Bioinformatics	Condor, PBS, Web services
GridFlow	Yes	XML	General	MPI/PVM
XCAT	Yes	XML	General	Web services
Geodise	Yes	XML	Design search optimization	Globus, Condor, Web service, Matlab
GridAnt	Yes	XML	Grid services deployment	GT2/GT3
Symphony	Yes	XML	General	JavaBeans
Discovery	Yes	XML/DPML	Data analysis/knowledge discovery	Web services
P-GRADE	Yes	No	General	Globus/Condor

7.6 CHAPTER SUMMARY

In this chapter, we have studied workflow technologies. A workflow system provides users with the ability to efficiently build service-oriented applications by simply plugging and playing services together.

Workflow standard bodies, such as WfMC, are attempting to define a standard specification for different WFMSs to interoperate. WfMC defines a reference model to specify the components that a WFMS should follow and the functionality of these components. However, workflow systems following the WfMC reference model are currently focused on business processing.

Web services, upon which OGSA is based, are emerging as a promising computing platform for heterogeneous distributed systems. Services composition is an important feature of Web services technology. Workflow languages such as WSFL, XLANG, WSCI, BPEL4WS and BPML have been implemented for service composition. Among them, BPEL4WS seems to have gained most attention, from both Web services and Grid communities. However, flow languages used for Web services composition cannot be used directly for Grid services composition because of the additional complexity of the Grid. Several flow languages, such as GSFL, SWFL, GWEL and GALE, have been implemented for Grid applications composition. Among them, GSFL and SWFL are based on WSFL; GWEL is based on BPEL4WS; GALE is purely based on XML. However, a standard flow language needs to be investigated, which may be a subset of a Web services flow language such as BPEL4WS to provide a standard way for services composition for the Grid.

Currently many Grid projects, as reviewed in the chapter, can provide some kinds of support for Grid workflow management. We found that almost all of the projects use an XML-based vocabulary for workflow description. However, each project has its own workflow vocabulary for workflow definition; whereas a standard vocabulary should be defined so that each workflow system would use the same language, which would make workflows interoperable across workflow systems. In addition, the WfMC reference model could be applied in designing interoperable workflow management systems for the Grid.

While workflow systems can efficiently improve the productivity in the construction of application, they are complex in nature, especially for the Grid. Many standards need to be investigated and specified by a Grid standard body such as GGF.

7.7 FURTHER READING AND TESTING

XML plays a critical role in defining workflows. All the flow languages reviewed in this chapter are XML-based. In this chapter, we reviewed XML-based flow languages for Web services composition. However, non-XML-based languages, such as PSL [54] and PIF [55] have been widely used for business process modelling, mainly in Business Process Re-Engineering (BPR) systems.

For further testing, we'd suggest readers visit IBM's Web site to download BPWS4J for experiencing with Web services-oriented workflow applications.

7.8 KEY POINTS

- A workflow management system provides users with the ability to quickly build Grid applications via the composition of services.
- The main components in a service-oriented workflow system include a GUI for discovering services available and visually building service control/data flows, a language for representing workflows, a workflow engine for managing the run-time behaviour and execution of services instances.
- WfMC provides a reference model for implementing interoperable workflow management systems.
- XML plays a critical role in defining workflows.
- WSFL, XLANG, WSCI, BPEL4WS and BPML are flow languages that can be used for composing Web services with WSDL interfaces.
- GSFL, SWFL, GWEL are flow languages for composing Grid applications. They are based on Web services flow languages, e.g. GSFL and SWFL are based on WSFL; GWEL is based on BPEL4WS.
- Many projects have been carried out to manage workflow on the Grid. However, they have adopted or defined different workflow definition languages. A standard workflow vocabulary is urgently needed to define a standard workflow definition language for the Grid.

7.9 REFERENCES

[1] WfMC, http://www.wfmc.org/.
[2] Workflow Management Coalition: Terminology & Glossary. Document Number *WfMC-TC-1011*, February 1999.
[3] Cichocki, A., Helal, A.S., Rusinkiewicz, M. and Woelk, D. (1998). *Workflow and Process Automation: Concepts and Technology*, Boston, MA, London. Kluwer.
[4] XPDL, www.wfmc.org/standards/docs/TC-1025_10_xpdl_ 102502.pdf.
[5] Wf-XML, www.wfmc.org/standards/docs/Wf-XML-1.0.pdf.
[6] Bolcer, G.A. and Kaiser, G.E. (1999). Collaborative Work: SWAP: Leveraging the Web to Manage Workflow. *IEEE Internet Computing*, 3(1): 85–88.
[7] Keith, D.S. ASAP/Wf-XML 2.0 Cookbook. In Layna Fischer (ed.), *The Workflow Handbook 2004*, Lighthouse Point, FL, USA. Future Strategies Inc.
[8] XLANG, http://www.ebpml.org/xlang.htm.

7.9 REFERENCES

[9] WSFL, http://www-3.ibm.com/software/solutions/webservices/pdf/WSFL.pdf.
[10] WSCI, www.w3.org/TR/wsci/.
[11] BPEL4WS, http://www-106.ibm.com/developerworks/library/ws-bpel.
[12] XSLT, http://www.w3.org/TR/xslt.
[13] XPath, http://www.w3.org/TR/xpath.
[14] XML Schema, http://www.w3.org/XML/Schema.
[15] BPML, http://www.bpmi.org/.
[16] Peltz, C. (2003). Web Services Orchestration. HP Report.
[17] Aalst, W., Dumas, M. and Hofstede, A. (September 2003). *Web Service Composition Languages: Old Wine in New Bottles?* Proceedings of the EUROMICRO 2003, Belek near Antalya, Turkey. CS Press.
[18] Staab, S., Aalst, W., Benjamins, V.R., Sheth, A.P., Miller, J.A., Bussler, Maedche, C.A., Fensel, D. and Gannon, D. (2003). Web Services: Been There, Done That? *IEEE Intelligent Systems*, 18(1): 72–85.
[19] BizTalk, http://www.microsoft.com/biztalk/.
[20] Collaxa Orchestration Server, http://www.collaxa.com/home.index.jsp.
[21] IBM BPWS4J, http://alphaworks.ibm.com/tech/bpws4j.
[22] Krishnan, S., Wagstrom, P. and Laszewski, G. (2002). GSFL: A Workflow Framework for Grid Services. Preprint ANL/MCS-P980-0802, Argonne National Laboratory, 9700 S. Cass Avenue, Argonne, IL 60439, USA.
[23] Huang, H. and Walker, D. (June 2003). *Extensions to Web Service Techniques for Integrating Jini into a Service-Oriented Architecture for the Grid.* Proceedings of the International Conference on Computational Science 2003 (ICCS '03), Melbourne, Australia. Lecture Notes in Computer Science, Springer-Verlag.
[24] Cybok, D. (March 2004). A Grid Workflow Infrastructure, Presentation in GGF-10, Berlin, Germany.
[25] Beiriger, J.I., Johnson, W.R., Bivens, H.P., Humphreys, S.L. and Rhea, R. (2000). *Constructing the ASCI Computational Grid.* Proceedings of the Ninth International Symposium on High Performance Distributed Computing (HPDC 2000), Pittsburgh, Pennsylvania, USA. CS Press.
[26] BioPipe, http://www.biopipe.org/.
[27] myGrid, http://www.mygrid.org.uk/.
[28] Goble, C.A., Pettifer, S., Stevens, R. and Greenhalgh, C. (2003). Knowledge Integration: In Silico Experiments in Bioinformatics, in *The Grid 2: Blueprint for a New Computing Infrastructure*, Ian Foster and Carl Kesselman (eds), 2nd edition, November, San Francisco, CA. Morgan Kaufmann.
[29] Taverna, http://taverna.sourceforge.net.
[30] Talisman, http://talisman.sourceforge.net/.
[31] SoapLab, http://industry.ebi.ac.uk/soaplab/.
[32] Scufl, http://taverna.sourceforge.net/scuflfeatures.html.
[33] Innovation Workflow, http://www.it-innovation.soton.ac.uk/mygrid/workflow/.
[34] Taylor, I., Shields, M., Wang, I. and Rana, O. (2003). Triana Applications within Grid Computing and Peer to Peer Environments. *Journal of Grid Computing*, 1(2): 199–217. Kluwer Academic.
[35] Huang, Y. (2003). JISGA: A JINI-BASED Service-Oriented Grid Architecture. *International Journal of High Performance Computing Applications*, 17(3): 317–327. SAGE Publication.

[36] ICENI, http://www.lesc.ic.ac.uk/iceni/.
[37] Aalst, W., Aldred, L., Dumas, M. and Hofstede, A. (2004). *Design and Implementation of the YAWL system.* Proceedings of the 16th International Conference on Advanced Information Systems Engineering (CAiSE '04), Riga, Latvia. Lecture Notes in Computer Science, Springer-Verlag.
[38] Bausch, W., Pautasso, C., Schaeppi, R. and Alonso, G. (2002). *BioOpera: Cluster-Aware Computing.* Proceedings of the IEEE International Conference on Cluster Computing (Cluster 2002), Chicago, Illinois, USA. CS Press.
[39] Cao, J., Jarvis, S.A., Saini, S. and Nudd, G.R. (2003). *GridFlow: Workflow Management for Grid Computing.* Proceedings of 3rd International Symposium on Cluster Computing and the Grid (CCGRID '03), Tokyo, Japan. CS Press.
[40] Cao, J., Jarvis, S.A., Saini, S., Kerbyson, D.J. and Nudd, G.R. (2002). ARMS: An Agent-based Resource Management System for Grid Computing. *Scientific Programming*, Special Issue on Grid Computing, 10(2): 135–148. Wiley.
[41] Spooner, D.P., Cao, J., Turner, J.D., Keung, H.N., Jarvis, S.A. and Nudd G.R. (2002). *Localised Workload Management Using Performance Prediction and QoS Contracts.* Proceedings of 18th Annual UK Performance Engineering Workshop (UKPEW 2002), Glasgow, UK.
[42] Nudd, G.R., Kerbyson, D.J., Papaefstathiou, E., Perry, S.C., Harper, J.S. and Wilcox, D.V. (2000). PACE – A Toolset for the Performance Prediction of Parallel and Distributed Systems. *International Journal of High Performance Computing Applications*, Special Issues on Performance Modeling, Part I, 14(3): 228–251. SAGE Publication.
[43] CCA, http://www.cca-forum.org/.
[44] XCAT, http://www.extreme.indiana.edu/xcat/.
[45] Geodise, http://www.geodise.org.
[46] Xu, F. and Cox, S. (2003). *Workflow Tool for Engineers in a Grid-Enabled Matlab Environment.* Proceedings of the UK e-Science All Hands Meeting 2003, Nottingham, UK.
[47] GridAnt, http://www-unix.globus.org/cog/projects/gridant/.
[48] Apache Ant, http://ant.apache.org/.
[49] Lorch, M. and Kafura, D. (2002). *Symphony – A Java-Based Composition and Manipulation Framework for Computational Grids.* Proceedings of 2nd IEEE/ACM International Symposium on Cluster Computing and the Grid (CCGRID '02), Berlin, Germany. CS Press.
[50] Discovery Net, http://www.discovery-on-the.net.
[51] Kacsuk, P., Dózsa, G., Kovács, J., Lovas, R., Podhorszki, N., Balaton, Z. and Gombás, G. (2003). P-GRADE: A Grid Programming Environment. *Journal of Grid Computing*, 1(2): 171–197.
[52] Pegasus, http://pegasus.isi.edu/.
[53] GriPhyN, http://www.griphyn.org/.
[54] PSL, http://www.mel.nist.gov/psl/.
[55] PIF, http://ccs.mit.edu/pif/.

8

Grid Portals

LEARNING OUTCOMES

In this chapter, we will study Grid portals, which are Web-based facilities that provide a personalized, single point of access to Grid resources that support the end-user in one or more tasks. From this chapter, you will learn:

- What is a Grid portal and what kind of roles will it play in the Grid?
- First-generation Grid portals.
- Second-generation Grid portals.
- The features and limitations of first-generation Grid portals.
- The features and benefits of second-generation Grid portals.

CHAPTER OUTLINE

8.1 Introduction
8.2 First-Generation Grid Portals
8.3 Second-Generation Grid Portals
8.4 Chapter Summary
8.5 Further Reading and Testing

8.1 INTRODUCTION

The Grid couples geographically dispersed and distributed heterogeneous resources to provide various services to users. We can consider two main types of Grid users: system developers and end users. System developers are those who build Grid systems using middleware packages such as Globus [1], UNICORE [2] or Condor [3]. The end users are the scientists and engineers who use the Grid to solve their domain-specific problems perhaps via a portal. A Grid portal is a Web-based gateway that provides seamless access to a variety of backend resources. In general, a Grid portal provides end users with a customized view of software and hardware resources specific to their particular problem domain. It also provides a single point of access to Grid-based resources that they have been authorized to use. This will allow scientists or engineers to focus on their problem area by making the Grid a transparent extension of their desktop computing environment. Grid portals currently in use include XCAT Science Portal [4], Mississippi Computational Web Portal [5], NPACI Hotpage [6], JiPANG [7], The DSG Portal [8], Gateway [9], Grappa [10] and ASC Grid Portal [11].

In this chapter, we will study Grid portals; the technologies they employ and the mechanisms that they use. So far, Grid portal development can be broadly classified into two generations. First-generation Grid portals are tightly coupled with Grid middleware such as Globus, mainly Globus toolkit version 2.x (GT2) written in C. The second generation of Grid portals are those that are starting to emerge and make use of technologies such as portlets to provide more customizable solutions.

This chapter is organized as follows. In Section 8.2, we describe technologies involved in the development of first-generation Grid portals. We first present the three-tiered architecture adopted by most portals of this generation. We then introduce some tools that can provide assistance in the construction of these portals. Finally we give a summary on the limitations of first-generation Grid portals. In Section 8.3, we present the state-of-the-art development of second-generation Grid portals. We first introduce the concept of portlets and describe why they are so important for building personalized portals. We then give three portal frameworks that can be used to develop and deploy portlets. We conclude the chapter in Section 8.4 and provide further reading material about portals in Section 8.5.

8.2 FIRST-GENERATION GRID PORTALS

In this section, we will study the first-generation Grid portals from the points of view of architecture, services, implementation techniques and integrated tools. Most Grid portals currently in use belong to this category.

8.2.1 A three-tiered architecture

The first generation of Grid portals mainly used a three-tier architecture as shown in Figure 8.1. As stated in Gannon *et al.* [12], they share the following characteristics:

- A three-tiered architecture, consisting of an interface tier of a Web browser, a middle tier of Web servers and a third tier of backend services and resources, such as databases, high-performance computers, disk storage and specialized devices.
- A user makes a secure connection from their browser to a Web server.
- The Web server then obtains a proxy credential from a proxy credential server and uses that to authenticate the user.

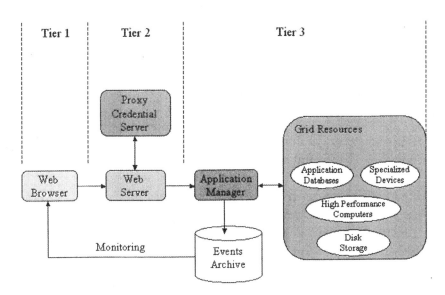

Figure 8.1 The three-tiered architecture of first-generation Grid portals

- When the user completes defining the parameters of the task they want to execute, the portal Web server launches an application manager, which is a process that controls and monitors the actual execution of Grid task(s).
- The Web server delegates the user's proxy credential to the application manager so that it may act on the user's behalf.

In some systems, the application manager publishes an event/message stream to a persistent event channel archive, which describes the state of an application's execution and can be monitored by the user through their browser.

8.2.2 Grid portal services

First-generation Grid portals generally provide the following Grid services.

- *Authentication*: When users access the Grid via a portal, the portal can authenticate users with their usernames and passwords. Once authenticated, a user can request the portal to access Grid resources on the user's behalf.
- *Job management*: A portal provides users with the ability to manage their job tasks (serial or parallel), i.e. launching their applications via the Web browser in a reliable and secure way, monitoring the status of tasks and pausing or cancelling tasks if necessary.
- *Data transfer*: A portal allows users to upload input data sets required by tasks that are to be executed on remote resources. Similarly the portal allows results sets and other data to be downloaded via a Web browser to a local desktop.
- *Information services*: A portal uses discovery mechanisms to find the resources that are needed and available for a particular task. Information that can be collected about resources includes static and dynamic information such as OS or CPU type, current CPU load, free memory or file space and network status. In addition, other details such as job status and queue information can also be retrieved.

8.2.3 First-generation Grid portal implementations

Most portals of this generation have been implemented with the following technologies:

- A dynamic Graphical User Interface (GUI) based on HTML pages, with JSP (Java Server Pages) or JavaScript. Common Gateway Interface (CGI) and Perl are also used by some portals. CGI is an alternative to JSP for dynamically generating Web contents.
- The secure connection from a browser to backend server is via Transport Layer Security (TLS) and Secure HTTP (S-HTTP).
- Typically, a Java Servlet or JavaBean on the Web server handles requests from a user and accesses backend resources.
- MyProxy [13] and GT2 GSI [14] are used for user authentication. MyProxy provides credential delegation in a secure manner.
- GT2 GRAM [15] is used for job submission.
- GT2 MDS [16] is used for gathering information on various resources.
- GT2 GSIFTP [17] or GT2 GridFTP [18] for data transfer.
- The Java CoG [19] provides the access to the corresponding Globus services for Java programs.

The first-generation Grid portals mainly use the GT2 to provide Grid services. One main reason for this is that Globus provides a complete package and a standard way for building Grid-enabled services.

8.2.3.1 MyProxy

MyProxy is an online credential management system for the Grid. It is used to delegate a user's proxy credential to Grid portals, which can be authenticated to access Grid resources on the user's behalf. Storing your Grid credentials in a MyProxy repository allows you to retrieve a proxy credential whenever and wherever you need one. You can also allow trusted servers to renew your proxy credentials using MyProxy, so, for example, long-running

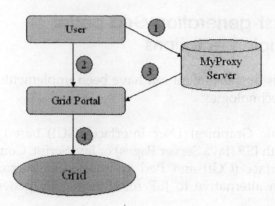

Figure 8.2 The use of MyProxy with a Grid portal

tasks do not fail due to an expired proxy credential. Figure 8.2 shows the steps to securely access the Grid via a Grid portal with MyProxy.

1. Execute `myproxy_init` command on the computer where your Grid credential is located to delegate a proxy credential on a MyProxy server. The delegated proxy credential normally has a lifetime of one week. The communication between the computer and the MyProxy server is securely managed by TLS. You need to supply a username and pass phrase for the identity of your Grid credential. Then you need to supply another different MyProxy pass phrase to secure the delegated proxy credential on the MyProxy server.
2. Log into the Grid portal with the same username and MyProxy pass phrase used for delegating the proxy credential.
3. The portal uses `myproxy_get_delegation` command to retrieve a delegated proxy credential from the MyProxy server using your username and MyProxy pass phrase.
4. The portal accesses Grid resources with the proxy credential on your behalf.
5. The operation of logging out of the portal will delete your delegated proxy credential on the portal. If you forget to log off, then the proxy credential will expire at the lifetime specified.

The detailed information about credentials and delegation can be found in Chapter 4, Grid Security.

8.2.3.2 The Java CoG

The Java Commodity Grid (CoG) Kit provides access to GT2 services through Java APIs. The goal of the Java CoG Kit is to provide Grid developers with the advantage to utilize much of the Globus functionality, as well as, access to the numerous additional libraries and frameworks developed by the Java community. Currently GT3 integrates part of Java CoG, e.g. many of the command-line tools in GT3 are implemented with the Java CoG.

The Java CoG has been focused on client-side issues. Grid services that can be accessed by the toolkit include:

- An information service compatible with the GT2 MDS implemented with Java Native Directory Interface JNDI [20].
- A security infrastructure compatible with the GT2 GSI implemented with the iaik security library [21].
- A data transfer mechanism compatible with a subset of the GT2 GridFTP and/or GSIFTP.
- Resource management and job submission with the GT2 GRAM Gatekeeper.
- Advanced reservation compatible with GT2 GARA [22].
- A MyProxy server managing user credentials.

8.2.4 First-generation Grid portal toolkits

In this section, we introduce four representative Grid portal toolkits: GridPort 2.0, GPDK, the Ninf Portal and GridSpeed. These toolkits provide some sort of assistance in constructing the first-generation Grid portals.

8.2.4.1 GridPort 2.0

The GridPort 2.0 (GP2) [23] is a Perl-based Grid portal toolkit. The purpose of GP2 is to facilitate the easy development of application-specific portals. GP2 is a collection of services, scripts and tools, where the services allow developers to connect Web-based interfaces to backend Grid services. The scripts and tools provide consistent interfaces between the underlying infrastructure, which are based on Grid technologies, such as GT2, and standard Web

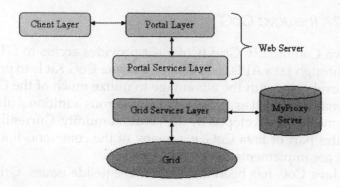

Figure 8.3 The architecture of GP2

technologies, such as CGI. Figure 8.3 shows the architecture of GP2. Its components are described below.

Client layer
The client layer represents the consumers of Grid portals, typically Web browsers, PDAs or even applications capable of pulling data from a Web server. Clients interact with a GP2 portal via HTML-form elements and use secure HTTP to submit requests.

Portal layer
The portal layer consists of portal-specific codes. Application portals run on standard Web servers and handle client requests and provide responses to those requests. One instance of GP2 can support multiple concurrent application portals, but they must exist on the same Web server where they share the same instance of the GP2 libraries. This allows the application portals to share portal-related user and account data and thereby makes possible a single-login environment. GP2 portals can also share libraries, file space and other services.

Portal services layer
GP2 and other portal toolkits or libraries reside at the portal services layer. GP2 performs common services for application portals including the management of session state, portal accounts and Grid information services with GT2 MDS.

Grid services layer
The Grid services layer consists of those software components and services that are needed to handle user requests to access the Grid. GP2 employs simple, reusable middleware technologies

8.2 FIRST-GENERATION GRID PORTALS

e.g. GT2 GRAM for job submission to remote resources; GT2 GSI and MyProxy for security and authentication; GT2 GridFTP and the San Diego Supercomputer Center (SDSC) Storage Resource Broker (SRB) for distributed file collection and management [24, 25]; and Grid Information Services based primarily on proprietary GP2 information provider scripts and the GT2 MDS.

GP2 can be used in two ways. The first approach requires that GT2 be installed because GP2 scripts wrap the GT2 command line tools in the form of Perl scripts executed from cgi-bin. GT2 GRAM, GSIFTP, MyProxy are used to access backend Grid services. The second approach does not require GT2, but relies on the CGI scripts that have been configured to use a primary GP2 Portal as a proxy for accessing GP2 services, such as user authentication, job submission and file transfer. The second approach allows a user to quickly deploy a Web server configured with a set of GP2 CGI scripts to perform generic portal operations.

8.2.4.2 Grid Portal Development Kit (GPDK)

GPDK [26] is another Grid portal toolkit that uses Java Server Pages (JSPs) for portal presentation and JavaBeans to access backend Grid resources via GT2. Beans in GPDK are mostly derived from the Java CoG kit. Figure 8.4 shows the architecture of GPDK. Grid service beans in GPDK can be classified as follows. These beans can be used for the implementation of Grid portals.

Security
The security bean, *MyproxyBean*, is responsible for obtaining delegated credentials from a MyProxy server. The *MyproxyBean* has a method for setting the username, password and designated lifetime of a delegated credential on the Web server. In addition, it allows delegated credentials to be uploaded securely to the Web server.

User profiles
User profiles are controlled by three beans: *UserLoginBean, UserAdminBean* and the *UserProfileBean*.

- The *UserLoginBean* provides an optional service to authenticate users to a portal. Currently, it only sets a username/password

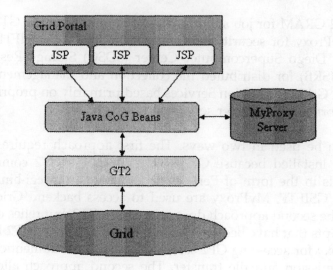

Figure 8.4 The GPDK architecture

and checks a password file on the Web server to validate user access.

- The *UserAdminBean* provides methods for serializing a *UserProfileBean* and validating a user's profile.
- The *UserProfileBean* maintains user information including preferences, credential information, submitted job history and computational resources used. The *UserProfileBean* is generally instantiated with session scope to persist for the duration of the user's transactions on the portal.

Job submission
The *JobBean* contains all the necessary functions used in submitting a job including memory requirements, name of executable code, arguments, number of processors, maximum wall clock or CPU time and the submission queue. A *JobBean* is passed to a *JobSubmissionBean* that is responsible for actually launching the job. Two varieties of the *JobSubmissionBean* currently exist. The *GramSubmissionBean* submits a job to a GT2 GRAM gatekeeper which can either run the job interactively or submit it to a scheduling system if one exists. The *JobInfoBean* can be used to retrieve a job-related time-stamped information including the job ID, status and outputs. The *JobHistoryBean* uses multiple *JobInfo* beans to provide a history of information about jobs that have been submitted. The history information can be stored in the user's profile.

8.2 FIRST-GENERATION GRID PORTALS

File transfer
The *FileTransferBean* provides methods for transferring files. Both *GSIFTPTranferBean* and the *GSISCPTransferBean* can be used to securely copy files from source to destination hosts using a user's delegated credential. The *GSISCPTransferBean* requires that GSI-enabled SSH [27] be deployed on machines to which file transfer via the GSI-enhanced "scp". The *GSIFTPTransferBean* implements a GSI-enhanced FTP for third-party file transfers.

Information services
The *MDSQueryBean* provides methods for querying a Lightweight Directory Access Protocol (LDAP) server by setting and retrieving object classes and attributes such as OS type, memory and CPU load for various resources. LDAP is a standard for accessing information directories on the Internet. Currently, the *MDSQueryBean* makes use of the Mozilla Directory SDK [28] for interacting with an LDAP server.

8.2.4.3 The Ninf Portal

The Ninf Portal [29] facilitates the development of Grid portals by automatically generating a portal front-end that consists of JSP and Java Servlets from a Grid application Interface Definition Language (IDL) defined in XML. The Ninf Portal then utilizes a Grid RPC system, such as Ninf-G [30] to interact with backend Grid services. Figure 8.5 shows the architecture of Ninf Portal. The Ninf Portal uses Java CoG to access a MyProxy server for the management of user credentials.

JSP
The portal user interface, which consists of JSPs and Java Servlets, can be automatically generated in the Ninf Portal. The JSP are used to interact with users and display messages on the client-side. They can also retrieve metadata from a data handling Servlet, which is used to read uploaded data, execute a Grid application and generate a result output page.

Ninf-G
Ninf-G is the Grid version of the Ninf system that runs on top of the GT2, offering network-based numerical library functionality via the use of RPC technology. Ninf-G supports asynchronous communications between Ninf-G clients and Ninf-G servers.

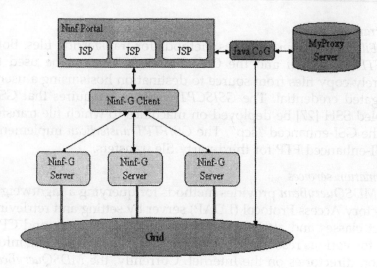

Figure 8.5 The Ninf Portal architecture

8.2.4.4 GridSpeed

GridSpeed [31], an extension of the Ninf Portal, is a toolkit for building Grid portals. It provides a Grid application portal-hosting server that automatically generates and publishes a customized Web interface for accessing the backend Grid services. The main aim of GridSpeed is to hide the complexity of the underlying infrastructure from Grid users. It allows developers to define and build their Grid application portals on the fly. GridSpeed focuses on the generation of portals for specific applications that provide services for manipulating complex tasks on the Grid. Figure 8.6 shows the architecture of GridSpeed. The main components are briefly described below.

Access Controller
Based on GT2 GSI, the Access Controller is used for user authentication and authorization. User credentials are managed by a MyProxy server and accessed via the Java CoG kit.

Descriptors
There are three kinds of descriptors: user, application and resource. A user descriptor contains information regarding a user's account information, a list of generated application portals and the location of the MyProxy server that is used to retrieve the user's credentials.

8.2 FIRST-GENERATION GRID PORTALS

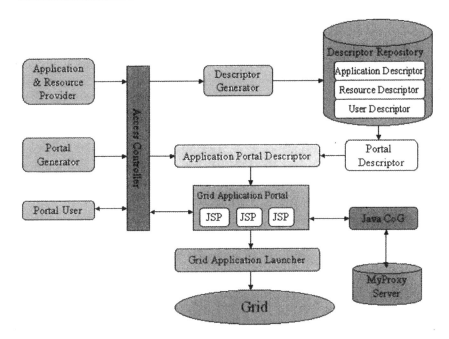

Figure 8.6 The architecture of GridSpeed

A resource descriptor contains information related to how to access a resource. An application descriptor contains information related to application information, such as parameters, template files and tasks. Each descriptor consists of an XML document defined by a GridSpeed XML Schema.

Descriptor Repository
The Descriptor Repository is used for searching, storing and editing all registered descriptors.

Application Portal Generator
The Application Portal Generator is the core component of the GridSpeed toolkit. It generates an application portal interface from a set of required descriptors that are dynamically loaded from the Descriptor Repository. The generator retrieves the necessary XML documents, which are then marshalled into Java objects via Castor [32], an open-source data binding framework for Java that can generate Java objects from XML descriptions. The generator produces a JSP file from the Java objects, which implement the actual application portal page.

8.2.5 A summary of the four portal tools

As shown in Table 8.1, the four toolkits provide various levels of support for portal developers to build Grid portals. Apart from GP2, which uses HTML pages for the portal–user interface, the other three toolkits use JSP technology. Grid portals can be grouped into two categories: user portals and application portals. A user portal provides a set of fundamental services for portal users, which includes single sign-on, job submission and tracking, file management, resource selection and data management. An application portal provides application-related services for users, e.g. to construct a domain application for the Grid. In this context, GP2 and GPDK are Grid user portal toolkits, and the Ninf Portal and GridSpeed are Grid application portal toolkits.

From the portal support point of view, GP2 provides a portal template and some CGI scripts in Perl for portal construction; GPDK provides a set of Java Beans for portal construction; the Ninf Portal can automatically generate a portal–user interface; GridSpeed can automatically generate a whole portal. When designing a Grid portal, the Ninf Portal allows portal developers to specify how to generate a portal via an application descriptor; GridSpeed provides a comprehensive mechanism supporting application, resource and user descriptors; GP2 and GPDK do not support this feature. Apart from GP2, the other three portal toolkits use Java CoG to access Grid resources. To provide secure access, all

Table 8.1 A comparison of portal tool kits

	GridPort 2.0	GPDK	The Ninf Portal	GridSpeed
Portal pages	HTML	JSP	JSP	JSP
Portal support	User portal	User portal	Application portal	Application portal
Portal construction	Perl/CGI	JavaBeans	Portal JSP generation	Portal generation
Portal descriptor	Not supported	Not supported	Application level	Application/ resource/ user level
Use of Java CoG	No	Yes	Yes	Yes
Use of MyProxy	Yes	Yes	Yes	Yes
Use of Globus	Yes	Yes	Yes	Yes
Portal customization	No	No	No	Being supported

the four portal toolkits use MyProxy for the management of user credentials. All the four toolkits access backend Grid resources via GT2 or earlier versions of Globus.

Whereas the four portal toolkits can provide some sort of assistance in building Grid portals, they are mainly used by portal developers instead of portal users, who cannot easily modify an existing portal to meet their specific needs. Portals developed at this stage are not customizable by the users. The GridSpeed development team is currently working on the issue.

8.2.6 A summary of first-generation Grid portals

First-generation Grid portals have been focused on providing basic task-oriented services, such as user authentication, job submission, monitoring and data transfer. However, they are typically tightly coupled with Grid middleware tools such as Globus. The main limitations of first-generation portals can be summarized as follows.

Lack of customization
Portal developers instead of portal users normally build portals because the knowledge and expertise required to use the portal toolkits, as described in this chapter, is beyond the capability of most Grid end users. When end users access the Grid via a portal, it is almost impossible for them to customize the portal to meet their specific needs, e.g. to add or remove some portal services.

Restricted Grid services
First-generation Grid portals are tightly coupled with specific Grid middleware technologies such as Globus, which results in restricted portal services. It is hard to integrate Grid services provided by different Grid middleware technologies via a portal of this generation.

Static Grid services
A Grid environment is dynamic in nature with more and more Grid services are being developed. However, first-generation portals can only provide static Grid services in that they lack a facility to easily expose newly created Grid services to users.

While there are limitations with first-generation Grid portals and portal toolkits, the experiences and lessons learned in developing Grid portals at this stage have paved the way for the development of second-generation Grid portals.

8.3 SECOND-GENERATION GRID PORTALS

In this section, we discuss the development of second-generation Grid portals. To overcome the limitations of first-generation portals, portlets have been introduced and promoted for use in building second-generation Grid portals. Currently, portlets are receiving increasing attention from both the Grid community and industry. In this section we review the current status of portlet-oriented portal construction. First we introduce the concepts of portlets and explain the benefits that they could provide.

8.3.1 An introduction to portlets

8.3.1.1 What is a portlet?

From a user's perspective, a portlet [33] is a window (Figure 8.7) in a portal that provides a specific service, e.g. a calendar or news feed. From an application development perspective, a portlet is a software component written in Java, managed by a portlet container, which handles user requests and generates dynamic contents. Portlets, as pluggable user interface components, can pass information to a presentation layer of a portal system. The content

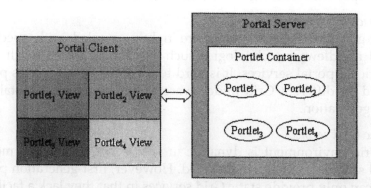

Figure 8.7 A portal with four portlets

generated by a portlet is also called a fragment. A fragment is a chunk of markup language (e.g. HTML, XHTML) adhering to certain rules and can be aggregated with other fragments to form a complete document. The content of a portlet is normally aggregated with the content of other portlets to form the portal page. A portlet container manages the life cycle of portlets in a portal.

8.3.1.2 Portlet container

A portlet container provides a run-time environment in which portlets are instantiated, executed and finally destroyed. Portlets rely on the overall portal infrastructure to access user profile information, participate in window-and-action events and communicate with other portlets, access remote content, lookup credentials and store persistent data. A portlet container manages and provides persistent storage mechanisms for portlets.

A portlet container is not a standalone container like a Java Servlet container; instead, it is implemented as a layer on top of the Java Servlet container and reuses the functionality provided by the Servlet container.

Figure 8.8 shows a Web page with two portlets. A portlet on a portal has its own window, a portlet title, portlet content (body) which can be rendered with `portlet.getContent()` method, and some actions to close, maximize or minimize the portlet.

8.3.1.3 Portlets and Java Servlets

Portlets are a specialized and more advanced form of Java Servlets. They run in a portlet container inside a servlet container which is a layer that runs on top of an application server. Like Java Servlets, portlets process HTTP requests and produce HTML output, e.g. with JSP. But their HTML output is only a small part of a Web page as shown in Figure 8.8. The portal server fills in the rest of the page with headers, footers, menus and other portlets.

Compared with Java Servlets, portlets are administered in a dynamic and flexible way. The following updates can be applied without having to stop and restart the portal server.

- A portlet application, consisting of several portlets, can be installed and removed using the portal's administrative user interface.

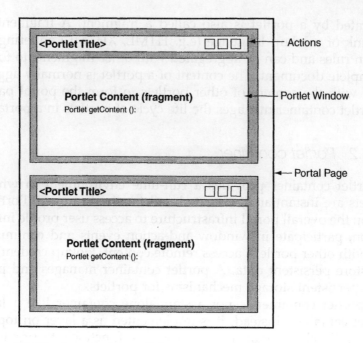

Figure 8.8 The layout of a portlet

- An administrator with the appropriate access rights can change the settings of a portlet.
- Portlets can be created and deleted dynamically.

8.3.1.4 Advantages of portlets over Java Servlets

Portlets also have many standard features that are not available to Java Servlets. One key feature is the built-in support to automatically use different JSP interfaces with different user devices. This allows users to write portlets that work on many devices, such as desktop computers with modern Web browsers, palmtop computers with limited Web browsers, alternatively Personal Digital Assistants (PDAs) or Web-enabled wireless phones. Users do not need to provide portability via the lowest common denominator. By reusing the same underlying business logic, the portal server will choose the most appropriate rendering for each client. Users can even have multiple portlet controllers which allows different page/action sequences to be used for each device type.

8.3.1.5 Portlet presentation

A portlet window consists of:

- A Title bar with the title of the portlet.
- Decorations, including buttons to change the window state of the portlet, such as maximize or minimize, the portlet or ones to change the mode of a portlet, such as show help or edit predefined portlet settings.
- Content produced by the portlet.

8.3.1.6 Portlet life cycle

The basic life cycle of a portlet includes the following three parts:

- *Initialization*: Using the *init* class to initialize a portlet and put it into service.
- *Request handling*: Processing different kinds of actions and rendering content for different clients.
- *Termination*: Using the *destroy* class to remove a portlet from a portal page.

The portlet receives requests based on the user interaction with the portlet or portal page. The request processing is divided into two phases:

- *Action processing*: If a user clicks on a link in a portlet, an action is triggered. The action processing must be finished before any rendering of the portlets on the page is started. In the action phase, the portlet can change the state of the portal.
- *Rendering content*: In the rendering phase, the portlet produces its markup content to be sent back to the client. Rendering should not change any state of the portlet. It refreshes a page without modifying the portlet state. Rendering multiple portlets on a page can be performed in parallel.

8.3.1.7 Access Web services via portlets

Figure 8.9 shows how to access Web services from a Web portal via a portlet. When a Web portal receives a servlet request, it generates

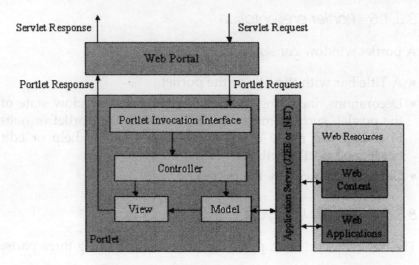

Figure 8.9 Accessing Web services in a Web portal via a portlet

and dispatches events to the portlet using parameters in the request and then invokes the portlet to be displayed through the Portlet Invocation Interface (PII). The portlet's internal design normally follows the Model-View-Controller (MVC) model [34] which splits the portlet functionality into a controller receiving incoming requests from the portlet PII, invoking commands operating on a model that encapsulates application data and logic to access backend Web content or applications and finally calling views for presentation of the results.

8.3.1.8 Events to access a Web page with portlets

The typical sequence of events to access a Web page via portlets is given below.

- A client (e.g. a Web browser) after being authenticated makes an HTTP request to a portal;
- The portal receives the request;
- The portal determines if the request contains an action targeted to any of the portlets associated with the portal page;
- If there is an action targeted to a portlet, the portal requests the portlet container to invoke the portlet to process the action;

- A portal invokes portlets, through the portlet container, to obtain content fragments that can be included in the resulting portal page;
- The portal aggregates the output of the portlets in the portal page and sends the portal page back to the client.

8.3.2 Portlet specifications

It is important for portlets developed from independent vendors to interoperate with each other. There is an urgent need to have a standard Portlet API for developing portlets. Currently, there are two groups that are working on the standardization of portlets. One is OASIS (Organization for the Advancement of Structured Information Standards) [35]. The other is JCP (Java Community Process) [36].

8.3.2.1 OASIS WSRP

The OASIS is a worldwide consortium that drives the development, convergence and adoption of a variety of e-business standards. The consortium has more than 400 corporate and individual members in over 100 countries. Web Services for Remote Portlets (WSRP) is an OASIS specification that will provide the "plug and play" of portlets, intermediary content aggregation applications and integration with applications from different sources. WSRP will allow applications to consume and/or produce Web services. These Web services will incorporate presentation elements and information that allow portal administrators to select and display portlets that originate from virtually anywhere on the Web without the need for further integration code. The WSRP producers and consumers may be implemented on different platforms, such as a J2EE or .NET.

The current goals of WSRP are to:

- Allow interactive Web services to be plugged into standards-compliant portals;
- Let anybody create and publish their contents and applications as Web services;
- Let administrators browse directories of WSRP services to plug into their portals with minimal programming effort;

- Let portals publish portlets so that they can be consumed by other portals without further programming;
- Make the Internet a marketplace of visual Web services, ready to be integrated into portals.

8.3.2.2 JSR 168

The JCP is an open organization of international Java developers and licensees whose charter is to develop and revise Java technology specifications, reference implementations and technology toolkits. Java Specification Requests (JSRs) are the actual descriptions of proposed and final specifications for the Java platform.

JSR 168 [37] is a JCP portlet specification that defines a set of Java APIs that permits interoperability between portals and portlets. It defines portlets as Java-based Web components, managed by a portlet container that process requests and generate dynamic content. Portals use portlets as pluggable user interface components that provide a presentation layer to information systems. This specification defines a Portlet API for portal composition addressing the areas of aggregation, personalization, presentation and security.

The current goals of JSR 168 are to:

- Define the run-time environment or the portlet container, for portlets;
- Define the Portlet API between portlet containers and portlets;
- Provide mechanisms to store transient and persistent data for portlets;
- Provide a mechanism that allows portlets to include servlets and JSPs;
- Define the packaging of portlets to allow easy deployment;
- Allow binary portlet portability among JSR 168 portals;
- Run JSR 168 portlets as remote portlets using the WSRP protocol.

8.3.2.3 WSRP and JSR 168

Although they are being governed by different standards bodies and review processes, WSRP and JSR 168 are complementary specifications. While JSR 168 defines a standard Portlet API that is specific to Java-based portals, WSRP defines a universal API that allows portals of any type to consume portlets of any type. They can be used together in the following two ways:

- Portlets written with the Java Portlet API may be wrapped as WSRP services and published in UDDI directories.
- WSRP services can be exposed as portlets with the Java Portlet API to aggregate them in portals.

Whereas JSR 168 defines a set of Java APIs that allows portlets to run on any compliant portals, WSRP allows Web services to be exposed as portlets in a plug-and-play fashion.

8.3.3 Portal frameworks supporting portlets

In this section, we introduce three representative and popular portal frameworks, namely Jetspeed, WebSphere's Portal and GridSphere. They have been widely used for building Web portals with portlets. We first describe the three frameworks and then compare them in terms of JSR 168 support, easy to use, availability and pre-built portlets.

8.3.3.1 Jetspeed

Jetspeed [38] is an open-source project from the Apache Software Foundation for building portals with portlets in Java. It executes with Tomcat Web server and uses the Cocoon [39], an XML publishing framework for processing XML information via XSLT. Jetspeed is the original source of the JSR 168. Jetspeed supports the RSS (Really Simple Syndication) [40] and OCS (Open Content Syndication) [41] formats. RSS is an XML format used for syndicating Web headlines. The OCS format describes multiple-content channels, including RSS headlines.

Whilst it comes with built-in portlets for OCS, RSS and for embedding HTML sources, creating new portlets requires Java programming. Modifying the look and feel of a portal from the Jetspeed default also requires JSP or XSLT programming. Jetspeed makes connections to external data and content feeds to retrieve and display the data. Users can implement a portal and access it from a Web browser or a wireless device, such as a WAP [42] phone or Palm device. Jetspeed supports built-in services for user interface customization, caching, persistence and user authentication, eliminating the need to implement these services. Some of the high-level features of Jetspeed include:

- A basis for standardizing the Java Portlet API specification (JSR 168)
- Template-based layouts including those for JSP and Velocity [43]
- Supports remote XML content feeds via OCS
- Custom default home page configuration
- Database user authentication
- In-memory cache for fast page rendering
- RSS support for syndicated content
- Integration with Cocoon, WebMacro [44] and Velocity to allow development with the latest XML/XSL technologies
- Wireless Markup Language (WML) support [45]
- XML-based configuration registry of portlets
- Web application development infrastructure
- Local caching of remote content
- Synchronization with Avantgo [46]
- Integrated with Turbine [47] modules and services
- A Profiler Service to access portal pages based on user, security (groups, roles), media types and language
- Persistence services available to all portlets to provide store state per user, page and portlet
- Interface skins so that users can choose colours and display attributes
- Customizer for selecting portlets and defining layouts for individual pages
- Portlet Structure Markup Language (PSML) can be stored in a database
- User, group, role and permission administration via Jetspeed security portlets
- Role-based security access to portlets.

Figure 8.10 shows the architecture of Jetspeed.

8.3 SECOND-GENERATION GRID PORTALS

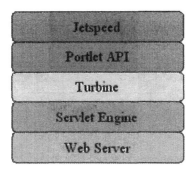

Figure 8.10 The architecture of Jetspeed.

Jetspeed is built on top of Turbine, a servlet-based framework, which is also part of the Jakarta Apache Project. Turbine handles user authentication and page layout as well as scheduling. Jetspeed can run on a number of servlet engines and databases. Jetspeed is bundled with Thomas Mueller's Hypersonic SQL database [48]. Tables are already created and populated with user data in Hypersonic SQL. Hypersonic SQL runs in process to Jetspeed (and Tomcat), so no additional configuration is necessary. To use a different database, such as Oracle, DB2, Sybase, mySQL or PostgresSQL [49], it is necessary to set up the database using the SQL scripts included with the Jetspeed source code. In addition, you must configure the `TurbineResources.properties` file that Jetspeed and Turbine use to point to the new database server.

Turbine
Turbine is a Java Servlet-based framework that allows Java developers to build secure Web applications. Turbine supports the MVC pattern that separates presentation from backend application/business logic. Turbine is made up of five different modules: Page, Action, Layout, Screen and Navigation. The invocation sequence in manipulating pages with the fives modules is shown in Figure 8.11.

The Page module is the first module in the chain of execution for the Page generation. It is considered to be the module that contains the rest of the modules (Action, Layout, Screen and Navigation). The Page module checks to see if there has been an Action defined in the request. If so, it attempts to execute that Action. After the Action has been executed, it asks the set Screen object for

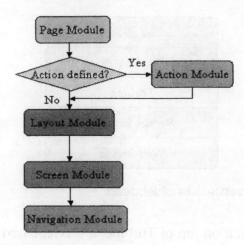

Figure 8.11 The sequences in the execution of Turbine components.

its Layout. The Page module then attempts to execute the Layout object returned by the Screen module.

The Layout module is called from the Page module. This module defines the physical Layout of a Web page. It generally defines the location of the Navigation portion (i.e. the top and bottom part of the Web page) as well as the location of where the body (or Screen) of the page is. The Layout module executes the Screen module to build the body of the Web page. It executes the Navigation modules to build the portions of the Web page that defines the navigation for the Web site.

PSML
Portlets are registered manually with Jetspeed using PSML. PSML informs Jetspeed about what portlets are available and registered with it. The configuration file for portlets is *jetspeed-config.jcfg* in the *WEB-INF/conf* directory. These default configuration files are called *default.psml* and *defaultWML.psml* and reside in *WEB-INF/psml*. Once a user is created, each user has configuration files associated with them: *homeHTML.psml* and *homeWML.psml*. These files are stored in *WEB-INF/psml/<username>* for each user. PSML is composed of two markups: registry and site markup.

Apache Cocoon
Apache Cocoon is a Web application development framework that implements the notion of "component pipelines", where each component in a pipeline specializes in a particular operation. This

8.3 SECOND-GENERATION GRID PORTALS

makes it possible to use a building block approach for constructing Web applications, by assembling components into pipelines without any programming.

WebMacro
WebMacro is a pure Java open-source template language and a viable alternative to JSP, PHP [50] and ASP [51].

Velocity
Velocity is a Java-based template engine. Velocity provides an alternative scripting language to JSP for developing portal pages.

ECS
The Element Construction Set (ECS) [52] is a Java API for generating elements for various markup languages including HTML 4.0 and XML, but can be extended to create tags for any markup language. ECS contains Java classes for the representation of HTML or XML tags.

Implementation of portlets with Jetspeed
Figure 8.12 shows a canonical example of the "Hello World" portlet using Jetspeed. In Jetspeed, all portlets inherit from `AbstractPortlet` and must provide a `getContent()` method to supply the portlet information. The "Hello World" example, as shown in Figure 8.12, uses `StringElement`, an ECS class, which can translate to WML or HTML, whichever is appropriate. The `RunData` object that gets passed in to `getContent()` is a Turbine object that stores various Servlet objects, e.g. `ServletContext`, `ServletRequest` or `HttpSession` as well as other Turbine objects including Users.

```
import org.apache.jetspeed.portal.portlets.AbstractPortlet;
import org.apache.turbine.util.RunData;
import org.apache.ecs.ConcreteElement;
import org.apache.ecs.StringElement;

public class HelloWorldPortlet extends AbstractPortlet
{
   public ConcreteElement getContent (RunData runData)
   {
     return (new StringElement ("Hello World!"));
   }
}
```

Figure 8.12 A "Hello World" portlet implemented with Jetspeed

The next step necessary for the deployment of a portlet is to add it to the Jetspeed Portlet Registry. A registry fragment file must be created that provides the portlet name, a description and the classname. The "Hello World" example registry file is shown in Figure 8.13.

8.3.3.2 IBM WebSphere Portal 5.0

The WebSphere Portal [53] is a J2EE application that runs on WebSphere Application Server. Its main function is to serve the WebSphere Portal framework to desktops and mobile devices of portal users. The WebSphere Portal creates an environment that provides the connectivity, administration and presentation services required. It has a portal toolkit that provides the capability to customize and manage an enterprise portal and create, test, debug and deploy individual portlets and their content. Templates allow developers to create their own portlets. Debugging and deployment tools are available to help shorten the development cycle. Sample portlets that demonstrate best programming practices are also provided.

WebSphere Portal 5.0 features
WebSphere Portal 5.0 has the following main features:

- A Portal framework that allows a user to integrate applications and data sources, and also perform administrative tasks, such as controlling portal membership.

```
<?xml version="1.0" encoding="UTF-8"?>
<registry>
   <portlet-entry name="Hello World" hidden="false"
        type="instance" application="false">
     <meta-info>
        <title>Hello World</title>
        <description>Hello World</description>
     </meta-info>
        <classname>Hello WorldPortlet</classname>
        <media-type ref="html"/>
   </portlet-entry>
</registry>
```

Figure 8.13 The "HelloWorldPortlet" descriptor in Jetspeed

8.3 SECOND-GENERATION GRID PORTALS

- Presentation services are used to create portal–user interface with a graphical user interface that can be customized to match the user's specific needs. A user can also configure a portal to render content on different client device types, such as laptops and mobile phones.
- Connectivity services allow a portal to access different data sources.
- Customization controls allow users to select which applications and content they view and how this information is organized on their portal pages.
- Portlets are included for accessing applications such a email, calendars, collaboration and syndicated news. Portlets are the integration codes that connect applications and data to the portal.
- A portlet API allows users to create new portlets and to add new applications and data sources to a portal as required. A toolkit is provided to simplify this task.
- Authenticated "single sign-on" allows authorized individuals to logon to multiple portal applications all at once without the need to remember multiple username and passwords and to sign on to each application individually.
- Administration or the delegation of administration of portal users, portlets and pages.

WebSphere Portal 5.0 is based on the IBM Portlet API, which is fundamentally similar to the Portlet API from JSR 168 with some differences. Stephan Hepper provides a comprehensive comparison between IBM WebSphere Portlet API and JSR 168 Portlet API in [54].

Implementation of portlets with WebSphere
Figure 8.14 shows a "Hello World" portlet implemented with WebSphere. Unlike the Jetspeed "Hello World" portlet example, there are no references to Turbine objects in WebSphere. In addition, the portlet extends the PortletAdaptor rather than AbstractPortlet, and offers an API very similar to the servlet API. Resources are initialized in the **init** method and the service method is used to handle client requests. Figure 8.15 shows the portlet registry file used to register the "Hello World" portlet in WebSphere.

```
package com.ibm.wps.samplets.helloworld:

import org.apache.jetspeed.portlet.*;
import org.apache.jetspeed.portlets.*;
import java.io.*;
public class Hello World extends PortletAdaptor
{
  public void init(PortletConfig portletConfig) throws UnavailableException
  {
    super.init( portletConfig );
  }
  public void service(PortletRequest portletRequest,
                      PortletResponse portletResponse)
                      throws PortletException, IOException
  {
    PrintWriter writer = portletResponse.getWriter();
    writer.println("<p>Hello World!</p>");
  }
}
```

Figure 8.14 A "Hello World" portlet implemented with WebSphere

8.3.3.3 GridSphere

GridSphere [55] is an open-source research project from the GridLab project [56]. It provides a portlet implementation framework based upon the IBM Portlet API and an infrastructure for supporting the development of re-usable portlet services. GridSphere allows developers to create and package third-party portlet-based Web applications that can be executed and administered within the GridSphere portlet container.

GridSphere includes a set of core portlets and services, which provide the basic infrastructure needed for developing and administering Web portals. A key feature of GridSphere is that it builds upon the Web Application Repository (WAR) deployment model to support third-party portlets. In this way, developers can distribute and share their work with other projects that use GridSphere to support their portal development.

GridSphere features
GridSphere has the following features:

- The Portlet API implementation in GridSphere is almost fully compatible with IBM WebSphere Portal version 4.2 or higher.
- Support for the easy development and integration of "third-party" portlets that can be plugged into the GridSphere portlet container.

8.3 SECOND-GENERATION GRID PORTALS

```xml
<?xml version="1.0" encoding="UTF-8"?>
<!DOCTYPE portlet-app-def PUBLIC "-//IBM//DTD Portlet Application 1.1//EN" "portlet_1.1.dtd">
<portlet-app-def>
<portlet-app uid="com.ibm.samples.HelloWorld.4943.1"> <portlet-app-name>Hello World Portlet
<portlet href="WEB-INF/web.xml#Servlet_439329280" id="Portlet_439329280">
    <portlet-name>HelloWorld</portlet-name>
    <cache><expires>0</expires><shared>no</shared></cache>
    <allows><minimized/></allows>
    <supports><markup name="html"><view/></markup></supports>
</portlet>
</portlet-app>
<concrete-portlet-app uid="640682430">
<portlet-app-name>Concrete Hello World - Portlet Sample#1</portlet-app-name>
<context-param>
    <param-name>Portlet Master</param-name>
    <param-value>yourid@yourdomnain.com</param-value>
</context-param>
<concrete-portlet href="#Portlet_439329280">
    <portlet-name>Hello World</portlet-name>
    <default-locale>en</default-locale>
    <language locale="en_US">
        <title>Hello World - Sample Portlet #1</title><title-short>Hello-World</title-short>
        <keywords>portlet hello world</keywords>
    </language>
</concrete-portlet>
</concrete-portlet-app>
</portlet-app-def>
```

Figure 8.15 The "Hello World" descriptor in WebSphere

- A high-level model for building complex portlets using visual beans and the GridSphere User Interface (UI) tag library.
- A flexible XML-based portal presentation description that can be modified to create customized portal layouts.
- A built-in support for Role-Based Access Control (RBAC) [57] in which users can be guests, users, administrators and super users.
- A portlet service model that allows for creation of "user services", where service methods can be limited according to user rights.
- Persistence of data provided using Hibernate for RDBMS database [58] support.
- Integrated Junit [59] and Cactus [60] unit tests for complete server-side testing of portlet services including the generation of test reports.
- GridSphere core portlets offer base functionality including login, logout, user and access control management.
- Localization support in the Portlet API implementation and GridSphere core portlets that support English, French, German, Czech, Polish, Hungarian and Italian.

The current GridSphere release provides a portal, a portlet container and a core set of portlets including user and group management, as well as layout customization and subscription.

Implementations of portlets with GridSphere
Figure 8.16 shows a "Hello World" portlet with GridSphere. All portlets need to extend the **AbstractPortlet** class. Figure 8.17 shows the "Hello World" portlet descriptor.

```
package portlets.examples;
public class Hello World extends AbstractPortlet
{
    public void doView(PortletRequest request, PortletResponse response)
        throws PortletException, IO Exception
    {
        PrintWriter out = response.getWriter();
        out.println("<h1>Hello World</h1>");
    }
}
```

Figure 8.16 The "Hello World" portlet in GridSpeed

```xml
<portlet-app-collection>
  <portlet-app-def>
  <portlet-app id="portlets.examples.HelloWorld">
    <portlet-name>Hello World Portlet Application</portlet-name>
    <servlet-name>HelloWorld</servlet-name>
    <portlet-config>
      <param-name>Portlet Master</param-name>
      <param-value>yourid@yourdomain.com</param-value>
    </portlet-config>
    <allows> <maximized/> <minimized/> <resizing/> </allows>
    <supports> <view/> <edit/> <help/> <configure/> </supports>
  </portlet-app>
  <concrete-portlet-app id="portlets.examples.HelloWorld.1">
    <context-param>
      <param-name>foobar</param-name> <param-value>a value</param-value>
    </context-param>
    <concrete-portlet> <portlet-name>Hello World</portlet-name>
      <default-locale>en</default-locale>
      <language locale="en_US"> <title>Hello World</title> <title-short>Hello World</title-short>
        <description>Hello World - Sample Portlet #1</description>
        <keywords>portlet hello world</keywords> </language>
      <config-param>
        <param-name>Portlet Master</param-name>
        <param-value>yourid@yourdomain.com </param-value>
      </config-param> </concrete-portlet></concrete-portlet-app>
  </portlet-app-def>
</portlet-app-collection>
```

Figure 8.17 The "Hello World" portlet descriptor in GridSphere

8.3.4 A Comparison of Jetspeed, WebSphere Portal and GridSphere

IBM WebSphere Portal, Jetspeed and GridSphere are portal frameworks that can be used to build Web portals with portlets. While they focus on portlets construction, they differ in their implementations and features. In this section, we give a comparison.

JSR 168 support
JSR 168 is becoming the standard Portlet API for building portable portlets. The portlet API from Jetspeed is the original source of JSR 168. The future Jetspeed 2 supports JSR 168. Jetspeed 2 uses Apache Pluto [61] container, which is the reference implementation of the JSR 168. Although the portlet API from IBM WebSphere Portal and the API in JSR 168 differ in some aspects, they share fundamental features in building portlets. The recently released version WebSphere Portal V5.02 supports JSR 168. The portlet API from GridSphere is JSR 168 compliant too.

Ease of use
Compared with Jetspeed and GridSphere, IBM WebSphere Portal provides an Integrated Development Environment (IDE) for building portlets.

Availability
Jetspeed and GridSphere are open source-based frameworks; IBM WebSphere Portal is not open source.

Pre-built Portlets
Jetspeed includes:

- An RSS portlet for rendering RSS documents as HTML pages.
- A FileServer portlet for providing static HTML pages.
- The Cocoon Portlet for taking a style sheet and a URL as parameters and transforming, and then returning the content to the user.
- A PortletViewer for providing additional information about a portlet including its configuration options, URL and properties.

IBM WebSphere Portal includes:

- A FileServer portlet for providing static HTML pages.
- A ServletInvoker portlet for invoking a servlet as a portlet, a JSP Portlet for generating JSP.

8.3 SECOND-GENERATION GRID PORTALS

- A CSVViewer portlet for displaying a file with data arranged in comma-separated values format.
- An RSS Portlet for rendering RSS documents as HTML pages.

GridSphere includes:

- Login/Logout Portlets for users to log in or out of a portal.
- An AccountRequest portlet for creating a new portal user to request an account.
- An AccountManagement portlet for managing users' accounts.
- A PortletSubscription portlet for users to add and remove portlets from their workspace.

8.3.5 The development of Grid portals with portlets

Portlet technology is gaining attention from the Grid community for building second-generation Grid portals to overcome problems encountered in first-generation Grid portal development frameworks and toolkits. A portlet in a Grid portal is not just a normal portlet that can be plugged into a portal; it is also associated with a backend Grid service. We define a portlet associated with a Grid service to be called a Grid Portlet. Figure 8.18 shows how to access a Grid service from a Grid portal via a Grid Portlet. The model is that a Grid Portlet interacts with a Grid service provided by Grid middleware such as Globus to access backend resources. Since Grid services provided by difference service providers using different Grid middleware technologies can be exposed as standard portlets, portals built from portlets are loosely coupled with Grid middleware technologies. Portal frameworks such as Jetspeed, WebSphere Portal and GridSphere have been widely used for building Web portals with portlets. They are being integrated with Grid services for constructing Grid portals with Grid Portlets. Currently no such framework exists that can provide an IDE in which a Grid portal can be visually built with Grid Portlets that are associated with backend Grid services.

With funding from the National Science Foundation Middleware Initiative (NMI), from the USA, the Open Grid Computing Environments (OGCE) project [62] was established in Fall 2003 to foster

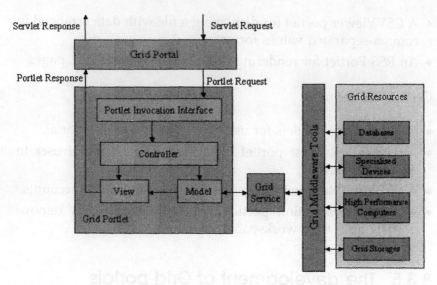

Figure 8.18 Access Grid resources via Grid Portlets

collaborations and sharable components with portal developers worldwide. Tasks include the establishment of a Grid Portal Collaboratory, a repository of portlets and portal service components, an online forum for developers of Grid Portals and the building of reusable portal components that can be integrated in a common portal container system.

The development of OGCE is based on the following projects:

- Java CoG Kit sponsored by SciDAC and NSF Alliance.
- The CHEF Project [63] of the University of Michigan.
- The Grid Portals Information Repository (GPIR) [64] and GridPort of the Texas Advanced Computing Center.
- The Alliance Portal Expedition project [65], including NCSA, Indiana University's Extreme Labs and the Community Grids Lab at Indiana University.

The Alliance Portal is an ongoing project that focuses on building second-generation Grid portals. It is based on Jetspeed and targeted at the construction of Grid portals using Grid Portlets. Currently the Alliance Portal can provide the following Grid Portlets that are leveraged from existing Grid services.

- *A Proxy Manager*: The Proxy Manager portlet is a utility that allows users to load GSI proxy credentials into their account via MyProxy.
- *An LDAP Browser*: The LDAP Browser portlet is an interface to access the contents of the LDAP servers.
- *A GridFTP Client*: The GridFTP Client portlet provides the basic client functions of Grid FTP with a user-friendly interface.
- *Gram Job Launcher*: The Gram Job Launcher portlet allows a user to submit jobs to a Grid environment using the Globus GRAM protocol. For this the user must have a valid GSI Proxy Certificate which can be loaded through the Proxy Manager portlet.
- *Grid utilities*: These include the "Grid Ping" utility and the "Grid Job Submission" utility. The GridPing portlet can ping a resource in a Grid environment and determine if a user has the access to it or not. The Grid Job Submission portlet is similar to the Gram Job Launcher portlet except that this portlet will not return till the job has been completed. The output and the error of the job is displayed by the portlet.
- *OGSA Browser*: The OGSA Browser portlet allows users to query a Grid service for its Service Data Elements (SDE). Users can query a SDE by using the name of the SDE as the query string. Users can also obtain a list of SDEs by using "serviceDataName" as the query string. Once users get a list of all the SDEs, they can click on each SDE to query it.

8.3.6 A summary on second-generation Grid portals

Second-generation Grid portals will be produced from pluggable Grid Portlets. Running inside a portlet container, portlets can be added into or removed from a portal, thus providing users with the ability to customize Grid services at a portal level. Grid Portlets are independent components that are based on existing Grid services. A Grid portal built from Grid Portlets can provide users with the ability to integrate services provided by different Grid-enabling technologies. Second-generation Grid portals with portlets have the following benefits compared with first-generation Grid portals.

- *Portal customization*: Users instead of a Grid system developer can construct their personalized portals out of the available portlets to meet their specific needs. Portlets can be easily added or removed from a portal.
- *Extensible Grid services*: Portals built from portlets are loosely coupled with Grid middleware technologies since Grid services can be exposed as standard portlets. A portal constructed from portlets provides users with the ability to integrate services from different Grid service providers.
- *Dynamic Grid services*: New services and components are being developed for the Grid. A Grid portal should be able to provide users with the ability to access dynamic Grid services in a Grid environment. To this end, a mechanism can be provided to expose Grid services as individual portlets that can be published and accessed via a portal.

8.4 CHAPTER SUMMARY

In this chapter, we have studied Grid portal-related technologies. Grid portals are Web-based interfaces that provide a single access point to the Grid. To make the Grid a reality, portals will play a critical role as they can provide user-friendly interfaces for the majority of end users to access the Grid, who may know little about the Grid infrastructure and services. The development of Grid portals has evolved from first to second generation, which are currently emerging. While first-generation portals provide services such as job submission, resource monitoring and data transfer, they mainly lack the ability to be customized. Portals of this generation were mainly built by portal system developers who had expertise in the Grid. Existing portal toolkits such as GPDK, GP2, the Ninf Portal and GridSpeed can provide some levels of assistance in building first-generation Grid portals. However, it is hard for end users who know little about Grid technologies to use these toolkits to build portals to meet their application-specific requirements.

Portlet technology is gaining increasing attention from the Grid community for building second-generation portals. Portals of this generation use portlets, which are pluggable software components that run inside a portlet container within a portal server. The portlet container manages the run-time behaviour of portlets. Grid end users can build their own portals by choosing portlets available.

While existing portal toolkits such as GridPort will support portlets in its future version GridPort 3.0, new portal frameworks such as Jetspeed, IBM WebSphere Portal and GridSphere have been developed for portlet construction. While these portal frameworks were originally focused on building Web portals with portlets, they are being integrated with Grid services to produce Grid Portlets. JSR 168 is becoming a standard for defining a Portlet API to build portable portlets. Currently IBM WebSphere V5.02 and GridSphere are JSR 168 compliant. The future Jetspeed 2.0 will be JSR 168 compliant too.

OGSA, as promoted by the Globus group and the Global Grid Forum, is the *de facto* standard in building service-oriented Grid systems. To this end, the development of portlets should take OGSA into account. Future portlets should be OGSA compliant, which means that these portlets should be associated with Grid services developed and deployed via OGSA compliant middleware. The work is currently being integrated into GridPort 3.0 and GridSphere.

8.5 FURTHER READING AND TESTING

JSP and Java Servlets have been widely used to generate dynamic Web pages for portals. There are many books on JSP and Java servlets. You also need knowledge on Globus, MyProxy and the Java CoG before you start your portal work. Detailed information can be found on their Web sites.

For testing purpose, you can start with your second-generation Grid portal development using portlets. You can choose open source-based Jetspeed or GridSphere as the portal framework. It will be easier to build portals with portlets using IBM WebSphere Portal because it provides a GUI-based integrated development environment.

8.6 KEY POINTS

- A Grid portal provides a Web page-based user interface as a single access point to the Grid.
- MyProxy has been widely used for the management of user credentials.

- JSP and Java Servlets are used for dynamically generating portal pages.
- Grid portals can be broadly classified into first- and second-generation portals.
- Grid portals mainly use JSP and JavaBeans to communicate with the Java CoG to interact with backend Globus-based services, specifically GT2.
- Existing Grid portal tools such as GPDK, GridPortal, the Ninf Portal and GridSpeed can provide some kinds of assistance in building first-generation Grid portals.
- First-generation portals are tightly coupled with Grid middleware technologies and can only provide static and restricted Grid services.
- First-generation Grid portals lack the ability to be customized in that portals can only be built by Grid system developers instead of users. It is difficult for end users to modify an existing portal of this generation to meet their specific needs.
- Portlet technology is gaining attention from the Grid community and being used to build second-generation Grid portals.
- Second-generation Grid portals are focused on portlets that support user customizability in that Grid users can build their personalized portals. Portals of this generation can provide extensible and dynamic Grid services.
- The Portlet API from JSR 168 is the portlet standard for writing portable portlets.

8.7 REFERENCES

[1] Globus, http://www.globus.org.
[2] UNICORE, http://www.unicore.de.
[3] Condor, http://www.cs.wisc.edu/condor/.
[4] Krishnan, S., Bramley, R., Gannon, D., Govindaraju, M., Indurkar, R., Slominski, A., Temko, B., Alameda, E., Alkire, R., Drews, T. and Webb, E. 2001. *The XCAT Science Portal*. Proceedings of Super Computing 2001 (SC '01), Denvor, Colorado, USA. CS Press.
[5] Haupt, T., Bangalore, P. and Henley, G. 2001. *A Computational Web Portal for the Distributed Marine Environment Forecast System*. Proceedings of the 9th International Conference on High-Performance Computing and Networking (HPCN), June 2001, Amsterdam, Netherlands. Lecture Notes in Computer Science, Springer-Verlag.
[6] The Hotpage Portal, http://hotpage.npaci.edu/.

8.7 REFERENCES

[7] Suzumura, T., Matsuoka, S. and Nakada, H. 2001. *A Jini-based Computing Portal System*. Proceedings of Super Computing 2001 (SC '01), Denvor, Colorado, USA. CS Press.

[8] The DSG Portal, http://159.dsg.port.ac.uk/projects/dsgportal/.

[9] Haupt, T., Akarsu, E., Fox, G. and Youn, C. 2000. The Gateway System: Uniform Web based Access to Remote Resources. *Concurrency – Practice and Experience*, 12(8): 629–642.

[10] Grappa, http://grid.uchicago.edu/grappa/.

[11] Allen, G., Daues, G., Foster, I., Laszewski, G., Novotny, J., Russell, M., Seidel, E. and Shalf, J. 2001. *The Astrophysics Simulation Collaboratory Portal: A Science Portal Enabling Community Software Development*. Proceedings of the 10th IEEE International Symposium on High Performance Distributed Computing 2001 (HPDC '01), San Francisco, California, USA. CS Press.

[12] Gannon, D. *et al.* 2002. Programming the Grid: Distributed Software Components, P2P and Grid Web Services for Scientific Applications. *Cluster Computing*, 5(3): 325–336.

[13] Novotny, J., Tuecke, S. and Welch, V. 2001. *An Online Credential Repository for the Grid: MyProxy*. Proceedings of the 10th IEEE International Symposium on High Performance Distributed Computing 2001 (HPDC '01), San Francisco, California, USA. CS Press.

[14] Foster, I., Kesselman, C., Tsudik, G. and Tuecke, S. 1998. *A Security Architecture for Computational Grids*. Proceedings of the 5th ACM Conference on Computer and Communications Security 1998, San Francisco, California, USA. ACM Press.

[15] Czajkowski, K., Foster, I., Karonis, N., Kesselman, C., Martin, S., Smith, W. and Tuecke, S. 1998. *A Resource Management Architecture for Metacomputing Systems*. Proceedings of the 12th International Parallel Processing Symposium & 9th Symposium on Parallel and Distributed Processing (IPPS/SPDP) Workshop on Job Scheduling Strategies for Parallel Processing, Orlando, Florida, USA. CS Press.

[16] Czajkowski, K., Fitzgerald, S., Foster, I. and Kesselman, C. August 2001. *Grid Information Services for Distributed Resource Sharing*. Proceedings of the 10th IEEE International Symposium on High-Performance Distributed Computing (HPDC-10), San Francisco, California, USA. CS Press.

[17] GSIFTP, http://www.globus.org/dataGrid/deliverables/gsiftp-tools.html.

[18] GridFTP, http://www.globus.org/dataGrid/Gridftp.html.

[19] Laszewski, G., Foster, I., Gawor, J. and Lane, P. 2001. A Java Commodity Grid Kit. *Concurrency and Computation: Practice and Experience*, 13(8–9): 643–662.

[20] JNDI, http://java.sun.com/products/jndi/.

[21] iaik security library, http://jce.iaik.tugraz.at/download/evaluation/index.php.

[22] Foster, I., Roy, A. and Sander, V. 2000. *A Quality of Service Architecture that Combines Resource Reservation and Application Adaptation*. Proceedings of the 8th International Workshop on Quality of Service, Westin William Penn, Pittsburgh, USA.

[23] GridPort, http://Gridport.npaci.edu/.

[24] Baru, C., Moore, R., Rajasekar, A. and Wan, M. 1998. *The SDSC Storage Resource Broker*. Proceedings of the CASCON'98 Conference, Toronto, Canada.

[25] SRB project, http://www.npaci.edu/SRB.

[26] Novotny, J. 2002. The Grid Portal Development Kit. *Concurrency and Computation: Practice and Experience,* 14(13–15): 1129–1144.
[27] GSI SSH, http://grid.ncsa.uiuc.edu/ssh/.
[28] Mozilla Directory, http://www.mozilla.org/directory/.
[29] Suzumura, T. *et al.* November 2002. *The Ninf Portal: An Automatic Generation Tool for the Grid Portals.* Proceedings of Java Grande 2002, Seattle, Washington, USA. ACM Press.
[30] Ninf-G, http://ninf.apGrid.org/.
[31] GridSpeed, http://grid.is.titech.ac.jp/gridspeed-www/.
[32] Castor, http://castor.exolab.org/.
[33] What is a Portlet? http://www.javaworld.com/javaworld/jw-08-2003/jw-0801-portlet.html.
[34] Krasner, G. and Pope, S. 1998. A Cookbook for Using the Model-View-Controller User Interface Paradigm in Smalltalk-80. *Journal of Object-Oriented Programming,* 1(3): 27–49.
[35] OASIS, http://www.oasis-open.org.
[36] JCP, http://www.jcp.org.
[37] JSR 168, http://www.jcp.org/jsr/detail/168.jsp.
[38] Jetspeed, http://jakarta.apache.org/jetspeed.
[39] Cocoon, http://cocoon.apache.org/.
[40] RSS, http://www.webreference.com/authoring/languages/xml/rss/intro/.
[41] OCS, http://internetalchemy.org/ocs/.
[42] WAP Forum, http://www.wapforum.org/.
[43] Velocity, http://jakarta.apache.org/velocity/.
[44] WebMacro, http://www.webmacro.org/.
[45] WML, http://www.oasis-open.org/cover/wap-wml.html.
[46] Avantgo, http://www.avantgo.com/doc/pylon/desktop_guide/AppFLookups6.html.
[47] Turbine, http://jakarta.apache.org/turbine/.
[48] Thomas Mueller's Hypersonic SQL database, http://hsql.sourceforge.net/.
[49] PostgresSQL, http://www.postgresql.org/.
[50] PHP, http://www.php.net/.
[51] ASP, http://msdn.microsoft.com/asp.
[52] ECS, http://jakarta.apache.org/ecs/.
[53] IBM WebSphere Portal, http://www.ibm.com/websphere.
[54] WebSphere Portlet API and JSR 168 Portlet API, http://www-106.ibm.com/developerworks/websphere/library/techarticles/0312_hepper/hepper.html.
[55] GridSphere, http://www.Gridsphere.org.
[56] GridLab, http://www.gridlab.org.
[57] RBAC, http://csrc.nist.gov/rbac/.
[58] Hibernate DBMS, http://tm4j.org/hibernate-backend.html.
[59] Junit, http://www.junit.org.
[60] Cactus, http://jakarta.apache.org/cactus/.
[61] Apache Pluto, http://jakarta.apache.org/pluto/.
[62] OGCE, http://www.ogce.org.
[63] CHEF, http://chefproject.org/portal.
[64] GPIR, http://www.tacc.utexas.edu/projects/gpir/.
[65] The Alliance Portal, http://www.extreme.indiana.edu/xportlets/project/index.shtml.

Part Four
Applications

9

Grid Applications – Case Studies

LEARNING OBJECTIVES

In this chapter, we will introduce Grid applications that have applied the core technologies presented in the previous chapters. This chapter will help show:

- Where and how to apply Grid technologies?
- The problem domains that the Grid can be applied to.
- The benefits the Grid can bring to distributed applications.

CHAPTER OUTLINE

9.1 Introduction
9.2 GT3 Use Cases
9.3 OGSA-DAI Use Cases
9.4 Resource Management Case Studies
9.5 Grid Portal Use Cases
9.6 Workflow Management – Discovery Net Use Cases
9.7 Semantic Grid – myGrid Use Case
9.8 Autonomic Computing – AutoMate Use Case
9.9 Conclusions

The Grid: Core Technologies Maozhen Li and Mark Baker
© 2005 John Wiley & Sons, Ltd

9.1 INTRODUCTION

In the previous chapters, we have discussed and explored core Grid technologies, such as security, OGSA/WSRF, portals, monitoring, resource management and scheduling and workflow. We have also reviewed some projects related to each area of these core technologies. Basically the projects reviewed in the previous chapters are focused on the Grid infrastructure, not applications. In this chapter, we present some representative Grid applications that have applied or are applying the core technologies discussed earlier and describe their make-up and how they are being used to solve real-life problems.

The reminder of this chapter is organized as follows. In Section 9.2, we present GT3 applications in the areas of broadcasting, software reuse and bioinformatics. In Section 9.3, we present two projects that have employed OGSA-DAI. In Section 9.4, we present a Condor pool being used at University College London (UCL) and introduce three use cases of Sun Grid Engine (SGE). In Section 9.5, we give two use cases of Grid portals. In Section 9.6, we present the use of workflow in Discovery Net project for solving domain-related problems. In Section 9.7, we present one use case of myGrid project. In Section 9.8, we present AutoMate for self-optimizing oil reservoir applications.

9.2 GT3 USE CASES

As highlighted in Chapter 2, OGSA has become the *de facto* standard for building service-oriented Grids. Currently most OGSA-based systems have been implemented with GT3. The OGSA standard introduces the concepts of Grid services, which are Web services with three major extensions as follows:

- Grid services can be transient services implemented as instances, which are created by persistent service factories.
- Grid services are stateful and associated with service data elements.
- Notification can be associated with a Grid service, which can be used to notify clients of the events they are interested in.

9.2 GT3 USE CASES

Compared with systems implemented with distributed object technologies, such as Java RMI, CORBA and DCOM, services-oriented Grid systems can bring the following benefits:

- Services can be published, discovered and used by a wide user community by using WSDL and UDDI.
- Services can be created dynamically, used for a certain time and then destroyed.
- A service-oriented system is potentially more resilient than an object-oriented system because if a service being used fails, an alternative service could be discovered and used automatically by searching a UDDI registry.

In this section, we present GT3 applications from two areas, one related to broadcasting large amount of data and the other involving software reuse.

9.2.1 GT3 in broadcasting

The multi-media broadcasting sector is a fast evolving and reactive industry that presents many challenges to its infrastructure, including:

- The storage, management and distribution of large media files. As mentioned in Harmer *et al.* [1], a typical one-hour television programme requires about 25 GB of storage and this could be 100–200 GB in production. In the UK, the BBC needs to distribute approximately 1 PB of material per year to satisfy its broadcasting needs. In addition, the volume of broadcast material is increasing every year.
- The management of broadcast content and metadata.
- The secure access of valuable broadcast content.
- A resilient infrastructure for high levels of quality of service.

A Grid infrastructure can meet these broadcasting challenges in a cost-effective manner. To this end, the BBC and Belfast e-Science Centre (BeSC) have started the GridCast project [2] which involves the storage, management and secure distribution of media files.

GT3 has been applied in the project to define broadcast services that can integrate existing BBC broadcast scheduling, automation and planning tools in a Grid environment. A prototype has been built with 1 Gbps connections between the BBC North Ireland station at Belfast, BBC R&D sector at London and BeSC. Various GT3 services have been implemented:

- For the transport of files between sites,
- The management of replicas of stored files,
- The discovery of sites and services on GridCast.

A services-oriented design with GT3 fits the project well because the broadcast infrastructure is by its nature service oriented.

9.2.2 GT3 in software reuse

GT3 can be used to execute legacy codes that normally execute on one computer as Grid services that can be published, discovered and reused in a distributed environment. In addition, the mechanisms provided in GT3 to dynamically create a service, use it for a certain amount of time and then destroyed it are suitable for making these programs as services for hire. In this section, we introduce two projects that are wrapping legacy codes as GT3-based Grid services.

9.2.2.1 GSLab

GSLab [3] is a toolkit for automatically wrapping legacy codes as GT3-based Grid services. The development of GSLab was motivated by the following aspects:

- Manually wrapping legacy codes as GT3-based Grid services is a time-consuming and error-prone process.
- To wrap a legacy code as a Grid service, the legacy code developer also needs expertise in GT3, which may typically be beyond their current area of expertise.

Two components have been implemented in GSLab: the GSFWrapper and the GSFAccessor. The GSFWrapper is used to automatically wrap legacy codes as Grid services and then deploy them in a container for service publication. The GSFAccessor is used to discover Grid services and automatically generate clients to access the discovered services wrapped from legacy codes via GSFWrapper. To improve the high throughput of running a large number of tasks generated from a wrapped Grid service, SGE version 5.3 has been employed with GSLab to dispatch the generated tasks to a SGE cluster. The architecture of GSLab is shown in Figure 9.1.

The process of wrapping legacy codes as Grid services involves three stages: service publication, discovery and access:

- *Publication*: GSFWrapper takes a legacy code as an input (step 1) and generates all the code needed to wrap the legacy application as a Grid Service Factory (GSF) and then deploy the wrapped GSF into a Grid service container for publishing (step 2). Once the Grid service container is started, the wrapped GSF will be automatically published in an SGE cluster environment and the jobs generated by the GSF will be scheduled in the SGE cluster.

- *Discovery*: A user browses the GSFs registered in a Grid service container via GSFAccessor (step 3) and discovers a GSF to use.

- *Access*: The user submits a job request to GSFAccessor via its GUI (step 4). Once the GSFAccessor receives a user job submission request, it will automatically generate a Grid service

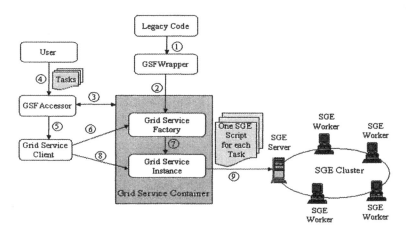

Figure 9.1 The architecture of GSLab

client (step 5) to request a GSF (step 6) to create a Grid service instance (step 7). Then the Grid service client will access the created instance (step 8) to generate tasks in the form of SGE scripts, which will then be used by an SGE server (step 9) through which to dispatch the tasks to an SGE cluster. One SGE script will be generated for each task in GSLab.

A case study, based on a legacy code called q3D [4], has been used to test GSLab. q3D is a C code for rendering 3D-like frames using either 2D geometric shapes or raster images as input primitives, which are organized in layers called *cels*. q3D has basic 3D features such as lighting, perspective projection and 3D movement. It can handle hidden-surface elimination (*cel* intersection) when rendering *cels*. Figure 9.2 shows four frames taken from an animation rendered by q3D. In the animation, the balloon moves gradually approaching the camera and the background becomes darker. Each frame in the animation has two *cels*: a balloon *cel* and a lake *cel*. Each frame is rendered individually from an input file called *stack* that contains the complete description of the frame such as the 3D locations of the *cels* involved. These *stack* files are generated by *makeStacks* from a script that describes the animation such as the camera path, *cels* path and lighting. *makeStacks* is a C code developed for q3D.

To wrap a legacy code as a Grid service, a user needs to provide the parameters to execute the legacy code in the GSFWrapper GUI, as shown in Figure 9.3. Then the GSFWrapper will automatically generate related codes and then deploy the service into a GT3 Grid service container.

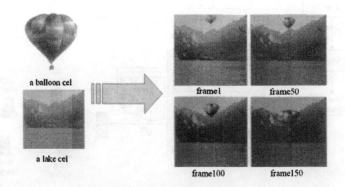

Figure 9.2 Four frames rendered by q3D using two cels

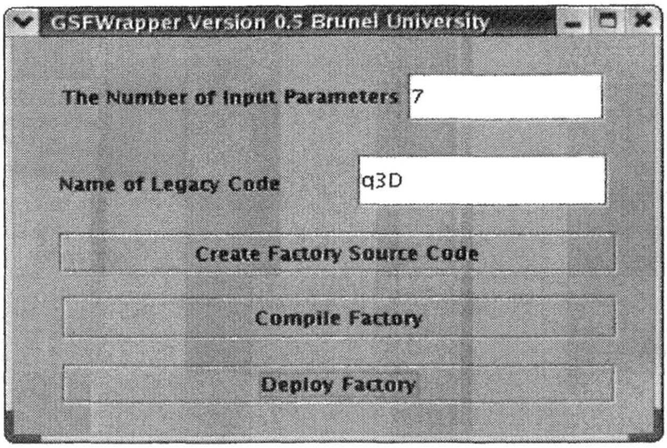

9.2 GT3 USE CASES

Figure 9.3 The GSFWrapper GUI

Once a service is published, the client uses the GSFAccessor GUI, as shown in Figure 9.4, to specify the parameters needed to execute the legacy code, e.g. the input data file name, number of jobs to run and output data file name. Once invoked, the GSFAccessor will generate the related code to call the Grid service that is deployed in

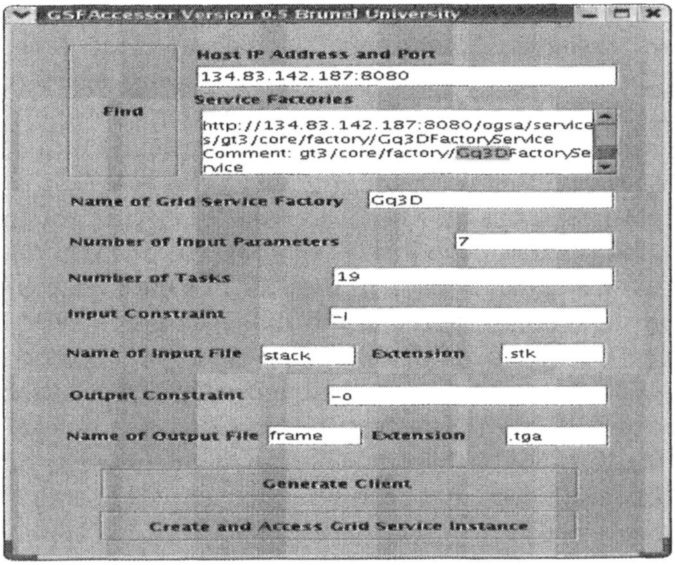

Figure 9.4 The GSFAccessor GUI

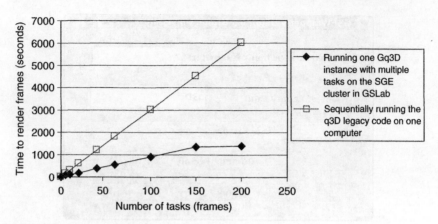

Figure 9.5 The performance of GSLab

an SGE-managed cluster and request its services. Figure 9.5 shows the performance of GSLab in wrapping the q3D legacy code as a Grid service accessed in an SGE cluster with five nodes, each of which has a Pentium IV 2.6-GHz processor and 512 MB RAM, running Redhat Linux.

9.2.2.2 GEMLCA

The Grid Execution Management for Legacy Code Architecture (GEMLCA) [5] provides a solution for wrapping legacy codes as GT3-based Grid services without re-engineering the original codes. The wrapped GT3 services can then be deployed in a Condor-managed pool of computers.

To use GEMLCA, a user needs to write a Legacy Code Interface Description (LCID) file, which is an XML file that describes how to execute the legacy code, e.g. the name of the legacy code and its main binary files, and the job manager (e.g. UNIX fork or Condor). Once deployed in GEMLCA, the legacy code becomes a Grid service that can be discovered and reused. A job submission is based on GT3 MMJFS as described in Chapter 2. A legacy code called MadCity [6], a discrete time-based microscopic simulator for traffic simulations, has been wrapped as a GT3 service and its performance has been demonstrated as a GEMCLA application.

The GEMLCA client has been integrated within the P-GRADE portal [7] to provide a GUI that supports workflow enactment.

Each legacy code deployed in GEMLCA [5, 8] can be discovered in the GUI and multiple published legacy codes can be composed to form another composite application.

9.2.3 A GT3 bioinformatics application

The Basic Local Alignment Search Tool (BLAST) [9] has been widely used in bioinformatics to compare a query sequences to a set of target sequences, with the intention of finding similar sequences in the target set. However, BLAST searches are computationally intensive. Bayer *et al.* [10] present a BLAST Grid service based on GT3 to speed up the search process, in which the BLAST service interacts with backend ScotGRID [11] computing resources. ScotGRID is a three-site (LHC Tier-2) centre consisting of an IBM 200 CPU Monte Carlo production facility run by the Glasgow Particle Physics Experimental (PPE) group [12] and an IBM 24 TByte data store and associated high-performance server run by EPCC [13]. A 100-CPU farm is based at Durham University Institute for Particle Physics Phenomenology (IPPP) [14]. Once deployed as a Grid service, the BLAST service can be accessed by a broad range of users.

9.3 OGSA-DAI USE CASES

A number of projects have adopted OGSA-DAI [15], in this section, we introduce eDiaMoND and ODD-Genes.

9.3.1 eDiaMoND

The eDiaMoND project [16] is a collaborated project between Oxford University, IBM, Mirada Solutions Ltd and a group of clinical partners. It aims to build a Grid-based system to support the diagnosis of breast cancer by facilitating the process of breast screening. Traditional mammograms (film) and paper records will be replaced with digital data. Each mammogram image is a size of 32 MB and about 250 TB data will need to be stored every year. OGSA-DAI has been used in the eDiaMoND project to access the large data sets, which are geographically distributed. The work carried out so far has shown the flexibility of OGSA-DAI and the granularity of the task that can be written.

9.3.2 ODD-Genes

ODD-Genes [17] is a genetics data analysis application built on SunDCG [18] and OGSA-DAI running on Globus. ODD-Genes allows researchers at the Scottish Centre for Genomic Technology and Informatics (GTI) in Edinburgh, UK, to automate important micro-array data analysis tasks securely and seamlessly using remote high-performance computing resources at EPCC. ODD-Genes performs queries on gene identifiers against remote, independently managed databases, enriching the information available on individual genes. Using OGSA-DAI, the ODD-Genes application supports automated data discovery and uniform access to arbitrary databases on the Grid.

9.4 RESOURCE MANAGEMENT CASE STUDIES

In Chapter 6, we have introduced resource management and scheduling systems, namely, Condor, SGE, PBS and LSF. In this section, we first introduce a Condor pool running at University College London (UCL). Then we introduce three SGE use cases.

9.4.1 The UCL Condor pool

A production-level Condor pool has currently been running at UCL since October 2003 [19]. In August 2004, the pool had 940 nodes on more than 30 clusters within the University. Roughly 1 500 000 hours of computational time have been obtained from Windows Terminal Service (WTS) workstations since October with virtually no perturbation to normal workstation usage. An average of 20 000 jobs are submitted on a monthly basis. The implementation of the Globus 2.4 toolkit as a gatekeeper to UCL-Condor allows users to access the pool via Globus certificates and the e-minerals mini-grid [20].

9.4.2 SGE use cases

9.4.2.1 SGE in Integrated Circuit (IC) design

Based in Mountain View, California, Synopsys [21] is a developer of Integrated Circuit (IC) design software. Electronic product technology is evolving at a very fast pace. Millions of transistors (billions in the near future) reside in ICs that once housed only thousands. But this increasing silicon complexity can only be harnessed with sophisticated Electronic Design Automation (EDA) tools that let design engineers produce products that otherwise would be impossible to design. With an SGE-managed cluster of 180 CPUs, the regression testing that used to take 10–12 hours now takes 2–3 hours.

9.4.2.2 SGE in financial analysis and risk assessment

Founded in 1817, BMO Financial Group [22] is one of the largest financial service providers in North America. With assets of about $268 billion as of July 2003, and more than 34 000 employees, BMO provides a broad range of retail banking, wealth management and investment banking products and solutions. Monte Carlo computationally intensive simulations have been used for risk assessment. To speed up the simulation process, an SGE-managed cluster has been built from a Sun Fire 4800 and V880 server, along with StorEdge 3910 server for storing data. The Monte Carlo simulations and other relevant risk-management computations are executed using this cluster. Results are fused and reports are prepared by 9:00 am the next business day, which used to take one-week time.

9.4.2.3 SGE in animation and rendering

Based in Toronto, Ontario in Canada, Axyz Animation [23] is a small- to mid-sized company that produces digital special effects. An SGE cluster has been built to speed up the animation and rendering process. With the help of the SGE cluster, the company has dramatically reduced time to do animations or render frames from overnight to 1–2 hours, eliminating bottlenecks from animation process and increasing server utilization rates to almost 95%.

9.5 GRID PORTAL USE CASES

9.5.1 Chiron

Chiron [24] is a Grid portal that facilitates the description and discovery of virtual data products, the integration of virtual data systems with data-intensive applications and the configuration and management of resources. Chiron is based on commodity Web technologies such as JSP and the Chimera virtual data system [25].

The Chiron portal was partly motivated by the Quarknet project [26] that aims to educate high school students about physics. Quarknet brings physicists, high school teachers and students to the frontier of the 21st century research about the structure of matter and the fundamental forces of nature. Students learn fundamental physics as they analyse live online data and participate in inquiry-oriented investigations, and teachers join research teams with physicists at local universities or laboratories. The project involves about 6 large physics experiments, 60 research groups, 120 physicists, 720 high school teachers and thousands of high school students. Chiron allows students to launch, configure and control remote applications as though they are using a local desktop environment.

9.5.2 Genius

GENIUS [27] is a portal system developed within the context of the EU DataGrid project [28]. GENIUS follows a three-tiered architecture as described in Chapter 8:

- A client running a Web browser;
- A server running Apache Web Server, the Java/XML framework EnginFrame [29];
- Backend Grid resources.

GENIUS provides secure Grid services such as job submission, data management and interactive services. All Web transactions are executed under the Secure Sockets Layer (SSL) via HTTPs. MyProxy is used to manage user credentials.

GENIUS has been used to run ALICE [30] simulation on the DataGrid testbed. In addition, GENIUS has also been used for performing ATLAS [31] and CMS [32] experiments in the context of the EU DataTAG [33] and US WorldGrid [34] projects.

9.6 WORKFLOW MANAGEMENT – DISCOVERY NET USE CASES

Discovery Net [35] is a services-oriented framework to support the high throughput analysis of scientific data based on a workflow or pipeline methodology. It uses the Discovery Process Markup Language (DPML) to represent and store workflows. Discovery Net has been successfully applied in the domains of Life Sciences, Environmental Monitoring and Geo-hazard Modelling. In particular, Discovery Net has been used to perform distributed genome annotation [36], Severe Acute Respiratory Syndrome (SARS) virus evolution analysis [37], urban air pollution monitoring [38] and geo-hazard modelling [39].

9.6.1 Genome annotation

The genome annotation application is data and computationally intensive and requires the integration of a large number of data sets and tools that are distributed across the Internet. Furthermore, it is a collaborative application where a large number of distributed scientists need to share data sets and interactively interpret and share the analysis of the results. A prototype of the genome annotation was successfully demonstrated at the Super Computing conference in 2002 (SC2002) [40] in Baltimore. The annotation pipelines were running on a variety of distributed resources including high performance resources hosted at the London e-Science center [41], servers at Baltimore and databases distributed around Europe and the USA.

9.6.2 SARS virus evolution analysis

In 2003, SARS spread rapidly from its site of origin in Guangdong Province, in Southern China, to a large number of countries throughout the world. Discovery Net has been used for the analysis of the evolution of the SARS virus to establish the relationship between observed genomic variations in strains taken from different patients, and the biology of the SARS virus. Similar to the genome application, discussed previously, the SARS analysis application also requires the integration of a large number of data sets and tools that are distributed across the Internet. It also needs the collaboration of distributed scientists and requires interactivity in the analysis of the data and in the interpretation of the generated results.

The SARS analysis workflows built with Discovery Net have been mostly automated and performed on the fly, taking on average 5 minutes per tool for adding the components to the servers at run time, thus increasing the productivity of the scientists. The main purpose of the workflows presented was to combine the sequence variation information on both genomic and proteomic levels; and to use the available public annotation information to establish the impact of those variations on the SARS virus development.

The data used consists of 31 human patient samples, 2 strains sequenced from palm civet samples which were assumed to be the source of infection and 30 sequences that were committed to Genbank [42] at the time of the analysis, including the SARS reference sequence (NC004718). The reference nucleotide sequence is annotated with the variation information from the samples, and overlaps between coding segments and variations are observed. Furthermore, individual coding segments are translated into five proteins that form the virus (Orf1A, Orf1B, S, M, E, N) and analysis is performed comparing the variation in these proteins in different strains.

All the samples were aligned in order to find the variation points, insertions and deletions. This is a time-consuming process, and with the help of the Grid, the calculation time went from three days on a standard desktop computer up to several hours.

9.6.3 Urban air pollution monitoring

Discovery Net is currently being used as knowledge discovery environment for the analysis of air pollution data. It is providing an infrastructure that can be used by scientists to study and understand the effects of pollutants such as Benzene, SO_2, NO_x or Ozone on human health. Sensors have been deployed to collect data. A sensor grid is being developed in Discovery Net to address the following four issues.

- *Distributed sensor data access and integration*: On one hand, it is essential to record the type of pollutants measured (e.g. Benzene, SO_2 or NO_x) for each sensor. On the other hand, it is essential to record the location of the sensor at each measurement time as the sensors may be mobile.
- *Large data set storage and management*: Each GUSTO (Generic Ultraviolet Sensors Technologies and Observations) sensor generates in excess of 8 GB of data each day, which must be stored for later analysis.
- *Distributed reference data access and integration*: Whereas the analysis of spatiotemporal variation of multiple pollutants in respect to one another can be directly achieved over archived data, more often it is their correlation with third-party data, such as weather, health or traffic data that is more important. Such third-party data sets (if available) typically reside on remote databases and are stored in a variety of formats. Hence, the use of standardized and dynamic data access and integration techniques to access and integrate such data is essential.
- *Intensive and open data analysis computation*: The integrated analysis of the collected data requires a multitude of analysis components, such as statistical, clustering, visualization and data classification tools. Furthermore, the analysis needs high-performance computing resources that utilize large data sets to allow rapid computation.

A prototype has been built to analyse the air pollution in the area around Tower Hamlets and Bromley areas in East London.

The simulated scenario is based on a distribution of 140 sensors in the area collecting data over a typical day from 8:00 am until 6:00 pm at two-second intervals; monitoring NO_x and SO_2. The simulation of the required data has taken into account known atmospheric trends and the likely traffic impact. Workflows built on the simulation results can be used to identify pollution trends.

9.6.4 Geo-hazard modelling

The Discovery Net infrastructure is being used to analyse cosmic shifts of earthquakes using cross-event Landsat-7 ETM+ images [43]. This application is mainly characterized by the high computational demands for the image mining algorithms used to analyse the satellite images (execution time for simple analysis of a pair of images takes up to 12 hours on 24 fast UNIX systems). In addition, the requirement to construct and experiment with various algorithms and parameter settings has meant that the provenance of the workflows and their parameter settings becomes an important aspect to the end-user scientists.

Using the geo-hazard modelling system, the remote sensing scientists have analysed data from an Ms 8.1 earthquake that occurred in 14 November 2001 in an uninhabitable area along the eastern Kunlun Mountains in China. The scientific results of their study provided the first ever 2D measurement of the regional movement of this earthquake and revealed illuminating patterns that were never studied before on the co-seismic left-lateral displacement along the Kunlun fault in the range of 1.5–8.1 m.

9.7 SEMANTIC GRID – MYGRID USE CASE

We have briefly introduced myGrid in Chapters 3 and 7. It is a UK e-Science pilot project, which is developing middleware infrastructure specifically to support *in silico* experiments in biology. myGrid provides semantic workflow registration and discovery. In this section, we briefly describe the application of myGrid to the study of Williams–Beuren Syndrome (WBS) [44].

WBS is a rare, sporadically occurring micro-deletion disorder characterized by a unique set of physical and behavioural features [45]. Due to the repetitive nature of sequence flanking in the WBS critical region (WBSCR), sequencing of the region is incomplete leaving documented gaps in the released sequence. myGrid has been successfully applied in the study of WBS in a series of experiments to find newly sequenced human genomic DNA clones that extended into these "gap" regions in order to produce a complete and accurate map of the WBSCR.

- On one hand, sequencing of the region is more complete. Six putative coding sequences (genes) were identified; five of which were identified as corresponding to the five known genes in this region.
- On the other hand, the study process on WBS has been speeded up. Manually, the processes undertaken could take at least 2 days, but the workflows developed in myGrid for WBS can achieve the same output in approximately an hour. This has a significant impact on the productivity of the scientist, especially when considering these experiments are often undertaken weekly, enabling the experimenter to act on interesting information quickly without being bogged down with the monitoring of services and their many outputs as they are running. The system also enables the scientists to view all the results at once, selecting those, which appear to be most promising and then looking back through the results to identify areas of support.

9.8 AUTONOMIC COMPUTING – AUTOMATE USE CASE

We have briefly introduced AutoMate in Chapter 3 as a framework for autonomic computing. Here, we briefly describe the application of AutoMate in the support of autonomic aggregations, compositions and interactions of software components and enable an autonomic self-optimizing oil reservoir application [46].

One of the fundamental problems in oil reservoir production is the determination of the optimal locations of the oil production and injection wells. As the costs involved in drilling a well and extracting oil is rather large (in millions of dollars per well), this is typically done in a simulated environment before the actual deployment in the field. Reservoir simulators are based on the numerical solution of a complex set of coupled non-linear partial differential equations over hundreds of thousands to millions of grid-blocks. The reservoir model is defined by a number of model parameters (such as permeability fields or porosity) and the simulation proceeds by modelling the state of the reservoir and the flow of the liquids in the reservoir over time, while dynamically responding to changes on the terrain. Such changes can, for example, be the presence of air pockets in the reservoir or responses to the deployment of an injection, or production oil well. During this process, information from sensors and actuators located on the oil wells in the field can be fed back into the simulation environment to further control and tune the model to improve the simulator's accuracy.

The locations of wells in oil and environmental applications significantly affect the productivity and environmental/economic benefits of a subsurface reservoir. However, the determination of optimal well locations is a challenging problem since it depends on geological and fluid properties as well as on economic parameters. This leads to a large number of potential scenarios that must be evaluated using numerical reservoir simulations. The high costs of reservoir simulation make an exhaustive evaluation of all these scenarios infeasible. As a result, the well locations are traditionally determined by analysing only a few scenarios. However, this *ad hoc* approach may often lead to incorrect decisions with a high economic impact.

Optimization algorithms offer the potential for a systematic exploration of a broader set of scenarios to identify optimum locations under given conditions. These algorithms together with the experienced judgement of specialists allow a better assessment of uncertainty and significantly reduce the risk in decision-making. However, the selection of appropriate optimization algorithms, the run-time configuration and invocation of these algorithms and the dynamic optimization of the reservoir remain a challenging problem.

The AutoMate oil reservoir application consists of:

1. Sophisticated reservoir simulation components that encapsulate complex mathematical models of the physical interaction in the subsurface, and execute on distributed computing systems on the Grid;
2. Grid services that provide secure and coordinated access to the resources required by the simulations;
3. Distributed data archives that store historical, experimental and observed data;
4. Sensors embedded in the instrumented oilfield providing real-time data about the current state of the oil field;
5. External services that provide data relevant to optimization of oil production or of the economic profit such as current weather information or current prices;
6. The actions of scientists, engineers and other experts, in the field, the laboratory and in management offices.

The overall oil production process described above is autonomic in that the peers involved automatically detect sub-optimal oil production behaviour at run time and orchestrate interactions among themselves to correct this behaviour. Further, the detection and optimization process is achieved using policies and constraints that minimize human intervention. The interactions between instances of peer services are opportunistic, based on run-time discovery and specified policies, and are not predefined.

9.9 CONCLUSIONS

In this chapter, we have introduced some representative Grid applications and described their make-up and how they are being used to solve real-life problems. These applications have applied or are applying the core technologies discussed in the previous chapters. We started this chapter by introducing GT3 applications such as in the areas of broadcasting and bioinformatics. GT3 has been used for building OGSI-based service-oriented Grid systems in which GT3 services can be published, discovered, and accessed by a broad user community. GSLab and GEMLCA projects have applied GT3 to leverage legacy codes as GT3 services to promote software reuse. OGSA-DAI is a middleware technology that can be used to access data from different data sources. There are a couple of projects that have employed OGSA-DAI. In this chapter, we focused on eDiaMoND to support the diagnosis of breast cancer by facilitating the process of breast screening, and ODD-Genes for genetics data analysis. For resource management, we introduced the UCL Condor pool and three SGE use cases. A cluster managed by Condor or SGE can be effectively used to solve computation intensive problems, e.g. using an SGE-managed cluster of 180 CPUs, the regression testing in integrated circuit design that used to take 10–12 hours now takes 2–3 hours. Grid portals are Web-based user interfaces that provide seamless access to a variety of backend resources. Many portal projects discussed in Chapter 8 have been focused on portal frameworks, i.e. how to build portals. In this chapter, we introduced Chiron and GENIUS for portal applications. Regarding workflow management, we described the application of Discovery Net to the areas of distributed genome annotation, SARS virus evolution analysis, urban air pollution monitoring and geo-hazard modelling. As one of the leading projects in Semantic Grid, myGrid has recently been applied to the study of WBS to speed up the process of the discovery of new genes or sequences. Finally we introduced AutoMate in the support of autonomic aggregations, compositions and interactions of software components and enable an autonomic self-optimizing oil reservoir application.

The Grid is still evolving. Hopefully in a couple of years, we will have a fully developed Grid environment which will be running across many virtual organizations located in different countries. In the near future, we should be able to easily access Grid resources including computing resources, software resources, data resources, storage resources, instrumentation resources without knowing where the resources come from. That is the final goal of the Grid and the right direction upon which the Grid community is currently moving towards.

9.10 REFERENCES

[1] Harmer, T.J., Donachy, P., Perrott, R.H., Chambers, C., Craig, S., Mallon, B. and Wright, C. *GridCast – Using the Grid in Broadcast Infrastructures*. Proceedings of UK e-Science All Hands Meeting 2003 (AHM '03), 2003, Nottingham, UK.
[2] GridCast, http://www.qub.ac.uk/escience/projects/gridcast/.
[3] Li, M., Yu, B. and Qi, M. *GSLab: A Toolkit for Automatically Wrapping Legacy Codes as GT3 Based Grid Services*. Technical Report, June 2004, Brunel University.
[4] Qi, M. and Willis, P. *Quasi3D Cel based Animation*. Proceedings of Vision, Video and Graphics 2003 (VVG '03), July 2003, Bath, UK.
[5] Delaitre, T., Goyeneche, A., Kacsuk, P., Kiss, T., Terstyanszky, G. and Winter, S.C. *GEMLCA: Grid Execution Management for Legacy Code Architecture Design*. Proceedings of the 30th EUROMICRO Conference, Special Session on Advances in Web Computing, 2004, Rennes, France. CS Press.
[6] Gourgoulis, A., Terstyansky, G., Kacsuk, P. and Winter, S.C. *Creating Scalable Traffic Simulation on Clusters*. Proceedings of the 12th Euromicro Conference on Parallel, Distributed and Network based Processing, 2004, La Coruna, Spain. CS Press.
[7] Kacsuk, P., Dózsa, G., Kovács, J., Lovas, R., Podhorszki, N., Balaton, Z. and Gombás, G. P-GRADE: A Grid Programming Environment, *Journal of Grid Computing*, 1(2): 171–197 (2003).
[8] GEMLCA, http://www.cpc.wmin.ac.uk/ogsitestbed/GEMLCA/.
[9] BLAST, http://www.ncbi.nlm.nih.gov/BLAST/.
[10] Bayer, M., Campbell, A. and Virdee, D. *A GT3 based BLAST Grid Service for Biomedical Research*. Proceedings of UK All Hands Meeting, 2004, Nottingham, UK.
[11] ScotGRID, http://www.scotgrid.ac.uk/.

9.10 REFERENCES

[12] PPE, http://ppewww.ph.gla.ac.uk/.
[13] EPCC, http://www.epcc.ed.ac.uk/.
[14] IPPP, http://www.ippp.dur.ac.uk/.
[15] OGSA-DAI Projects, http://www.ogsadai.org.uk/projects/.
[16] e-DiaMoND, http://www.ediamond.ox.ac.uk/.
[17] ODD-Genes, http://www.epcc.ed.ac.uk/~oddgenes/.
[18] SunDCG, http://www.epcc.ed.ac.uk/sungrid/.
[19] UCL Condor Pool, http://grid.ucl.ac.uk/Condor.html.
[20] eMinerals, http://eminerals.org/.
[21] Synopsys, http://www.synopsys.com.
[22] BMO, http://www.bmo.com.
[23] AXYZ, http://www.axyzfx.com.
[24] Zhao, Y., Wilde, M., Foster, I., Voeckler, J., Jordan, T., Quigg, E. and Dobson, J. *Grid Middleware Services for Virtual Data Discovery, Composition, and Integration.* Proceedings of ACM/IFIP/USENIX 5th International Middleware Conference, October 2004, Toronto, Canada. ACM.
[25] Foster, I., Voeckler, J., Wilde, M. and Zhao, Y. *Chimera: A Virtual Data System for Representing, Querying, and Automating Data Derivation.* Proceedings of the 14th Conference on Scientific and Statistical Database Management, July 2002, Edinburgh, UK.
[26] The Quarknet Project, http://quarknet.fnal.gov.
[27] GENIUS, http://genius.ct.infn.it/.
[28] DataGrid, http://www.opensource.org/licenses/eudatagrid.php.
[29] EnginFrame, http://www.enginframe.com.
[30] ALICE, http://www.cern.ch/Alice.
[31] ATLAS, http://atlasexperiment.org/.
[32] CMS, http://cmsinfo.cern.ch/Welcome.html/.
[33] DataTAG, http://datatag.web.cern.ch/datatag/.
[34] WorldGrid, http://www.ivdgl.org/projinfo/.
[35] Discovery Net, http://www.discovery-on_the.net/.
[36] Rowe, A., Kalaitzopoulos, D., Osmond, M., Ghanem, M. and Guo, Y. *The Discovery Net System for High Throughput Bioinformatics.* Proceedings of the 11th International Conference on Intelligent Systems for Molecular Biology, July 2003, Brisbane, Australia.
[37] Curcin, V., Ghanem, M. and Guo, Y. *SARS Analysis on the Grid.* Proceedings of UK e-Science All Hands Meeting, September 2004, Nottingham, UK.
[38] Ghanem, M., Guo, Y., Hassard, J., Osmond, M. and Richards, R. *Sensor Grids for Air Pollution Monitoring.* Proceedings of UK e-Science All Hands Meeting, September 2004, Nottingham, UK.
[39] Liu, J.G. and Ma, J. *Imageodesy on MPI & GRID for Co-seismic Shift Study Using Satellite Optical Imagery.* Proceedings of UK e-Science All Hands Meeting, September 2004, Nottingham, UK.
[40] SC2002, www.sc-conference.org/sc2002.
[41] LeSC, http://www.lesc.ic.ac.uk/.
[42] GenBank, http://www.ncbi.nlm.nih.gov/Genbank/GenbankOverview.html.
[43] Landsat, http://www.landsat.org/.
[44] Stevens, R.D., Tipney, H.J., Wroe1, C.J., Oinn, T.M., Senger, M., Lord, P.W., Goble, C.A., Brass, A. and Tassabehji, M. Exploring Williams–Beuren Syndrome Using myGrid. Bioinformatics, 20(Suppl. 1): i303–i310 (2003).

[45] Morris, C. The Natural History of Williams Syndrome: Physical Characteristics. *Journal of Paediatrics*, 113: 318–326 (1988).
[46] Matossian, V., Bhat, V., Parashar, M., Peszynska, M., Sen, M., Stoffa, P. and Wheeler, M.F. Autonomic Oil Reservoir Optimization on the Grid. *Concurrency and Computation: Practice and Experience*, 17(1): 1–26, January 2005, Wiley.

Glossary

Term	Meaning	Description
Apache Ant		Ant is a Java-based software tool for automating the software build processes. It is similar to make, which automates the compilation of programs whose files are dependent on each other. URL: http://ant.apache.org/
Apache Axis		Axis is a SOAP engine for Web services to exchange messages with SOAP. URL: http://ws.apache.org/axis/
Apache Cocoon		Cocoon is a Web application development framework that implements the notion of "component pipelines", where each component in a pipeline specializes in a particular operation. URL: http://cocoon.apache.org/

Apache Jakarta Tomcat		Tomcat is a Java Servlet container for managing the run-time environment of servlets and Java Server Pages (JSPs). URL: http://jakarta.apache.org/tomcat/
Apache Jakarta Turbine		Turbine is a Java Servlet-based framework that allows developers to build secure Web applications. URL: http://jakarta.apache.org/turbine/
Apache Jakarta Velocity		Velocity provides an alternative scripting language to JSP for developing Web pages. URL: http://jakarta.apache.org/velocity/
Apache Jetspeed		Jetspeed is a framework for building Web portals with portlets. URL: http://portals.apache.org/jetspeed-1/
AppLeS	Application Level Scheduler	AppLeS is an adaptive application level scheduling system in which schedules are generated based on application-specific information.
ASAP	Asynchronous Service Access Protocol	ASAP is a protocol for asynchronous communications between clients and services.
Automatic Computing		Autonomic computing refers to an infrastructure that automatically adapts to meet the demands of the applications that are running in it. Such a system has the features of self-optimization, self-protection, self-healing and self-configuration.

BPEL4WS	Business Process Execution Language for Web Services	Based on XLANG and WSFL, BPEL4WS is both a block-structured and a graph-based workflow language for Web services composition. URL: http://www-128.ibm.com/developerworks/library/ws-bpel/
BPML	Business Process Modelling Language	BPML is a meta-language for modelling business processes.
BSFL	Block Structured Flow Language	A BSFL specifies a predefined order in executing Web services.
CA	Certification Authority	A CA is an authority that issues and manages security credentials and public keys for secure communications.
CCA	Common Component Architecture	CCA is a component model for high-performance computing applications. URL: http://www.cca-forum.org/
Checkpointing		Checkpointing is the means of saving an executing program's state so that in case of failure it may be restarted at the last saved checkpoint.
CIM	Common Information Model	CIM provides a common definition so that vendors can exchange management information between systems throughout a network environment. http://www.dmtf.org/standards/cim

ClassAd	Classified Advertisement	A ClassAd is a descriptive language used by Condor for matching resources.
Condor		Condor is a resource management and job scheduling system. URL: http://www.cs.wisc.edu/condor/
Condor-G		Condor-G is a system that can submit Condor jobs to a Globus environment.
CORBA	Common Object Request Broker Architecture	CORBA is a middleware technology for building distributed client/server applications in which clients and CORBA objects are independent of location, platform and programming language.
CVS	Concurrent Versions System	CVS is a version control system. URL: http://www.cvshome.org/
DAML	DARPA Agent Markup Language	Based on RDF, DAML is an XML-based ontology language developed for the Semantic Web. URL: http://www.daml.org/
DAML+OIL		DAML+OIL is an ontology language combining the features of DAML and OIL.
DAML-S		DAML-S is both a language and an ontology for annotating Web services with semantic capabilities.
DCE RPC	Distributed Computing Environment RPC	DCE RPC is an implementation of RPC from the Open Software Foundation (OSF).

DCOM	Distributed Component Object Model	DCOM is a middleware technology for building Windows-based distributed client/server applications in which clients and DCOM components are independent of location and programming language.
Deadline Constraint Scheduling		Deadline constraint scheduling suspends or terminates running jobs at a certain time.
Dedicated Scheduling		Dedicated scheduling means that jobs scheduled to dedicated resources cannot be preempted.
DER	Distinguished Encoding Rules	DER is a binary representation of an X.509 digital certificate.
DES	Data Encryption Standard	DES is a symmetric key method for data encryption.
DL	Description Logic	DL is a formal method for knowledge representation.
DMTF	Distributed Management Task Force	DMTF is standards body for enterprise and Internet technologies.
ECS	Element Construction Set	ECS is a Java API for generating elements for various markup languages including HTML 4.0 and XML.
EJB	Enterprise JavaBeans	EJB is the server-side technology in the J2EE framework.
Exclusive Scheduling		Exclusive scheduling gives a job an exclusive use of the host that it executes on.

FaCT	Fast Classification of Terminologies	FaCT is a reasoning system that uses DL classifier.
Fair Share Scheduling		The fair share scheduling divides the processing resources among users and groups to provide fair access to resources. A fair share policy can be configured at either the queue or host level.
GARA	Globus Architecture for Reservation and Allocation	GARA is part of the Globus Toolkit for resource reservation and allocation.
GBFL	Graph-Based Flow Language	GBFL is a graph-based workflow language that uses graphs to specify the data and control flows between services.
GGF	Global Grid Forum	GGF is a standards body for Grid technologies. URL: http://www.gridforum.org/
GIS	Grid Information Service	GIS is part of the Globus Toolkit used to manage resources information.
Globus Toolkit		The Globus Toolkit provides middleware technologies for building Grid systems. URL: http://www.globus.org
GMA	Grid Monitoring Architecture	The GMA is the GGF's informational recommendation for monitoring the Grid.
Goal-oriented SLA Scheduling		This helps users configure workloads so that user jobs are completed on time and reduces the risk of missed deadlines. An SLA defines how a service is

GLOSSARY

		delivered and the parameters for the delivery of a service.
GRAM	Globus Resource Allocation Manager	GRAM is part of the Globus Toolkit used for job submission.
Grid Portal		A Grid portal is a system in which Grid resources can be accessed via a Web-based user interface.
GridSphere		GridSphere is a framework for building Web portals with portlets. URL: http://www.gridsphere.org
GSFL	Grid Services Flow Language	GSFL is a WSFL-based workflow language for OGSA compliant Grid service composition.
GSH	Grid Service Handler	GSH is a globally unique URI for a Grid service or a Grid service instance.
GSI	Grid Security Infrastructure	GSI provide an infrastructure for secure communication in grid environments. It is based on public-key encryption, X.509 certificates and the SSL communication protocol.
GSR	Grid Service Reference	GSR is a reference associated with an implementation of a Grid service or a Grid service instance.
GT2	Globus Toolkit 2.x	GT2 is implemented in C language.
GT3	Globus Toolkit 3.x	GT3 is built on OGSI and Web services.

GT4	Globus Toolkit 3.9.x	GT4 is built on WSRF.
GWSDL		GWSDL is an extension of WSDL for describing Grid services.
IETF	Internet Engineering Task Force	IETF is a standard body for Internet technologies. URL: http://www.ietf.org/
IIOP	Internet-Inter ORB Protocol	IIOP is the protocol for CORBA clients to communicate with objects running in an Internet environment.
J2EE	Java 2 Platform Enterprise Edition	J2EE defines the standards for building enterprise applications in Java. URL: http://java.sun.com/j2ee/index.jsp
Java CoG Kit	Java Commodity Grid Kit	The Java CoG Kit is a software package for accessing Globus resources from Java.
JCE	Java Cryptography Extension	JCE is a set of software packages from Sun Microsystems that provides a framework for encryption, key generation and key agreement and Message Authentication Code (MAC) algorithms. URL: http://java.sun.com/products/jce/index.jsp
JCP	Java Community Process	JCP is a standard body for Java-based technologies. URL: http://www.jcp.org
JNDI	Java Native Directory Interface	JNDI is an interface for accessing information directories built with standard protocols such as LDAP.

Job Dependency Scheduling		Job dependency scheduling specifies that the execution of a job is dependent on the completion of another job, which can be specified in the job submission.
JRMP	Java Remote Method Protocol	JRMP is the communicate protocol used by RMI clients and objects.
JSP	Java Server Page	JSP is a server-side scripting language that can be used to dynamically generate Web pages.
JSR 168	Java Specification Request 168	JSR 168 is a JCP portlets specification for developing interoperable portlets.
LDAP	Lightweight Directory Access Protocol	LDAP is a set of protocols for accessing information directories on the Internet.
LMJFS	Local Managed Job Factory Service	LMJFS is part of the Globus Toolkit version 3 that is used for running a local user host environment for job submission.
LSF	Load Sharing Facility	LSF is a resource management and job scheduling system. URL: http://www.platform.com/products/LSF/
MDS	Monitoring and Directory Service	MDS is an information service provided by the Globus Toolkit.
MDS2		MDS2 is an LDAP-based information service provided by GT2.

MDS3		MDS3 is an OGSI-based information service provided by GT3.
MMJFS	Master Managed Job Factory Service	MMJFS is part of the Globus Toolkit version 3 for running a master user host environment for job submission.
MPI	Message Passing Interface	MPI is a high-level messaging API for peer-to-peer communications in a parallel environment.
MVC	Model-View-Controller	MVC is a model for user interface management systems with an aim to separate presentation logic from application logic.
MyProxy		MyProxy is an online credential management system for the Grid.
N1GE	N1 Grid Engine	N1GE is the commercially supported version of SGE. URL: http://wwws.sun.com/software/gridware/
.Net		.Net is a software framework for building Windows-based Web services applications. URL: http://www.microsoft.com/net/
NWS	Network Weather Services	NWS is a system for system performance prediction. URL: http://nws.cs.ucsb.edu/
OASIS	Organization for the Advancement of Structured Information Standards	OASIS is a standards body for e-Business technologies. URL: http://www.oasis-open.org

OCS	Open Content Syndication	OCS is an XML-based format for describing multiple-content channels, including RSS headlines.
OGSA	Open Grid Services Architecture	OGSA, based on Web services, is the *de facto* standard for building service-oriented Grid systems. URL: http://www.globus.org/ogsa/
OGSA-DAI	Open Grid Services Architecture-Data Access and Integration	OGSA-DAI is a middleware technology for accessing and integrating data from different data sources such as relational and XML databases, as well as file systems on the Grid. URL: http://www.ogsadai.org.uk/
OGSI	Open Grid Services Infrastructure	OGSI is a specification for implementing interfaces defined by OGSA. OGSI has been replaced by WSRF.
OIL	Ontology Inference Layer	OIL is an XML-based representation and inference layer for ontologies.
OMG	Object Management Group	OMG is a standards body for interoperable enterprise applications. URL: http://www.omg.org/
ONC RPC	Open Network Computing RPC	ONC RPC is an implementation of RPC from Sun Microsystems.
Ontology		An ontology provides a common vocabulary for a domain and defines the meaning of the terms and the relationships between them.

ORPC	Object Remote Procedure Call	ORPC is the protocol for DCOM clients to communicate with DCOM components.
OWL	Web Ontology Language	OWL is a revision of DAML+OIL. URL: http://www.w3.org/2001/sw/WebOnt/
OWL-S		OWL-S is derived from DAML-S and uses OWL as the ontology language to enrich Web services with semantic capabilities.
PBS	Portable Batch System	PBS is a resource management and job scheduling system. URL: http://www.openpbs.org/
PEM	Privacy Enhanced Mail	PEM is an ASCII form (Base64) representation of an X.509 digital certificate.
PIF	Process Interchange Format	PIF is an interchange format for different process representations.
PKI	Public Key Infrastructure	PKI enables users to securely and privately exchange data through a public and a private cryptographic key pairs, which can be obtained from a trusted authority.
Portlet		A portlet is a Java Servlet, conforming to JSR 168, for interacting with users in a specified part of a Web page.
Preemptive Scheduling		Preemptive scheduling lets a pending job with a high-priority job take resources away from a running job with a lower priority.
PSL	Process Specification Language	PSL defines a neutral representation for process interchange.

PVM	Parallel Virtual Machine	PVM is a software system for developing parallel applications. Using PVM, a heterogeneous collection of UNIX and/or Windows systems can work as a single virtual machine. URL: http://www.csm.ornl.gov/pvm/pvm_home.html
RDF	Resource Description Framework	RDF is an XML-based language for describing structured metadata.
RDFS	RDF Schema	RDFS is an extension of RDF with more modelling primitives.
RFT	Reliable File Transfer	RFT is part of the Globus Toolkit used for reliable file transfer.
RMI	Remote Method Invocation	RMI is a middleware technology for building Java-based distributed client/server applications in which clients and objects are independent of location and platform.
RPC	Remote Procedure Call	RPC is a middle technology for building distributed client/server applications in which a client calls a remote procedure as if it were local.
RSA	Rivest, Shamir and Adleman	RSA is an asymmetric key method for data encryption.
RSL	Resource Specification Language	RSL the description language used by the Globus Toolkit for job submission.
RSS	Really Simple Syndication	RSS is an XML-based format for syndicating Web headlines.

Semantic Grid		The Semantic Grid applies technologies used in the Semantic Web to the Grid to annotate resources and services with semantic meanings for an efficient services/resources discovery.
SGE	Sun Grid Engine	SGE is a resource management and job scheduling system. URL: http://gridengine.sunsource.net/
Semantic Web		The Semantic Web is an initiative to augment unstructured Web content as structured information and to improve the efficiency of Web information discovery and machine-readability.
SLA	Service Level Agreement	An SLA is an agreement between a service consumer and provider that defines the guarantees regarding the use of the services.
SNMP	Simple Network Management Protocol	SNMP is a standard protocol for network management.
SOA	Service Oriented Architecture	SOA is a model for developing loosely coupled distributed systems in which software components are exposed as services so that they can be published in a network environment and then discovered by clients.
SOAP		SOAP is an XML-based messaging protocol for communications between Web Services applications. URL:http://www.w3.org/TR/soap/

SSL	Secure Sockets Layer	SSL is a protocol for secure information transmission over Internet.
TLS	Transport Layer Security	TLS, a successor to SSL, is a protocol for secure communications across a public network through data encryption.
UDDI	Universal Description, Discovery, and Integration	UDDI is an industry standard for service registration and discovery in Web services. URL: http://www.uddi.org/
URI	Uniform Resource Identifier	A URI is a standard way for identifying a resource on the Web.
VO	Virtual Organisation	A VO is a dynamic environment that couples geographically distributed resources, which may run across multiple institutions. A VO may have rules as specifying how its resources can be securely accessed and shared by its members.
W3C	World Wide Web Consortium	The W3C is standard body for Web technologies. URL:http://www.w3.org/
WAP	Wireless Application Protocol	WAP is set of protocols for accessing Internet resources from mobile devices. URL: http://www.wapforum.org/
WBEM	Web-Based Enterprise Management	WBEM provides a set of standard Web-based technologies for managing enterprise environments. http://www.dmtf.org/standards/wbem/

Web Portal		A Web Portal is a system in which content can be accessed via Web-based user interfaces
WebSphere		WebSphere is a framework for building Web portals with portlets. URL: http://www-306.ibm.com/software/websphere/
WfMC	Workflow Management Coalition	WfMC is a standards body for workflow management systems. URL: http://www.wfmc.org/
WFMS	Workflow Management System	A WFMS is a system that manages the execution of workflows.
Workflow Engine		A workflow engine manages the run time of workflow processes.
Workflow Language		A workflow language is used to describe the functional relations of processes in a workflow.
WPDL	Workflow Process Definition Language	WPDL is a workflow language defined by WfMC.
WS	Web Services	WS are XML-based middleware for building service-oriented distributed applications. URL: http://www.w3.org/2002/ws/
WSCI	Web Services Choreography Interface	WSCI is a WSDL-based block-structured workflow language for Web services composition. URL: http://www.w3.org/TR/wsci/

WSDD	Web Services Deployment Descriptor	WSDD is an XML-based document describing how to deploying Web services.
WSDL	Web Services Description Language	WSDL is an XML-based language for describing Web services interfaces. URL: http://www.w3.org/TR/wsdl
WSFL	Web Services Flow Language	WSFL is a WSDL-based graph-structured workflow language for Web services composition.
WSIF	WS-Inspection Language	WSIF is an XML-based language for describing WS-Inspection documents. URL: http://www-106.ibm.com/developerworks/webservices/library/ws-wsilspec.html
WS-Inspection		WS-Inspection is an industry standard for service registration and discovery.
WSRF	Web Services Resource Framework	WSRF is a set of specifications that models stateful resources with Web services.
WSRP	Web Services for Remote Portlets	WSRP is an OASIS specification for accessing Web services via portlets.
WS-Security		WS-Security is a specification for secure message exchanging with SOAP.
X.509		X.509 is a standard for defining digital certificates, which are used to authenticate messages sent over a network.

XDR	eXternal Data Representation	XDR is a data representation approach for exchanging data between heterogeneous computing systems.
XLANG		XLANG is a WSDL-based block-structured workflow language for Web services composition.
XML-Encryption		XML-Encryption is a specification for encrypting messages in XML. URL: http://www.w3.org/Encryption/2001/
XML-Signature		XML-Signature is a specification for processing digital signatures in XML messages. URL: http://www.w3.org/Signature/
XPDL	XML Process Definition Language	XPDL is an XML-based workflow language defined by WfMC to replace WPDL.
XSD	XML Schema Definition	XSD is an XML-based language for describing and verifying XML documents.

Index

Access Control List 126
Apache Ant 51
Apache Axis 30
  Java2WSDL 30, 31
  WSDL2Java 30, 32
AppLes 291
Asynchronous
  communication 19
AtlasGrid 193
Autonomic computing 108
  self-configuring 110
  self-healing 109
  self-learning 110
  self-optimizing 109
  self-protection 109
Autonomic Grid services 113
Autopilot 164

BioOpera 326
BioPipe 323
BPEL4WS 315
BPML 317

Cactus 366
CCA 112, 327

Certificate Authority 132, 133
CHEF 370
CIM 4
ClassAd 266
CODE 168
Condor 254
Condor-G 267
CORBA 19
  IDL 19
  IIOP 19
Credential 146
  delegation 139
CrossGrid 193
Cryptography 127
  asymmetric cryptography 129
  symmetric cryptosystems 128

DAGMan 264
DAIS 59
DAML 85
DAML+OIL 84
DAML-S 89
  ServiceGrounding 90
  ServiceModel 90
  ServiceProfile 89–90

Data provenance 106, 107
DataGrid 193
DataTAG 193
DCE RPC 15
DCOM 18
  ORPC 18
Dedicated scheduling 266
DES 128
Description Logic 84
Digital signature 130
Discovery Net 329
DMTF 4

ECS 361
Equal-share scheduling 274

F-Logic 87
FaCT 85
Fault tolerance 161
Firewall 133

GALE 322
Ganglia 217
GARA 341
Genetic algorithms 296
Geodise 95, 327
GGF 4
GLUE 212
GMA 154
  consumer 155
  Directory Service 156
  producer 156–7
GPDK 343
GPIR 176, 370
GrADS 164, 293
Grid definition 2, 3
Grid service container 42
Grid service data 37
Grid service instance 35
GridAnt 327
GridFlow 326
GridICE 172
GridLab 202
GridMon 219

GridPort 341
GridRM 180
  jGMA 182
GridSpeed 346
GridSphere 364
GriPhyN 329
GRM/PROVE 220
GSFL 318
GSH 36
GSI 134
GSR 36
GSSAPI 205
GT3 40
  GT3 GRAM 44
  GT3 GWSDL 50
  GT3 Index Service 48
  GT3 LMJFS 45
  GT3 MJS 45
  GT3 MMJFS 45
  GT3 programming model 50
  GT3 RFT 49
GT4 69
GWEL 321

Hawkeye 185

ICENI 325
INFN-Grid 172
IT Innovation Workflow
  Enactment Engine 324

J2EE 30
  JAXB 30
  JAXP 30
  JAXR 30
  JAX-RPC 30
JAMM 189
Java CoG 339, 341
Java RMI 16
  JRMP 17
JCE 142
JCP 355
Jetspeed 357
  PSML 360
JIGSA 325

JNDI 341
Job checkpointing 264
Job flocking 265
Job scheduling architecture 245
   centralised scheduling 245
   distributed scheduling 246
   hierarchical scheduling 248
Job selection paradigms 253
   backfilling selection 254
   first come first serve 253
   priority-based selection 253
   random selection 253
JSR 168 356
Junit 366

Key 129
   private key 129
   public key 129

L-Bone 193
LCG 172
LDAP 170, 175, 190, 191, 193, 195, 196, 197, 209, 232
LSF 279

MapCenter 192
MDS3 196
Mercury 201
Metadata 78, 79
Monitoring 153
   cross-API Monitoring 161–2
   dynamic monitoring 160–1
   static monitoring 160
   workflow monitoring 161
MS.NETGrid 40
MVC model 101
MyGrid 94, 324
MyProxy 339

N1GE 269
Nagios 221
Naming Schema 181
NASA IPG 168
.Net 32

ASP 32
DISCO 32
MSXML 32
NetLogger 222
NetSolve 215
Nimrod/G 293
Ninf Portal 345
NMI 205
NPACI Grid 205
NWS 205

OASIS 4
OCS 357
OGCE 369
OGSA 5, 34, 69
OGSA-based Grid services 34
OGSA-DAI 53
OGSA-DAI portTypes 54
   DAIServiceGroupRegistry 56
   GDSPortType 54
   GridDataPerform 55
   GridDataServiceFactory 56
   GridDataTransport 55
OGSA portTypes 38
   Factory 38
   GridService 38
   HandleResolver 38
   NotificationSource/ NotificationSink 39
   Registration 38–9
OGSA-WG 12
OGSI 35, 40, 43, 66
OGSI::Lite 41
OGSI.NET 40
OGSI-WG 13
OIL 84
OilEd 88
OMG 19
ONC RPC 15
OntoEdit 87
Ontology 78, 79, 80
Ontology-based Resource Matching 93
Ontology languages 83

OpenSSL 146
OWL 86
  OWL DL 86
  OWL Full 86
  OWL Lite 86
OWL-S 90

PBS 274
P-GRADE 329
PIF 331
PlanetLab 193
Portal services 338
Portals 336
Portlet 350
Preemptive Scheduling 266
Protégé 88
PSL 331
Public-key certificate 130
Public Key Infrastructure 131
PyOGSI 41

RDF 81
  data model 82
RDFS 84
Resource reservation 290
R-GMA 209
RPC 15
  IDL 15
RRDtool 218
RSA 129
RSS 357

SAAJ 30
SCALEA-G 223
SDE 37
Security 123
  Assurance 125
  Auditability 126
  Authentication 125
    Mutual authentication 135
  Authorization 125
  Availability 125
  Confidentiality 125
  Integrity 125
  Non-repudiation 126
Semantic Grid 77
Semantic Grid portal 99
Semantic service annotation and
  adaptation 98
Semantic Web 78
Semantic workflow 94, 95
Service Data Element (SDE) 37
SGE 269
Simulated annealing 296
Single sign-on 139
SOA 21
SOAP 23
  Body 23
  Envelope 23
Socket programming 14
SSL 205
SWFL 321
Symphony 328
Synchronous communication 16,
  17, 19

Time-stamped data 159
The Alliance Portal 370
Triana 324
Turbine 359

UDDI 26
  Green page 26
  White Page 26
  Yellow Page 26
UDDI4J 30

Velocity 358
visPerf 214
VO 3
W3C 4
WBEM 4
Web services 21
  deployment model 29
  hosting environments 34
  programming model 29
  SOAP engine 30

INDEX

WebSphere 362
WfMC 303
WfMC reference model 305
WFMS 304
Workflow definition 304
Workflow engine 306
WPDL 308
WS-Addressing 63
WS-AtomicTransaction 6
WS-Coordination 6
WS-Federation 6
WS-GAF 6
WS-I 6
WS-Inspection 27
WS-Management 6
WS-Policy 6
WS-ReliableMessaging 5
WS-Security 43
WS-SecureConversation 6
WS-Trust 6
WSCI 313
WSDL 24
  binding 25
  data types 24
  message 24–5
  port 25

portType 25
service 25–6
WSFL 311
WSIF 34
WSRF 60, 66, 69
  WS-BaseFaults 62
  WS-Notification 61
    WS-BaseNotification 61
    WS-BrokeredNotification 61
    WS-Topics 61
  WS-RenewableReferences 62–3
  WS-Resource 63
  WS-ResourceLifetime 60
  WS-ResourceProperties 61
  WS-ServiceGroup 62
WSRP 355

X.509 PKI 131
XDR 205, 218
XLANG 311
XML-Encryption 43
XML-Signature 43
XPDL 308
XSD 24

YAWL 326